Environmental Flows, Ecological Quality and Ecosystem Services

Environmental Flows, Ecological Quality and Ecosystem Services

Editor

Carles Ibáñez

MDPI • Basel • Beijing • Wuhan • Barcelona • Belgrade • Manchester • Tokyo • Cluj • Tianjin

Editor
Carles Ibáñez
EURECAT, Technological Centre of Catalonia
Spain

Editorial Office
MDPI
St. Alban-Anlage 66
4052 Basel, Switzerland

This is a reprint of articles from the Special Issue published online in the open access journal *Water* (ISSN 2073-4441) (available at: https://www.mdpi.com/journal/water/special_issues/ Environmental_Flows_Ecological_Quality_Ecosystem_Services).

For citation purposes, cite each article independently as indicated on the article page online and as indicated below:

LastName, A.A.; LastName, B.B.; LastName, C.C. Article Title. *Journal Name* **Year**, *Volume Number*, Page Range.

ISBN 978-3-0365-2245-6 (Hbk)
ISBN 978-3-0365-2246-3 (PDF)

Cover image courtesy of Carles Ibáñez

Contents

About the Editor

Carles Ibáñez is Senior Researcher and Head of the Department of Climate Change at EURECAT, Technological Institute of Catalonia, as well as Scientific Director of the Climate Resilience Centre. Former Head of the Aquatic Ecosystems Program of IRTA (2005–2017), a public research institute of the Government of Catalonia. He received his PhD in Biology from the University of Barcelona and post-doc in the Laboratory of Fluvial System Ecology of CNRS (France). He was a visiting researcher at the Socio-ecological Synthesis National Centre of United States (University of Maryland). He has 30 years of research experience in the field of aquatic, coastal and wetland ecology, climate change impacts and adaptation, sustainable management of water resources and environmental management, with around 100 papers published in peer-reviewed international journals and 15 books and book chapters. He has directed 14 PhD Theses and many Master Theses. He is member of the Advisory Council for Sustainable Development of the Government of Catalonia and is Associated Editor of the journal Estuaries & Coasts.

Editorial

Special Issue: Environmental Flows, Ecological Quality, and Ecosystem Services

Carles Ibáñez

Department of Climate Change, Area of Sustainability, EURECAT, 43870 Amposta, Catalonia, Spain;
carles.ibanez@eurecat.org

Citation: Ibáñez, C. Special Issue: Environmental Flows, Ecological Quality, and Ecosystem Services. *Water* **2021**, *13*, 2760. https://doi.org/10.3390/w13192760

Received: 24 September 2021
Accepted: 26 September 2021
Published: 6 October 2021

Publisher's Note: MDPI stays neutral with regard to jurisdictional claims in published maps and institutional affiliations.

Global environmental change is greatly disturbing rivers and estuaries by a number of stressors, among which water withdrawal, damming, pollution, invasive species, and climate change are the most worrying. All of them have a direct connection to river hydrology and preserving flow regimes has become one of the key issues for protecting our future water resources and river ecosystems. Thus, the science and methods to establish adequate environmental flows (e-flows) is a keystone to implement an integrated management of water resources. This is even more critical in water scarce river basins, where the preservations of ecological functions and values of aquatic ecosystems (i.e., water quality, sediment dynamics, productivity, biodiversity, carbon cycling, etc.) critically depend on river flow regime. However, scientists and managers often find it very difficult to quantitatively link the ecological status and biodiversity of aquatic ecosystems and their services to specific e-flow regimes. This Special Issue includes papers investigating the links between river flow regime, the status of aquatic ecosystems, and the benefits they provide to our society either from the science or management perspective.

The main group of papers focuses on the conceptual, quantitative, and qualitative links of flow regime and e-flows with river ecosystem functions and values, such as the preservation and ecological status of inland wetlands [1], the functions and values of temporarily closed estuaries [2], the preservation of fish species richness [3], the impact on fish spawning period [4], the ecological quality, bird diversity and shellfish fisheries in a lowland river and its coastal area [5], and the ecosystem productivity of coastal areas [6]. Moreover, one of the papers [7] performs an in-depth review of ecohydrological links in the lower Ebro River and its Delta, which is one of the most studied cases regarding the impacts of flow regime alteration on socioecological functions and values. Finally, a couple of papers deal with management and policy issues, such as Chinese policies of hydropower projects to avoid negative environmental impacts [8] and the proposal of a new framework for managing ecological quality and ecosystem services in coastal waters [9].

Most of the papers highlight the idea that, besides promising progress in establishing and implementing e-flows in rivers and coastal areas, efforts are not sufficient to preserve the ecological integrity and health of river ecosystems. On the other hand, the multidimensional character of e-flow research and management suggests that a holistic socio-ecological approach is needed to successfully establish, test, and implement sound-flow regimes under an adaptive framework.

Funding: This research received no external funding.

Institutional Review Board Statement: Not applicable.

Informed Consent Statement: Not applicable.

Conflicts of Interest: The author declares no conflict of interest.

References

1. Nikitina, O.I.; Dubinina, V.G.; Bolgov, M.V.; Parilov, M.P.; Parilova, T.A. Environmental Flow Releases for Wetland Biodiversity Conservation in the Amur River Basin. *Water* **2020**, *12*, 2812. [CrossRef]
2. Adams, J.B.; Van Niekerk, L. Ten Principles to Determine Environmental Flow Requirements for Temporarily Closed Estuaries. *Water* **2020**, *12*, 1944. [CrossRef]
3. Driver, L.J.; Cartwright, J.M.; Knight, R.R.; Wolfe, W.J. Species-Richness Responses to Water-Withdrawal Scenarios and Minimum Flow Levels: Evaluating Presumptive Standards in the Tennessee and Cumberland River Basins. *Water* **2020**, *12*, 1334. [CrossRef]
4. Li, J.; Qin, H.; Pei, S.; Yao, L.; Wen, W.; Yi, L.; Zhou, J.; Tang, L. Analysis of an Ecological Flow Regime during the Ctenopharyngodon Idella Spawning Period Based on Reservoir Operations. *Water* **2019**, *11*, 2034. [CrossRef]
5. Belmar, O.; Ibáñez, C.; Forner, A.; Caiola, N. The Influence of Flow Regime on Ecological Quality, Bird Diversity, and Shellfish Fisheries in a Lowland Mediterranean River and Its Coastal Area. *Water* **2019**, *11*, 918. [CrossRef]
6. Cozzi, S.; Ibáñez, C.; Lazar, L.; Raimbault, P.; Giani, M. Flow Regime and Nutrient-Loading Trends from the Largest South European Watersheds: Implications for the Productivity of Mediterranean and Black Sea's Coastal Areas. *Water* **2019**, *11*, 1. [CrossRef]
7. Ibáñez, C.; Caiola, N.; Belmar, O. Environmental Flows in the Lower Ebro River and Delta: Current Status and Guidelines for a Holistic Approach. *Water* **2020**, *12*, 2670. [CrossRef]
8. Wu, M.; Chen, A.; Zhang, X.; McClain, M.E. A Comment on Chinese Policies to Avoid Negative Impacts on River Ecosystems by Hydropower Projects. *Water* **2020**, *12*, 869. [CrossRef]
9. Tagliapietra, D.; Povilanskas, R.; Razinkovas-Baziukas, A.; Taminskas, J. Emerald Growth: A New Framework Concept for Managing Ecological Quality and Ecosystem Services of Transitional Waters. *Water* **2020**, *12*, 894. [CrossRef]

Article

Ten Principles to Determine Environmental Flow Requirements for Temporarily Closed Estuaries

Janine Barbara Adams [1,2,*] **and Lara Van Niekerk** [2,3]

[1] Botany Department, Nelson Mandela University, Port Elizabeth 6031, South Africa

[2] DST/NRF Research Chair in Shallow Water Ecosystems, Institute for Coastal and Marine Research, Nelson Mandela University, Port Elizabeth 6031, South Africa; LvNieker@csir.co.za

[3] Council for Scientific and Industrial Research (CSIR), P.O. Box 320, Stellenbosch 7599, South Africa

[*] Correspondence: janine.adams@mandela.ac.za; Tel.: +27-41-504-2429

Received: 29 May 2020; Accepted: 6 July 2020; Published: 8 July 2020

Abstract: Temporarily closed estuaries require seasonal opening to tidal flows to maintain normal ecological processes. Each estuary has specific environmental flow (EFlow) requirements based on the relationship between freshwater inflow, coastal dynamics, rate of sandbar formation, and the open/closed state of the mouth. Key abiotic processes and ecosystem services linked to mouth state were highlighted. We reviewed completed EFlow requirement studies for temporarily closed estuaries in South Africa and found that the formulation of these requirements should consider the timing and magnitude of flows in relation to the morphology of an estuary, its mouth structure, catchment size, and climate. We identified ten key principles that could be adapted to similar systems in equivalent climatic settings. Principle 1 recognizes that each estuary is unique in terms of its EFlow requirements because size, scale, and sensitivity of core elements to freshwater inflow are specific for each system; EFlows cannot be extrapolated from one estuary to another. Principle 2 highlights the importance of baseflows in keeping an estuary mouth open because a small reduction in flow can cause the mouth to close and alter essential ecological processes. Principle 3 outlines the role of floods in resetting natural processes by flushing out large volumes of sediment and establishing the equilibrium between erosion and sedimentation. Principle 4 emphasizes the need for open mouth conditions to allow regular tidal flushing that maintains water quality through reducing retention times and preventing the onset of eutrophic conditions. Principle 5 advises artificial breaching to be practiced with caution because execution at low water levels encourages sedimentation that reduces the scouring effect of flushing. Principle 6 holds that elevated inflow volumes from wastewater treatment works or agricultural return flows can increase the frequency of mouth opening and cause ecological instability. Principle 7 states that water released from dams to supply the environmental flow cannot mimic the natural flow regime. Principle 8 specifies the need for short- and long-term data to increase the confidence levels of EFlow assessments, with data to be collected during the open and closed mouth states. Principle 9 advocates the implementation of a monitoring program to track the achievement of EFlow objectives as part of a strategic adaptive management cycle. Finally, Principle 10 recommends the adoption of a holistic catchment-to-coast management approach underpinned by collaboration with regulatory authorities and stakeholders across a range of sectors. These principles can be used to guide the formulation and management of EFlows, an essential strategy that links the maintenance of estuarine ecological integrity with social well-being.

Keywords: intermittently; lakes; lagoons; bar; microtidal; coastal; semi-closed; berm; water quality; ecosystem services

---

1. Introduction

The importance of freshwater inflow to estuaries and their response to altered flows are particularly relevant in semi-arid countries such as South Africa. Rainfall is variable and unpredictable with approximately 1000 mm per annum falling along the eastern subtropical coast and <200 mm per annum on the west coast. Modifications to freshwater inflow by human activities threaten the health and functioning of South African estuaries. While several studies have addressed responses to change in freshwater inflow [1–7], few have quantified the environmental flow (EFlow) required to maintain these ecosystems and their benefits in the face of competing water uses and regulated flows. Flow studies in estuaries pose unique challenges because the combined influence of the river and the sea generates spatial and temporal variability at a range of scales.

In South Africa, estuaries are defined as partially enclosed, permanent water bodies, either continuously or periodically open to the sea and extending landwards as far as the upper limit of tidal action, salinity penetration, or back-flooding under closed-mouth conditions. During floods, they can become river mouths where no seawater enters. When there is little or no fluvial input, they can be isolated from the sea by a sandbar and become fresh or hypersaline [8]. More than 90% of South African estuaries have restricted inlets brought about by low river runoff combined with strong coastal wave action and high rates of sediment movement [7,9]. Sand bars (berms) form in mouth regions causing >75% to intermittently close off from the sea [8,10,11]. Such estuaries have been classified globally as either temporarily closed estuaries (TCEs), temporarily open/closed estuaries (TOCEs), intermittently open lakes and lagoons (ICOLLS), or intermittently open/closed estuaries (IOCEs) that remain closed until sufficient inflow, often flood events, enables the berm to be breached [9,12,13]. They are also known collectively as blind estuaries, bar-built systems, coastal plain estuaries, barrier beaches, and estuarine embayments and are located in coastal plains where there is sufficient marine sediment to support the development of barrier beaches, dunes, or bars [14]. They are a feature of microtidal (<2m tidal range) and low-mesotidal (1.3 m mean tidal range) coastlines in the mid-latitudes, predominantly those with temperate climates [12]. These estuaries occur on the south-eastern coast of New Zealand, Australia, Brazil, and Uruguay; the south-western coast of India and Sri Lanka [15]; and are numerically dominant in microtidal Mediterranean climates. For example, of the 47 estuaries in south-western Australia, 82% are close to the sea [16]. Lagoons, creeks, and back dune lagoons in south-eastern Australia are also close to the sea during low flow conditions [17].

Artificial breaching of closed estuaries and coastal lagoons is a common practice worldwide, used to offset flow reduction, prevent flooding, and improve fisheries and water quality [18–21]. In Central and Southern California, many historically closed estuaries have been permanently opened using groins, levees, and regular dredging [22], practices that can have unforeseen negative environmental impacts. TCEs can have high biodiversity importance and productivity. Closed phases are associated with high abundance and biomass of benthic microalgae, submerged macrophytes, zooplankton, and hyper- and zoobenthos [23]. Peaks in food resources and available habitats coupled with stable physico-chemical conditions benefit estuarine-spawning fish that, in turn, support piscivorous avifauna. The estuaries function as important roosting and feeding sites for resident and migratory birds. Similarly, closed systems in south-eastern Australia and in California provide nursery habitats for juvenile fish and migratory shore birds [17,24,25].

This article focuses specifically on South African TCEs that are < 500 ha in size and seasonally or intermittently open with land–sea connectivity occurring at annual scales. This group differs from large (>500 ha) closed systems such as ICOLLS and coastal lagoons that often remain closed for several years at a time. Recently, the sizeable group of South African TCEs [10] was sub-divided into large and small systems based on a threshold estuarine habitat area of 15 ha [8]. Small TCEs are strongly event-driven and can experience rapid increases and decreases in freshwater inflow over a few hours. They are generally perched and when open (semi-closed), the outflow channel restricts tidal amplitude to 10–15 cm compared to 15–30 cm in large TCEs. The main forces that maintain open mouth conditions are river inflow and to a lesser extent tidal flows [8], while the main closing forces are wave energy and sediment availability from marine, fluvial, or aeolian sources (Figure 1). South charged into a high energy coastal environment [26]. They stay closed until their basins fill up and the sand berm is breached. Mouth opening may result in the removal of significant amounts of sediment. However, infilling from marine and fluvial sediment following flow reduction can be rapid. Groundwater inflows are an important additional source of freshwater input in arid and semi-arid environments for the development of refugia during droughts, and the maintenance of water level and salinity regimes [27]. However, surface water flow is the dominant driver in mouth behavior of TCEs given their small size and sensitivity to flow changes and are thus the focus of EFlow studies.

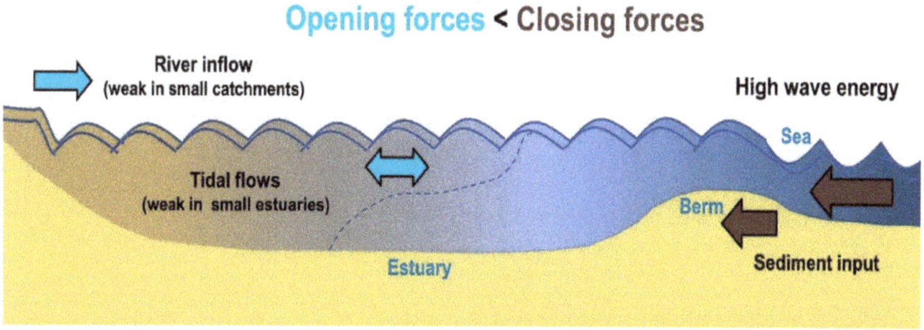

Figure 1. Longitudinal representation of a temporarily closed estuary showing the main opening and closing forces that influence mouth condition.

Estuaries require an adequate flow of freshwater to maintain a healthy state. EFlows describe the quantity and quality of freshwater flows in terms of timing, duration, frequency, and intensity necessary to sustain aquatic ecosystems. Managing EFlows to estuaries should account for these aspects of the annual and multi-year hydrograph. These EFlows support human cultures, economies, sustainable livelihoods, and well-being [28]. EFlow requirements were seldom considered in water resource planning and management [29–33] until methods were developed for estuaries in South Africa, Australia, and the USA (CA and TX) [33,34]. EFlow requirements have also been described for Mediterranean coastal ecosystems, such as the Ebro Delta, where anthropogenic pressures are magnified by the natural hydrological variability of aquatic ecosystems [35,36]. To ensure ecological sustainability and social well-being, innovation is needed in the emerging science of EFlows [28].

The aim of the study was to review EFlow assessments conducted for a range of these systems to identify and describe general principles that can be applied to this category of estuary worldwide. The exercise was based on EFlow studies conducted over the past 25 years in South Africa and included the outcome of research aimed at understanding the responses of TCEs to changes in freshwater inflow [15,23,31,32,37,38]. The discussion includes estuarine lakes and predominantly open estuaries where they illustrate a relevant phenomenon during the closed-mouth state. Processes under these conditions are comparable to that in TCEs. Reviews of TCE responses to changes in freshwater inflow are described in detail elsewhere [15,23]. The study presented in this paper updates these reviews and relates estuary response to changes in important ecosystem services [39]. Some of the key abiotic processes and biotic responses that occur when freshwater is abstracted and the mouth of the estuary closes more frequently to the sea were highlighted (Table 1).

Table 1. Response of temporarily closed estuaries (TCEs) to reduced freshwater inflow and increased duration and frequency of closed-mouth conditions in terms of resultant abiotic drivers, the main flow component driving the process, biotic responses, and impacts on ecosystem services.

Closed-Mouth Conditions and Abiotic Driver	Flow Component *	Biotic Response	Impact on Ecosystem Services
No tidal exchange	Reduced dry-season base flows and/or drought flows	Loss of intertidal habitat, salt marsh and nursery habitat. Reduced biological diversity (birds, macrofauna, macrophytes).	Loss of wetland purification capacity, erosion control, bank protection, and flood mitigation.
Loss of connectivity with sea	Reduced dry-season base flows and/or drought flows	Loss of invertebrate and fish recruitment. Interruption to faunal life cycles. Decline in salt tolerant biota.	Loss of nursery habitat and biodiversity. Reduced fisheries.
Loss of connectivity with the catchment	No dry-season base flows or drought flows, reduction in flood occurrence and magnitude	No recruitment of catadromous fauna (eels and freshwater mullet) that live in freshwater and breed in the sea.	Reduced food security and loss of cultural elements.
Increased water level (wet conditions)	Reduced dry-season base flows	Loss of intertidal habitat (intertidal salt marsh) and wading birds due to flooding.	Reduced tourism appeal. Less recreational bird watching opportunities. Flooding of adjacent properties leading to artificial breaching.
Decreased water level (dry conditions)	Reduced/no drought flows	Die-back of submerged plants. Loss of nursery habitat. Reduced foraging and nesting habitat for water birds.	Reduced bait and fisheries resources. Reduced ecotourism.
Eutrophication and low water transparency	Increase or decrease dry-season base flows, coupled with nutrient enrichment	Loss of submerged aquatic vegetation (seagrass). Occurrence of harmful algal blooms and invasive aquatic macrophytes. Fish kills.	Loss of waste assimilative capacity. Reduced nutrient cycling. Loss of fisheries. Reduced revenues from recreation and tourism. Decline in real estate values.
Increased retention of pollutants and human pathogens	Reduced dry-season base flows and/or reduction in flood occurrence and magnitude	Accumulation of pollutants in fish and shellfish.	Biota unsuitable for human consumption. Reduced food supply Loss of recreational facility (no swimming or boating). Negative impact on human health and wellbeing. Reduced biodiversity. Erosion and destabilization of vegetated banks.
Salinity extremes (hypersaline/hyposaline)	Reduced or no dry-season base flows or drought flows	Die-back of sensitive biota. Changes in species and community composition. Reduced biomass.	Loss in bank buffering capacity and flood control. Flooding of adjacent properties. Reduced aesthetic appeal and real estate value.

* Drought flows are generally assumed at 90% exceedance of mean monthly flows over a 60 to 80 year simulation period, seasonal low flows at 80–50% exceedance during the dry season, and floods greater than 95% exceedance. However, estuary sensitivity to flow can vary outside these ranges depending on estuary size, wave exposure, and catchment runoff.

2. EFlow Studies in South Africa

South Africa's National Water Act (Act 38 of 1998) recognizes water as a national asset and requires sufficient quality and quantity of water to be "reserved" for estuaries to ensure their functioning and health. EFlows must be allocated to meet environmental and basic human needs before water can be abstracted for other uses. In response, the Ecological Reserve Method to determine the Environmental Water Requirements (EWR) of estuaries was devised in 1999 by a core team of estuary specialists [40,41]. The method summarized in Figure 2 incorporates an ecosystem approach that requires an understanding of the effect of changes in river inflow on the abiotic components of an estuary (e.g., hydrodynamics, sediment dynamics, and water quality) and the response of its biotic elements (e.g., microalgae, macrophytes, invertebrates, fishes and birds) [32,40,41].

An estuarine health index is applied to measure the present state from a reference condition taken approximately 120 years ago, before large-scale urban and water resource development occurred in South Africa. The index rates estuaries from 0 to 100 as natural (91–100), largely natural (74–90), moderately modified (75–61), largely modified (60–39), and highly degraded (40–21) to extremely degraded (20–0) [42,43]. Changes in health state provide a measure of how resilient an estuary is to changes in EFlows. EFlows are set to maintain the estuary in the desired condition identified after taking different development and hydrological scenarios into account. The achievement of the desired state depends on the provision of minimum flows from the catchment that will keep the mouth open, flush the water column, scour sediments, and ensure the maintenance of ecosystem services such as the estuary nursery function.

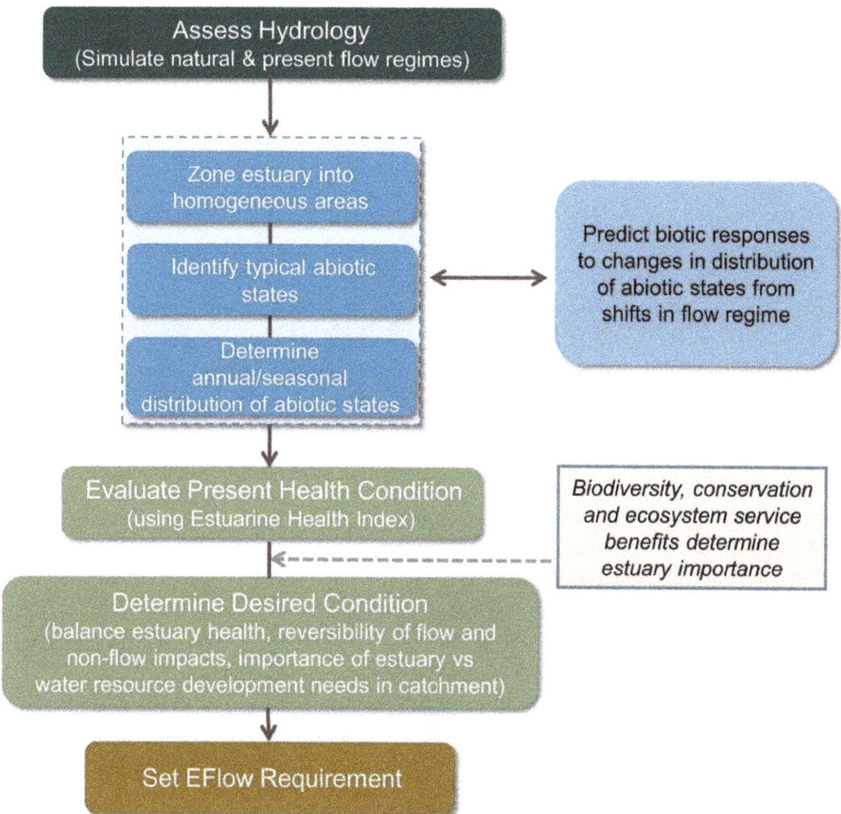

Figure 2. The method used to determine EFlows for estuaries (modified from [32] (Figure 2 Reprinted from Science of the Total Environment, 656, An environmental flow determination method for integrating multiple-scale ecohydrological and complex ecosystem processes in estuaries, 482–494, 2018, with permission from Elsevier)).

The South African EFlow method uses a "top down" approach, which defines freshwater inflows from the perspective of the natural flow regime using a scenario-based approach [28,29]. Because of the lack of measured river inflow data for smaller estuaries, rainfall-runoff models are used to simulate inflows. Either monthly and/or daily-flow time-series are generated with the temporal resolution determined by variability in river flow and its influence on salinity distribution and mouth conditions. However, hydrological data for long time-periods (e.g., 60–80 year simulation periods), which are more representative of South Africa's highly variable runoff regimes, are normally only available for monthly flow data. Drought flows are generally assumed at 90% exceedance of mean monthly flows over the simulation period, seasonal low flows at between 80 and 50% exceedance (with a focus on the dry season), and floods at a greater-than-95% exceedance. Estuary sensitivity to flow can vary depending on estuary size and associated tidal flows, wave exposure, and catchment runoff signal. While some estuaries only close under drought conditions, others may close under dry-season base flows. Correlations between mouth state and measured and/or simulated flows are therefore used to determine an estuary's responses to flow variations and superimpose that on the natural, present, and potential future flow regimes to quantify change (see [31,32] for more information on ecohydrological coupling).

EFlow requirement data for this review were available for 38% of TCEs in South Africa from studies commissioned by the government department responsible for water resource management or by regional catchment management agencies that needed the data to issue new water use licenses. Data were also obtained from assessments undertaken on behalf of local municipal authorities such as eThekweni Municipality that planned to release treated sewerage into the Mdloti and Mhlanga estuaries [44]. The company Richards Bay Minerals commissioned an EFlow study for the Siyaya Estuary in KwaZulu-Natal [45], and the Water Research Commission funded an assessment of the East Kleinemonde Estuary as part of a larger Resource Directed Measures project [46]. Studies that advanced the understanding of EFlows were identified in this review in an effort to document incremental learning. Applying the EFlow requirement method in South Africa has followed a "learning-by-doing" approach. The principles and lessons learnt were identified mainly from five estuary case studies (Great Brak, Uilkraals, Mhlanga, Goukamma, and East Kleinemonde estuaries). The location of these and other estuaries mentioned in the text are shown in Figure 3.

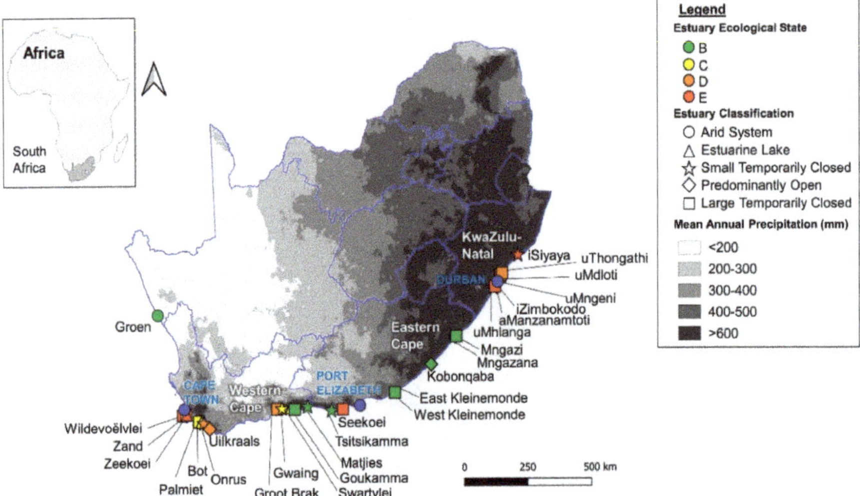

Figure 3. The South African coastline indicating estuaries mentioned in the text.

3. Understanding the Relationship between Freshwater Inflow and Ecosystem Services

This section updates our knowledge on the response of TCEs to changes in freshwater inflow and links these responses to changes to Ecosystem Services. Managing estuaries through EFlows requires that a balance be achieved between using estuary resources and sustaining their ability to continue to deliver goods and services. Table 1 outlines the responses of TCEs to reduced freshwater inflow and closed mouth conditions that influence their abiotic properties and causes changes in the biota, thus ultimately impacting on the provision of ecosystem services.

3.1. Maintaining Intertidal Habitat Is Key to Productivity

Freshwater input helps maintain an open estuary mouth that allows tidal exchange with nearshore marine water and supports the formation of intertidal habitat. South African estuaries experience three dominant hydrodynamic states defined by the state of the mouth, i.e., open, semi-closed and closed [47]. In the semi-closed state, the mouth is shallow and perched (elevated above mean sea level) with a narrow opening that restricts tidal exchange by allowing only a trickle of water out to sea. This leads to the loss of intertidal habitat, such as salt marsh, and bird species that feed in this area (Table 1). When this state persists and little tidal exchange occurs, the estuary becomes "fresh", benefiting biota that prefer fresh to brackish conditions, e.g., the submerged macrophyte *Ruppia cirrhosa* instead of *Zostera capensis*.

The loss of tidal action adversely affects the quantity and availability of intertidal benthic organisms that provide food for wading birds. Since many are Palaearctic migrants, the impact of persistent mouth closures on global bird populations is far reaching [48]. Other effects include the loss of shallow water habitats, favored by herons and flamingos, and the loss of islands that provide roosts and breeding sites safe from terrestrial predators [46]. In response, EFlows are set to maintain intertidal habitats as this provides ecosystem services such as water quality maintenance, erosion control, bank protection, and flood mitigation (Table 1).

3.2. Catchment and Marine Connectivity Sustain Biodiversity, Fisheries, and Genetic Diversity

Mouth closure leads to the loss of land–sea connectivity and nursery habitats, which affects recruitment of invertebrates and fish, and impacts fisheries (Table 1). Some migratory invertebrates, such as penaeid swimming prawns, anomuran mudprawns and mangrove crabs, have an obligatory marine phase in their life cycle and cannot re-enter estuaries under closed conditions [49]. Predominantly adult fish communities were recorded at the Swartvlei Estuary during closed periods because juveniles could not reach nursery areas [50]. Prolonged closure results in low recruitment to estuaries of juvenile marine fish and prevents the migration of adults back to the sea [51,52]. In the Bot Estuary, non-migrating resident fish were dominant after three years of closure [53]. Some species can, however, be recruited into closed systems by larval fish moving over the berm during overwash events, as recorded at East Kleinemonde Estuary, where marine species, e.g., *Rhabdosargus holubi*, were encountered in nursery areas [9,49,54]. During extended closures, fish populations decline rapidly due to predation by other fish, birds, and mammals.

EFlows are designed and set to ensure an estuary mouth is open during key recruitment periods, i.e., spring and summer for most migratory fish. More recently, they have been used to ensure genetic exchange between TCEs by ensuring simultaneous breaching in adjacent systems [55]. Freshwater species with estuarine or marine life-cycle phases, e.g., freshwater prawns and shrimps (*Macrobrachium* spp.), catadromous crabs (*Varuna litterata*) and *Anguillidae* eels, can be cut off from these estuaries if there is inadequate freshwater inflow [32]. Consequently, they depend on a catchment-to-coast management approach for their survival. Coastal connectivity is maintained through freshwater flow to the sea that carries nutrients, detritus, and sediment in the form of productive plumes or fronts. These serve as migration and spawning cues [5,32,56]. Lamberth et al. [56] showed the negative influence of freshwater inflow reduction on commercial line fish catchments documented for 40 km off the east coast of South Africa.

3.3. Duration and Extent of Water Level Fluctuations Act as Biological Resetting Events

When an estuary mouth closes, water levels can build up behind the sand bar and flood nearby habitats, leading to a loss of ecosystem services such as bird watching (Table 1). Macrophyte response to water level changes is summarized in Table 2. The ecosystems approach requires that the relationship between abiotic drivers and biotic responses is central to the EFlow assessment. Succulent salt marsh species are sensitive to prolonged inundation and die back after three months [57–59]. The impact of changes in water level on plants depends on the duration, intensity, and frequency of an event and the phenology of the plant at the time [60]. Large sediment seed reserves ensure macrophyte survival and persistence even after prolonged unfavorable conditions [61]. If evaporation exceeds freshwater inflow, water levels decrease exposing habitats to desiccation. Extensive submerged macrophyte beds (e.g., *Ruppia cirrhosa*) can form during mouth closures but desiccate and die back if permanently exposed; they recover rapidly, however, from a large seed bank [62]. In contrast, low water levels associated with floods and mouth opening act as natural stressors, causing extensive dieback of submerged macrophytes, and boom-and-bust population dynamics in the endemic red-listed fish, *Clinus spatulatus*, has been reported [55]. Submerged macrophyte die-back and fish kills were also reported in the West Kleinemonde Estuary where the water level naturally decreased by 1.65 m in 24 h when the mouth opened; thousands of fish became trapped in the littoral submerged macrophyte beds [63].

Table 2. Response of macrophytes in TCEs to changes in environmental conditions.

Condition	Macrophyte Response
Increase in freshwater inflow ↑ mouth breaching perched estuaries drain ↓ water level ↓ salinity	Submerged macrophytes die back, epiphytes on reeds and sedges become lost. Salt marsh, salt pans and mangroves decline.
Decrease in freshwater inflow ↑ duration mouth closure ↑ water level ↑ salinity ↑ sedimentation	Loss of intertidal salt marsh, reeds, sedges, and mangroves, submerged macrophytes increase. Submerged macrophyte species composition changes rapidly in response to altered salinity. Reed growth increases. Macroalgal and submerged macrophyte growth increase in response to low flow conditions.
Artificial mouth opening ↑ salinity ↑ tidal currents	Reed beds and swamp forest die-back. Submerged macrophytes die-back

Extreme water level fluctuations thus act as biological resetting events, preventing dominance of selected species and temporarily reducing estuary productivity and biodiversity. These events need to be identified, understood, and incorporated into the allocation of EFlows given that the management objective is to maintain biodiversity and the natural estuary functioning as best as possible. Increasing freshwater abstraction can mimic this stress in a system not previously subjected to it, causing disruption of natural processes. Freshwater abstraction caused the Kobonqaba Estuary in the Eastern Cape and Uilkraals Estuary in the Western Cape to close to the sea in 2010, the first time in recorded history [64]. High water levels in the Kobonqaba Estuary flooded the mangroves, causing die-back and tree loss [65] and die-back of salt marsh in the Uilkraals. In future, sea level rises in response to climate change could result in more frequent open mouth states and increased tidal exchange. However, an increase in the frequency and intensity of sea storms would increase sand delivery, thus raising the berm height and causing the estuary mouth to close. Predicted decreases in dry-season base flows and increases in droughts could also cause more frequent mouth closures, flooding, and die-back of salt marsh plants, or low water levels causing die-back of submerged macrophytes. These extreme conditions need to be incorporated into EFlow assessments to ensure natural variability and future resilience to climate change.

3.4. Water Quality Changes Impacts on Nutrient Cycling, Fisheries, and Cultural Values

TCEs are particularly vulnerable to water quality changes during the closed state when longer water residence times allow nutrients to accumulate and primary producers to grow. Micro- and macroalgal blooms have increased in South African estuaries, particularly in TCEs. Besides wastewater input, agricultural return flow also reduces water quality that leads to anoxic conditions and eutrophication. Oxygen extremes and fish kills are common. A recent assessment showed the largest wastewater volumes to be discharged to TCEs compared to other estuary types [66]. Input of sewage effluent is a major problem on the developed east coast around Durban, where this is often the only flow into small estuaries [31]. River flow into the Mdloti Estuary was found not to be a predictor of physico-chemical state as the system is driven by effluent inflows [67]. The minimum discharge standards of municipal effluents are extremely high (21 mg L^{-1} (ca. 1500 μM)) for inorganic nitrogen (DIN) and 10 mg L^{-1} (ca. 320 μM) for phosphorus (DIP) and cause estuary eutrophication [31,66]. Some estuaries (e.g., Wildevoëlvlei and Zeekoe) have transitioned to alternate stable states characterized by toxic cyanobacteria blooms resulting from wastewater inputs and increased residence times [66]. Nutrien assimilation and cycling function of estuaries are essential ecosystem services that need to be protected for estuaries to retain their resilience. Polluted estuaries lose their recreational and aesthetic value (Table 1). In particular, during the closed state, there is an increase in the retention of pollutants and human pathogens. Microbiological indicators used for contact recreation are the abundance of *Enterococci* and *Escherichia coli* [68].

3.5. Salinity Extremes Drive Biological Responses and Can Limit the Provision of Ecosystem Services

Depending on rainfall and salinity conditions, closed estuaries can become hyper- or hyposaline. Along the semi-arid west coast, hypersaline conditions occur regularly during closed phases, whereas hyposaline conditions are common on South Africa's subtropical east coast (Figure 3). However, many east coast estuaries have lost their marine connection to the sea owing to encroachment by infrastructure. EFlow assessments use the duration and extent of salinity penetration as a key indicator to evaluate estuary health state, and where possible, recommend mitigation measures that control salinity regimes to protect estuary biodiversity and productivity. Common to hypersaline estuaries is a low freshwater supply that persists for varying times, relatively long water residence times, high evaporative loss from the estuary basin, and temporary loss of connectivity to the sea [69]. South Africa's arid environment has resulted in a number of case studies that highlight the relationship between flow, salinity regimes, and biotic responses. In the hypersaline Groen Estuary (Figure 3), biotic recruitment from the marine environment follows mouth opening, but species disappear progressively as salinity increases and threshold salinity values are reached. Salinity ranged from 223 in the lower reaches to <10 upstream over a distance of <1 km [70]. In the Seekoei Estuary, fish kills occurred when salinity rose above 90. Mass mortality of estuarine fish is also associated with exposure to low salinity (<6) resulting from extended closed-mouth conditions [71]. The sand prawn (*Callianassa kraussi*) cannot breed in salinity less than 20, suggesting that infrequent mouth openings substantially reduce populations [72]. Dieback of vegetation due to salinity changes results in the destabilization and erosion of banks with potential flooding of surrounding properties (Table 1). Where no site-specific information is available, generic abiotic-biotic (driver-response) relationships are extrapolated to systems that share characteristics in EFlow studies.

4. Principles for the Determination of EFlows for Temporarily Closed Estuaries

Completed EFlow studies in South Africa were critically assessed for lessons learnt and to identify key principles (Table 3). These principles are specific to temporarily closed estuaries, with a bias towards smaller-sized classes by global standards (<500 ha in size). This unique subset of estuaries often requires additional considerations in the determination of EFlows and associated management mitigations, such as the importance of mouth state, and related sensitivity to water quality and water level change. EFlow determination for estuaries lags behind that of river assessments; these principles can be used globally to assist in the setting of EFlows for similar types of estuaries.

4.1. Principle 1: EFlows Are Unique for Every Estuary

The way in which estuaries are influenced by freshwater inflow is the result of flow patterns that occur over weeks or months. Strong longitudinal abiotic gradients develop and change in response to tides and freshwater inflow and, in turn, influence biotic composition and function. The size, shape (bathymetry and topography), and characteristics of the freshwater inflow regime of an estuary determines its EFlow requirements. Each one has unique elements that govern its sensitivity to freshwater inflow modifications. For example, large estuarine lakes experience changes in water level and salinity on annual time scales, whereas smaller closed estuaries experience these on a monthly scale. In general, the smaller an estuary, the more sensitive it will be to a modification in river flow [32]. While low flows are generally associated with closed-mouth conditions and high flows (particularly episodic flood events), with open mouth conditions, actual flow magnitudes are specific to each estuary and have little relevance to others. Table 3 indicates the river flow at which the mouth closes for different estuaries. This is not related to the MAR or the present ecological state of the estuary. The objective of EFlow studies in South Africa is to maintain the natural mouth phase regime of an estuary, whether it be predominantly open or closed most of the time, and thus supports natural functions and ecosystems services. Only in exceptional circumstances, e.g., as a result of historical poor land-use planning or management of effluent and storm water, will deviation from what is natural be deemed an appropriate management response.

Table 3. Key lessons from EFlow case studies that advanced the understanding of the relationship between flow and estuary response.

Great Brak Estuary [73–76]
• Releasing water from dams can supply the required EFlow but floods are still needed to flush the estuary of sediment and organic load (Principles 3, 7).
• Flow released in spring and summer promotes salt marsh growth and fish and mudprawn recruitment from the sea into the estuary (Principle 1).
• Artificial breaching increases sedimentation. Sand berm during closed phase increases over four months to reach maximum levels (Principle 5).
• Need to sample in closed and open phase to determine ecological health as tidal flushing by sea water "masks" pollution and leads to an "optimistic" evaluation of water quality status (Principles 4, 8).
• Macroalgal blooms indicate deterioration in water quality, a relationship not detected in water column nutrient measurements (Principle 8).
• Abiotic and biotic indicators need to be monitored representing a range of trophic groups. "Higher" organisms add cost to a monitoring program, but are good integrators of ecological health (Principle 8).
• Long-term time series are needed to provide information on event-scale responses, e.g., droughts, extended mouth closure, and extreme coastal storms (Principles 8, 10).

Uilkraals Estuary [77]
• Small reduction in baseflow can cause mouth closure (Principle 2).
• Dam construction and capture of low flows caused the mouth to close for the first time; it was previously permanently open (Principle 1).
• Significant inflow is needed to flush open the closed system as the estuary water surface area is large relative to the Mean Annual Runoff (Principle 1).
• Closed mouth conditions lead to nutrient enrichment from catchment inputs (Principle 4).

Mhlanga Estuary [31,44]
• Increased flow from wastewater treatment works increased the frequency of mouth opening and resulted in an unstable system (Principle 6).
• Decline in estuary health was caused by eutrophic conditions and unnatural, frequent mouth breaching. (Principle 6).
• Sand berm increased rapidly over seven days to reach its maximum height as a result of wave exposure along sediment-rich coastline (Principle 1).

Goukamma Estuary [73,78]
• Desktop study found the estuary to be in a good condition but subsequent monthly sampling exposed the negative impact of poor land-use and agricultural practices in the catchment that lead to eutrophic conditions (Principles 8, 9).
• There was a deterioration in water quality owing to surrounding dairy farming (Principle 6).
• Field- and long-term-monitoring data are needed for high-confidence EFlow assessments (Principles 8,9,10).

East Kleinemonde Estuary [58,79,80]
• This is a small estuary usually almost permanently closed and therefore not sensitive to baseflow reduction. Floods play a key role in maintaining/controlling the mouth regime (Principles 1, 3).
• Daily flow data and water level recordings are important to set EFlows. Relationships are established between abiotic drivers and biotic response (Principle 8).
• Salt marsh dies when the mouth closes and water level increases but colonizes rapidly when habitat becomes available (Principle 8).

The wave climate and tidal regime at an estuary's location, and associated cross-shore and longshore sediment transports, are key factors influencing mouth functioning. For example, the temporarily closed Mngazi Estuary drains a much larger catchment area than the predominantly open Mngazana Estuary, but closes to the sea as the mouth is exposed to high wave energy and sediment deposition. On the other hand, a rocky headland at Mngazana Estuary leads to increased turbulence and scouring of the channel, preventing sediment deposition and provides protection from wave action (Figure 4). In addition, the neap tidal volume of the estuary (170,000 m^3) is larger than the Mngazi Estuary (65,000 m^3), indicating that tidal flows assist in keeping the mouth open under higher wave conditions. The median flow (50th percentile) for the Mngazana Estuary is 0.3 m^3 s^{-1} compared to 0.6 m^3 s^{-1} in the Mngazi Estuary. Due to wave refraction and diffraction around the headland at Mngazana mouth, a rip current is often generated, which flows adjacent to the rocky shoreline towards the sea. This current assists in removing sediment from the mouth area. These factors contribute to keeping the Mngazana Estuary permanently open. This shows the need for site-specific data for EFlow assessments and shows that EFlow requirements cannot be extrapolated from one adjacent estuary to the others without considering site-specific variances.

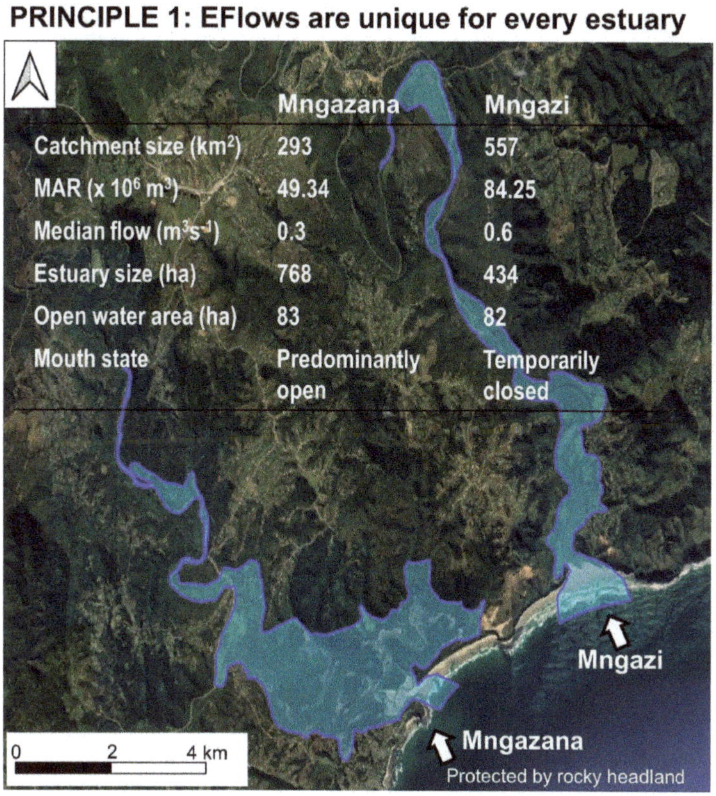

Figure 4. Characteristics of the predominantly open Mngazana Estuary compared to the temporarily closed Mngazi Estuary.

Along high-energy coastlines, the mouths of TCEs can close within days to weeks because flows from relatively small catchments rapidly decrease below the volumes needed to maintain an open mouth, while the mouths of more protected TCEs may only close after a few months. For example, at the larger Great Brak TCE (105 ha), tidal flows through an unrestricted (non-perched) mouth assist in maintaining open mouth conditions. Mouth closure occurs over neap tides and high wave conditions during low flow periods (Table 4). In other small estuaries, e.g., Little Amanzimtoti (10 ha) and Mbokodweni (18 ha), mouth opening and closing often occurs in less than a day during low-flow conditions. Such systems can be open for a few hours at a time depending on the rate at which the sand berm at the mouth is built by the sea. Aided by high wave energy and high volumes of suspended sediment, the height of the sand berm at the Mhlanga Estuary in KwaZulu-Natal (KZN) increased over seven days to reach its maximum [49], while the Great Brak Estuary took up to four months to reach similar levels [73].

Table 4. Relevant data on the percentage Mean Annual Runoff remaining, and flow required to maintain an open mouth in temporarily closed estuaries. Estuary size (ha) is provided as a proxy for tidal flows, while wave exposure is rated from very exposed to protected. Mouth constriction (perched above normal tidal action) is indicated and the resultant annual percentage open mouth conditions are provided as an indication of the interplay between runoff, tidal flows, and wave exposure. These data were extracted for each estuary from reports of the Department of Water Affairs. Estuaries are listed from west to east (Figure 2) and are classified as large and small temporarily closed, as per [8].

Estuary	Present Ecological State	Natural MAR ($\times 10^6$ m^3)	% of Natural MAR to Maintain Present State	River Flow at Which Mouth Closes (m^3 s^{-1})	Estuary size (ha)	Wave Exposure	Mouth Perched above Normal Tidal Action	% Mouth Open
Palmiet (large)	Moderately modified	255	63	0.3–1	33	Medium exposure	Yes	99–75
Onrus (small)	Largely modified	9.6	77	0.015	11	Medium exposure	Yes	50–25
Great Brak (large)	Largely modified	36.79	44	<0.3	105	Exposed	No	50–25
Gwaing (small)	Moderately modified	26.64	75	<0.2	9	Protected	No	99–75
Goukamma (large)	Near Natural	57.5	85	<0.5	18	Very Exposed	No	99–75
Matjies (small)	Near Natural	5.10	84	<0.03	3	Medium exposure	Yes	75–50
Tsitsikamma (small)	Near Natural	19.90	67	<0.05	7	Exposed	Yes	75–50
East Kleinemonde (large)	Near Natural	2.856	96	<0.03	59	Protected	Yes	50–25
Mngazi (large)	Near Natural	84	97	<0.3	17	Very Exposed	No	75–50
Little Amanzimtoti (small)	Highly degraded	2.8	232.5	<0.03	10	Exposed	Yes	75–50
Mbokodweni (large)	Highly degraded	31.5	169.8	<0.2	18	Very Exposed	Yes	99–75
Mhlanga (large)	Largely modified	12.4	158	<0.4	83	Very Exposed	Yes	50–25
Mdloti (large)	Largely modified	98.7	73	<0.3	58	Very Exposed	Yes	50–25
Tongati (large)	Largely modified	70.79	111.9	<0.4	37	Very Exposed	Yes	99–75
Siyaya (small)	Highly degraded	6.5	71	<0.3	10	Very Exposed	Yes	25–0

4.2. Principle 2: A Small Reduction in Baseflow Leads to Mouth Closure

A small reduction in baseflow can cause the mouth of an estuary to close. The relatively large 286 ha Uilkraals Estuary (Figure 5, Table 2) changed from a predominantly open to a temporarily closed estuary following the construction of the Kraaibosch dam 10 km upstream from the mouth. The estuary now receives 70% of its natural MAR [77] and the mouth closes at flows less than 0.03 m^3 s^{-1} (Figure 5). The EFlow study for this estuary recommended that invasive alien plants be removed from the catchment in an effort to restore dry season baseflow. While a relatively low flow rate is needed to keep the Uilkraals mouth open, which is protected from wave energy, much higher flow rates may be needed for estuaries located along high-energy beaches, such as those of the KwaZulu-Natal south coast (Figure 3). Changes in annual flow rates given as a percentage change in mean annual runoff mask the degree to which seasonal flows can be modified. For example, while the MAR may only be modified by 10%, the dry-season baseflows may be altered by as much as 50%, thus stressing the importance of baseflow regimes in EFlow allocations to ensure the flow of ecosystem benefits.

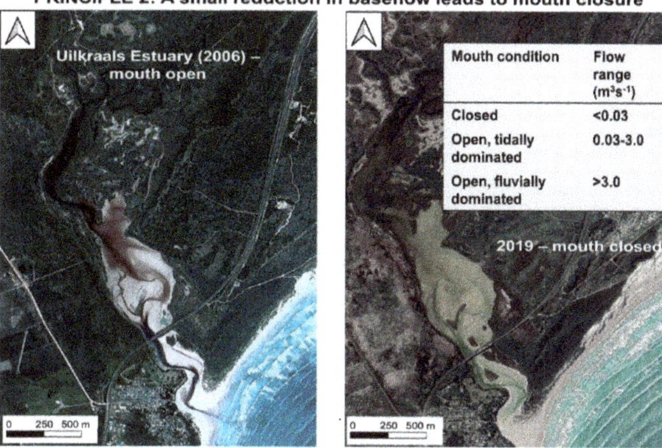

Mouth condition	Flow range ($m^3 s^{-1}$)
Closed	<0.03
Open, tidally dominated	0.03-3.0
Open, fluvially dominated	>3.0

Figure 5. The Uilkraals Estuary open in 2006 (**left**) and closed in 2019 (**right**).

4.3. Principle 3: Floods Flush and Reset Closed Estuaries

Floods are the most important natural means of eroding and transporting sediment from estuaries. Large volumes can be rapidly removed during major floods, with a return period of 1 in 20 years or more. Floods therefore determine the equilibrium between sedimentation and erosion (Figure 6). Smaller floods with return periods of 1–2 years can sometimes also have a significant influence [81]. The annual runoff of South African rivers is highly variable and unpredictable compared with larger Northern Hemisphere systems, and fluctuates between floods and extremely low to zero freshwater inflow [82,83]. Droughts and floods determine the envelope in which estuaries operate [84] and must be included in EFlow assessments. Accumulated sediment is both catchment-derived and transported into estuaries on flood tides by longshore coastal currents. Soil erosion in catchments poses a major threat to estuaries and can cause rapid infilling of small systems [24,85]. In the small Onrus Estuary (10 ha), flows of ~2 m^3 s^{-1} (Oct 2012) resulted in mouth opening but no sediment scouring (Figure 7a). Larger flows of between 4 and 12 m^3 s^{-1} (August to December 2013) were required to remove accumulated sediments and organic matter and to reset the sediment cycle and tidal conditions (Figure 7b).

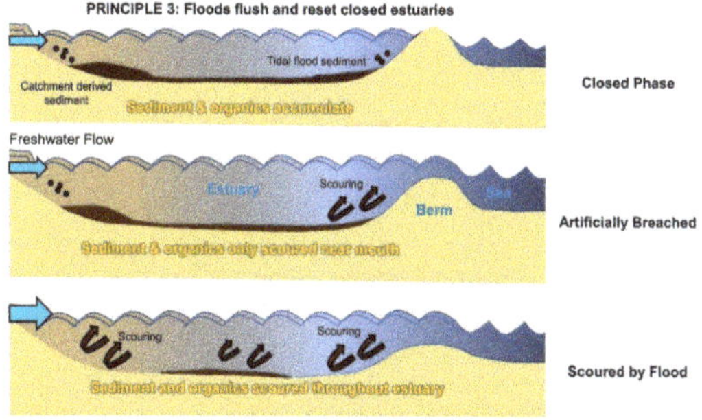

Figure 6. Longitudinal view of a temporarily closed estuary showing the influence of freshwater inflow on the accumulation of sediment and organic matter during closed, artificially breached, and flood-scoured phases.

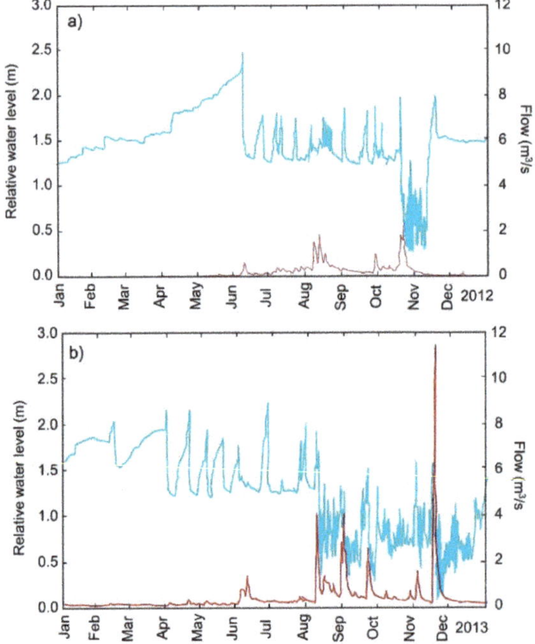

Figure 7. Relative water level (m) (blue line) and freshwater inflow ($m^3\ s^{-1}$) (red line) measured at the Onrus Estuary in 2012 (**a**) and 2013 (**b**).

4.4. Principle 4: Open Mouth Conditions Maintain Good Water Quality

When the mouth of a closed estuary opens and normal tidal exchanges occur, regular flushing improves water quality (Figure 8) [66]. Reduced flow and no tidal exchange leads to higher water retention times and limited mixing that creates eutrophic and polluted conditions, particularly in urbanized freshwater-dominated estuaries. To illustrate these cause-effect relationships, increased freshwater abstraction upriver of the Palmiet Estuary has increased the duration and frequency of closed mouth conditions and has resulted in the development of algal blooms and anoxic conditions, particularly in the stratified, deeper waters [32]. EFlow assessments can prescribe a minimum baseflow to maintain open mouth conditions and resultant tidal flushing [28]. Flow releases and mouth states should always mimic the natural conditions and maintain historically established mouth dynamics. In Southern California, temporarily closed estuaries have been changed to permanently open under the pretext of restoration. This has resulted in changes to habitat types because the closed state is essential to normal estuarine function. For an activity to be restorative, it should follow natural hydrological and sediment dynamic regimes [22].

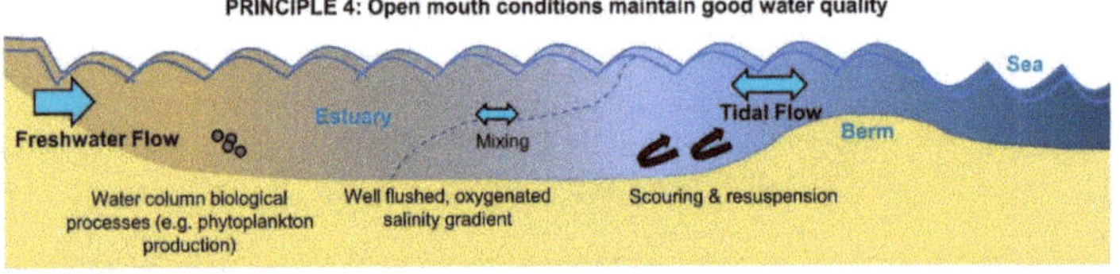

Figure 8. Longitudinal view of a temporarily closed estuary, illustrating how open mouth conditions ensure tidal flushing and good water quality.

In urban estuaries affected by low-lying development and nutrient enrichment, management objectives can also focus on the maintenance of water quality for socio-economic and human health reasons [86]. In these systems, artificial breaching of the mouth to flush out contaminants is seen as a solution to ameliorate poor water quality. The mouth of the Zand Estuary in South Africa has been actively managed to facilitate the penetration of seawater on spring tides, thus keeping the estuary well flushed in the low flow season. Reduced water residence times and elevated salinity limited the occurrence of harmful algal blooms, bacterial contamination, and fish kills [87]. However artificial breaching as a flow mitigation measure without supporting freshwater input from the catchment can result in reduced water column depth and increases stress on resident biota by exposing communities to low oxygen bottom waters that occur in enriched systems [88–90]. Artificial breaching is thus not advocated as a mitigation for poor water quality and this should rather be addressed at source. Restoration of morphological complexity including green infrastructure such as wetlands will help retain sediment and nutrient inputs, thus mitigating adverse water quality. In addition, artificial breaching of an estuary with poor water quality can pose a risk to near shore marine ecology and users (e.g., recreational beaches or aquaculture facilities). This aspect is less relevant for small estuaries (with small volumes) discharging in high energy coastal environments that assist with dilution of pollutants.

4.5. Principle 5: Artificial Breaching Causes Sedimentation

Artificial breaching as a flow mitigation measure needs to be practiced with caution because, besides the adverse ecological effects, it leads to sedimentation and thus reduced flushing in an estuary (Figure 9). Estuaries are frequently breached to lower water levels and reduce floods risks to low-lying development. Artificial breaching is usually advocated when an estuary is closed for a prolonged period. Flooding of the causeway for extended periods at the Seekoei Estuary, which has seen substantial flow reduction, provided the motivation for artificial breaching. However, its estuary management plan indicates that mouth management only would not restore estuary function, and freshwater inflow is essential to maintain estuary health [91].

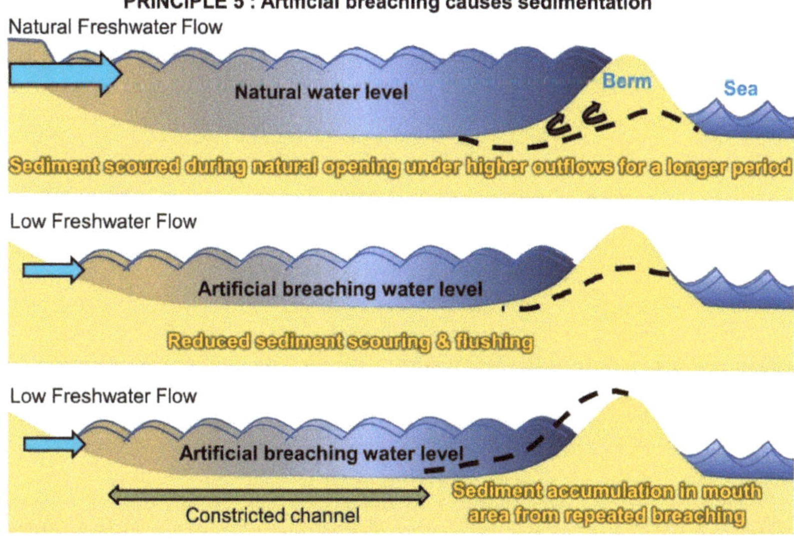

Figure 9. Longitudinal view of a temporarily closed estuary showing the influence of artificial breaching on sediment scouring and flushing.

When breaching occurs at natural levels (often 1–2 m higher than at present), a large volume of water flows out to sea over an extended period which, in turn, scours sediment from the lower and middle reaches of the estuary (Figure 9). The sediment flushing potential increases exponentially with increasing outflow velocities. When breaching takes place at lower water levels, flushing intensity is reduced and sediment accumulates. This practice also disrupts normal deposition-erosion cycles and reduces the scouring effectiveness of subsequent breaching. Consequently, the berm widens and estuary channels constrict [64].

Environmental Impact Assessment legislation in South Africa makes provision for the formulation of a Mouth Management Plan (captured in an operational Maintenance Management Plan) that enables artificial breaching to be based on sound scientific knowledge of estuary function. Timing of breaching must take ecological processes, such as fish and invertebrate recruitment, into account. Artificial breaching practices are not advocated as a mitigation measure for freshwater inflow reduction as it comes with its own long-term impacts, but where it is an established practice, it should be integrated into EFlow studies and freshwater allocated to support the mitigation where needed.

4.6. Principle 6: Wastewater Input and Agricultural Return Flow Can Cause Unstable Conditions

Increased inflow volumes from wastewater treatment works can be detrimental because increased baseflows and higher water levels cause more frequent mouth openings in TCEs (Figure 10a) causing ecological instability, low biotic diversity, and biomass. At the Mhlanga Estuary, sewage discharge and growing urbanization have resulted in freshwater inflow volumes increasing by 158% to a present MAR of 19.6×10^5 m^3 [92]. Consequently, the estuary has changed from being predominantly closed to predominantly open. For example, between March 2002 and March 2003, the mouth opened 23 times and the ecosystem became unstable with reduced microalgal biomass and low biodiversity [49]. Because the estuary is perched above mean sea level, it drains when the mouth opens. Water levels fall and much of the floodplain becomes permanently exposed, losing important invertebrate and fish habitats. Reed beds dry out and periphyton, which forms the base of the food chain, no longer grows. Limited reed productivity reduces detrital input to the ecosystem, affecting the entire food chain, especially detritivore fish [93]. In some smaller closed estuaries, such the Tongati (37 ha), input from the wastewater treatment works in the dry season exceeds the minimum flow by approximately 300% and keeps the mouth open. Eutrophic freshwater conditions that develop in small closed KwaZulu-Natal systems promote colonization by invasive aquatic plants such as water hyacinth [31,66]. EFlow assessments that aim to improve estuary ecological status will focus on returning the flow regime as close as possible back to natural, which may involve diversion or reuse of wastewater.

4.7. Principle 7: Water Released from Dams to Supplement the EFlow Cannot Replace the Natural Flow Regime

Releasing water from upstream dams to keep an estuary mouth open can be viewed as a mitigation strategy to reduce the impact of reduced freshwater inflow (Figure 10b). There is little published on this as dam releases are mostly used as a mitigation measure in river ecosystems. In South Africa, freshwater releases from the Wolwedans Dam upstream of the Great Brak Estuary facilitated the development of an in-depth understanding of estuarine EFlow requirements as an early case study (Table 2). Thirty years of monitoring and adaptive management involving scientists, government departments, and citizens refined the flow releases and subsequent management practices [81]. Planned releases in spring and summer allowed the mouth to remain open and improved its ecological health by supporting salt marsh growth, fish, and mudprawn recruitment [73]. However, while the dam releases maintained open mouth conditions, they did not ensure sustained good water quality. When closed, decomposition of residual organic material caused oxygen stress to fish, and the development of macroalgal blooms indicated deteriorating water quality that nutrient measurements alone did not reveal [38,73–76]. Floods are therefore needed to flush out organic material accumulated during the closed period. In addition, natural flood patterns act as a queuing factor to attract juvenile fish and invertebrates to the estuary as nursery habitats. Most South African dams were not, however, designed for flood releases and cannot meet this EFlow requirement to support critical ecosystem services.

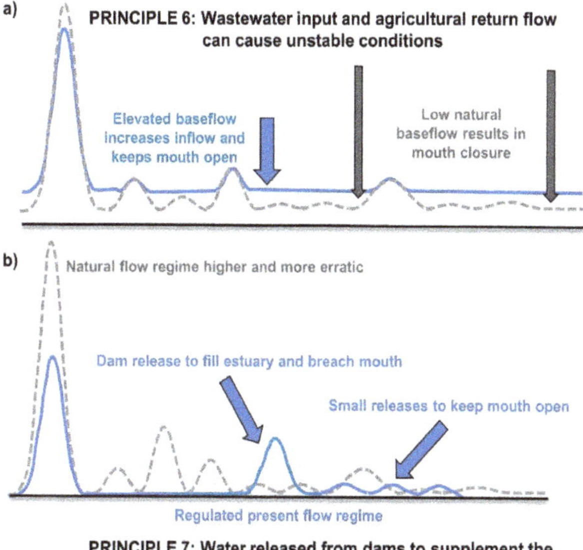

Figure 10. Graphics comparing (**a**) natural and elevated baseflows, highlighting their effect on estuary mouth condition, and (**b**) flow patterns from natural flow regimes and those of regulated dam releases.

4.8. Principle 8: Field and Long-Term Data Are Needed for High-Confidence EFlow Assessments

The availability of high-confidence data influences the reliability of EFlow study outcomes. Funds are often too limited to support the acquisition of sufficient detail, which leads to low-confidence results [32]. Confidence in hydrology data are particularly important. Essential relationships between flow, hydrodynamics, sediment dynamics, water quality, and biotic responses can only be determined from field data collected concurrently. Discipline-orientated data sets are often unrelated to influencing environmental factors [41]. Long-term water level data and mouth state observations allow for the correlation between mouth state and specific flow ranges in TCEs. For high-confidence studies and accurate assessments of estuary health, closed systems must also be sampled in the open and closed mouth states (Table 2).

Although the Great Brak Estuary was monitored for 12 years, an accurate assessment of ecological health was only achieved once the "higher trophic level" biotic components, particularly fish, were included [73]. The Goukamma Estuary, situated in a scenic protected area, was thought to be in good condition until regular sampling of the water column exposed eutrophic and anoxic conditions in highly stratified bottom waters caused by inputs from surrounding dairy farms ([78], Table 3). Artificial breaching of an estuary mouth can confuse the understanding of the relationship between freshwater inflow and mouth condition unless there are accurate long-term records. Field and long-term monitoring data are thus especially needed at modified systems to assess the environmental flow requirements of estuaries. Figure 11 (modified from [55]) indicates an excellent long-term data set for water level, fish abundance, and mouth state that was used to identify status and trends needed to set high-confidence EFlows.

PRINCIPLE 8: Field and long term data are needed for high confidence EFlow assessments

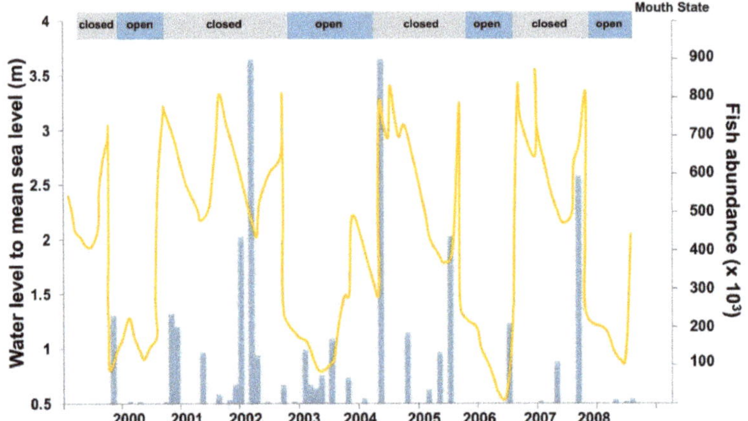

Figure 11. Graphic illustration of the long-term data for water levels (yellow line), fish abundance (blue bars), and mouth state used to identify trends and patterns on which high-confidence EFlow assessments were based (modified from [55]).

The EFlows method used in South Africa analyses changes in runoff on a monthly time scale. This scale is inadequate for small estuaries as they are highly variable and respond to daily flows [32]. Reliable EFlow requirement estimations sometimes need flow data on a daily basis to quantify and incorporate high variability. The installation of flow gauges and water level recorders at TCEs enable important relationships between abiotic drivers and biotic responses to be measured and incorporated. Such long-term monitoring data were used, for example, to link macrophyte response to changing mouth conditions in the unmodified East Kleinemonde Estuary (Table 4). Established flow-response relationships for critical ecosystem services can, in turn, be extrapolated to similar systems where observational data are absent, thus increasing confidence in EFlow assessments.

4.9. Principle 9: Monitoring Must Take Place in a Strategic Adaptive Management Cycle

Most EFlow studies emphasize the need for detailed resource monitoring such as water level recordings, continuous gauging of river inflow, mouth observations, longitudinal and vertical salinity distributions, and monitoring of nutrient concentrations and selected biotic components. Table 5 shows the data required for an EFlow assessment (Figure 2). Resource monitoring assesses whether the allocated EFlow is achieving the desired state (health category) and associated ecosystem services (as captured in ecological condition categories). Each EFlow assessment identifies the monitoring required for selected abiotic and biotic components and specifies the temporal and spatial scale required for interpretation. Optimal frequency and duration of measurement is indicated where appropriate (Table 5), however, because of the variability of estuaries' high-confidence studies would benefit from continuous monitoring and frequent measurements. This mostly does not take place because of limited resources (funding and human).

Implementation of EFlows and compliance monitoring must take place in an adaptive management cycle as natural systems can behave in complex manners and resource use changes over time, thus requiring modifications to EFlow allocations to achieve management objectives (Figure 12a). This "learning-by-doing" approach is well documented in ecological monitoring programs [94]. Finally, the national department responsible for water initiated a National Estuary Monitoring Programme in South Africa where the first tier focused on abiotic parameters and the second tier monitored biotic parameters [95]. Significant funding constraints has however halted progress with this program. This emphasizes the need to secure high-level government support and to embed monitoring in regional or national budgets.

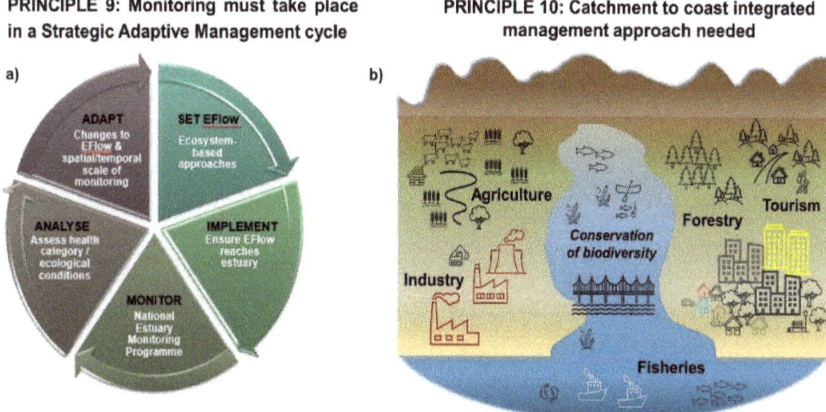

Figure 12. (**a**) The Strategic Adaptive Management cycle for the implementation of EFlows and compliance monitoring, and (**b**) Sectors involved in a catchment-to-coast integrated management approach.

4.10. Principle 10: Catchment to Coast Integrated Management Approach Needed

Small catchments mean TCEs are sensitive to local land use changes. This necessitates a holistic catchment-to-coast management approach that requires collaboration across government departments and sectors (Figure 12b). The Integrated Water Resource Management (IWRM) process captures some of this complexity and EFlows are an important component [30]. Estuaries are subjected to multiple pressures, e.g., artificial breaching and biological invasions [7], and need to be assessed as part of an estuary EFlow assessment to ensure that desired state objectives are achievable. Co-operative governance can address non-flow related impacts. For example, fishing during the closed mouth state in a small estuary can constitute "mining" of estuarine-dependent species. This requires fisheries management interventions to ensure stock recovery. Implementing EFlows is an important component of estuary restoration and a socio-ecological systems approach is necessary to achieve estuary health, ecosystem services, and human wellbeing objectives [86]. The Emerald Growth concept of Tagliapietra et al. [96] presents a new framework for managing ecological quality by analyzing socio-economic growth and human well-being as part of ecosystem services of transitional waters. The determination of EFlow allocations should be a cross-sectorial endeavor because reduced flows to estuaries have effects in sectors such as fisheries, tourism, and biodiversity [31]. Many of these sectors are regarded as "just another stakeholder" in water management instead of becoming an integral part of steering or oversight committees leading EFlow processes. Integration would allow broad discussions across all sectors, not only from a water supply perspective. In South African EFlow studies, international obligations (e.g., Ramsar, Convention on Biological Diversity) are often not achieved owing to an under-representation of sectors in decision-making. In order to ensure a sustainable flow of ecosystem services, EFlows should thus not be set in isolation from other sector objectives.

Table 5. Data and monitoring pre-requisites for EFlow studies (modified from Taljaard et al. 2003, [97]).

Component	Baseline Data and Monitoring Requirements for High-Confidence EFlow Studies
Hydrology	Primary catchment delineation.
	Measured rainfall data in the catchment (or a representative adjacent catchment) for 50 years.
	Hydrological parameters (evaporation rates, radiation rates, soil permeability, catchment land use) based on long-term studies.
	Measured river inflow data (gauging station) at the head of the estuary over 5–20 years.
	Record of flow reduction activities (e.g., abstraction, impoundment) and flow enhancement activities (e.g., discharges, transfer schemes).
	Flood hydrographs (hourly measurements at calibrated weir) for a range of flood size classes and preferably for a minimum of 50 years.
Bathymetry	Bathymetric/topographical surveys including berm height, cross sections at 10–50 m in the mouth region, cross section profiles at 500 to 1000 m intervals upstream of the mouth, and floodplain topography preferably done with Lidar. Repeated every 3 years and after episodic events like floods to record changes and erosion/deposition cycles.
Hydrodynamics	Continuous water level recordings near estuary mouth (but away from wave action). Minimum period of 5 to 20 years depending on frequency of breaching.
	Water level recordings at 2 to 6 stations along the length of the estuary over a spring and a neap tidal cycle (14 days). Data are used for model calibration. In large complex estuaries recorders to be left for ongoing monitoring.
	Daily mouth state observations (open/closed/overtopping), particularly those with a semi-closed mouth state. Minimum period of 5 to 20 years depending on frequency of breaching.
	Stationary camera observations of mouth behavior. Minimum period of 5 to 20 years depending on frequency of breaching.
	High resolution historical aerial photography and satellite imagery (1:10,000 scale), preferably rectified going as far back as historically possible (e.g., in South Africa 1920s). Old maps, anecdotal information and farm records are also used to determine natural process regimes.
	Data on wave conditions near mouth (minimum period of 5 to 20 years).
Sediments	Sediment samples collected along length of entire estuary at 500 to 1000 m intervals using a Van Veen or a Zabalocki-type Eckman grab (to characterize recent sediment movement) for particle size analyses. Samples preferably collected at 3-year intervals and after floods.
	Sediment core samples collected using a corer (for historical sediment characterization) at intervals similar to cross-section profiles (see bathymetry) or where considered appropriate by sediment specialist. Collected at 3-to-6-year intervals and after floods.
	Long-term suspended sediment load near head of estuary over 5–15 years (including detritus component—particulate carbon/loss on ignition), needed to compile sediment transport curve for river inflow.

Table 5. *Cont.*

Water quality	Quarterly/monthly longitudinal profiles (salinity, temperature, oxygen, pH, turbidity, nutrients) along the length of the estuary collected over a spring and neap tide during high and low tide with a focus on the end of low flow season and the peak of high flow season. Preferably for 5 to 20 years.
	If possible, long-term stationary in situ salinity and temperature loggers (minimum wet-dry cycle) placed in the lower, middle, and upper estuary reaches. Minimum period of 5 to 20 years depending on frequency of breaching.
	Water quality (e.g., system variables, nutrients, and toxic substances) measurements for river water entering at the head of the estuary and for nearshore seawater. Minimum period of 5 to 20 years, but ongoing preferable.
	Measurements of organic content and toxic substances (e.g., trace metals and hydrocarbons) in sediments along the length of the estuary for urban estuaries (once off).
Microalgae	Quarterly data on relative abundance of dominant phytoplankton groups, i.e., flagellates, dinoflagellates, diatoms, and blue-green algae, during typical high and low flow conditions preferably for a series of years, thereafter every 3 years.
	Quarterly chlorophyll-a measurements taken at the surface, 0.5 m and 1 m depths, under typically high and low flow conditions, preferably for a series of years, thereafter every 3 years.
	Quarterly intertidal and subtidal benthic chlorophyll-a measurements preferably for a series of years, thereafter every 3 years.
Macrophytes	Aerial photographs or high resolution satellite imagery of the estuary (ideally 1:5000 scale) reflecting the present state and reference condition (earliest year available). A GIS vegetation map indicating the present and reference condition distribution of the different macrophyte habitat types (e.g., salt marsh, mangroves, reeds, and sedges) to be repeated every 3 years.
	Number of macrophyte habitat types, identification and total number of macrophyte species, number of rare or endangered species, or those with limited populations documented during a field visit. The extent of anthropogenic impacts (e.g., trampling, mining) must be noted. To be repeated every 3 years.
	Permanent transects (fixed monitoring stations that can be used to measure change in salt marsh in response to changes in salinity and inundation patterns) set up along an elevation gradient. Measurements of percentage cover of each plant species in duplicate quadrats (1 m^2). Measurements of sediment salinity, water content, depth to water table and water table salinity. These data should be available for low- and high-flow periods and repeated every 3 years.

Table 5. *Cont.*

Invertebrates	Species and abundance of zooplankton, based on quarterly samples collected across the estuary at each of a series of stations along the estuary. Sampling stations must be representative of the substrate and salinity zones typical of a particular estuary, e.g., 0–10, 10–20, 20–30, >30.
	Benthic invertebrate species and abundance, based on subtidal grab samples and intertidal core samples at a series of stations up the estuary, and pump sampling or counts of hole densities. Sampling stations must be representative of the substrate and salinity zones characteristic of the estuary.
	Macrocrustacean species and abundance based on sampling at each station using a benthic sled with flow meter, prawn/crab traps, and appropriate gear for shoreline.
	Measures of sediment characteristics at each station. In situ water quality variables need to be collected at time of sampling (e.g., salinity, temperature, oxygen). These data should be available for summer and winter seasons of the year, or for low- and high-flow periods every 3 years.
Fish	Species and abundance data of fish, based on quarterly seine net and gill net sampling, with less gill net than seine samples. In small estuaries, these nets should only be used in the mouth, middle, and upper reaches. Sampling stations must be representative of the salinity zones characteristic of the particular estuary, i.e., 0–10, 10–20, 20–30, >30 (at least one station should be in this range). Within each salinity zone, representative habitats need to be sampled, such as submerged macrophytes, prawn beds, sand flats (representing different food sources). In situ water quality variables need to be collected at time of sampling (e.g., salinity, temperature, oxygen). The number of seine net stations in a small estuary (<5 km long) should not be <5, distributed along the length of the estuary. For larger estuaries, 10 to 15 seine net stations selected geographically along the entire length of the estuary. A rough estimate for setting the distance between stations is to divide the length of the estuary by 10 (i.e., if an estuary is 30 km long, the distance between stations should be about 3 km). These data should be available for four seasons of the year, or for low- and high-flow periods in a series of years. To be repeated every 3 years.
Birds	One year of monthly counts of all water-associated birds, by species, for the whole estuary, preferably separated into counting areas and/or a series of at least 10 years of summer and winter counts, in addition to historical data on the same.

5. Conclusions and Recommendations

Increasing human pressures on estuaries highlights the need for effective EFlow assessments for these systems. TCEs are threatened by relatively small levels of water abstraction and related changes in mouth condition. Similarly, small changes in water quality in high retention systems can lead to a rapid degradation in overall condition. This is further exacerbated by poor land-use, over exploitation of living resources, and artificial breaching. EFlows should always be assessed in the context of natural estuary functioning, present management actions, and restoration activities needed to maintain or improve conditions.

This article highlights 10 principles that underpin the determination of EFlow requirements (Table 6). Principles 1–4 address recommendations for EFlows, namely the importance of base flows, floods, and open mouth conditions and that each estuary is considered unique as EFlow requirements do not transfer well between systems. Principles 5–7 address some of the pressures such as additional flows, artificial breaching, and water releases from dams. These are mitigation and management actions that need to be addressed in EFlow studies. Principles 8–10 consider monitoring and management recommendations relating to long term monitoring, strategic adaptive management, and an integrated catchment-to-coast approach. These 10 principles can be adapted to TCEs in climatic settings similar to those in South Africa. Globally, studies have incorporated social, economic, and cultural issues into integrated water resources management frameworks. The emerging frontier of EFlows is an essential strategy that links the maintenance of ecological integrity with social wellbeing.

Table 6. Summary of the 10 principles to determine the EFlows for temporarily closed estuaries.

Principle 1	EFlows are unique for every estuary
Principle 2	A small reduction in baseflow leads to mouth closure
Principle 3	Floods flush and reset closed estuaries
Principle 4	Open mouth conditions maintain good water quality
Principle 5	Artificial breaching causes sedimentation
Principle 6	Wastewater input and agricultural return flows can cause unstable conditions
Principle 7	Water released from dams to supplement the EFlow cannot replace the natural flow regime
Principle 8	Field and long-term data are needed for high-confidence EFlow assessments
Principle 9	Monitoring must take place in a strategic adaptive management cycle
Principle 10	Catchment to coast integrated management approach needed

The 10 key principles identified from evaluating several completed studies can be used to guide the allocation of estuarine EFlows. Although EFlow implementation is often slow, there have been successes. EFlows have, for example, prevented the issuing of water use licenses in stressed catchments and in ecologically sensitive estuaries where EFlow requirements are high. They have guided plans to divert sewage effluent in an effort to restore its health. Water demand and/or the need for wastewater disposal in urban areas, however, means that the health of some systems will never be improved. The "learning-by-doing" approach would benefit from a globally based comparative assessment of successes and failures in implementing estuary EFlows so that lessons learnt can be shared.

Challenges facing the implementation of EFlows include those related to a changing climate, which will shift baselines for estuary management and prompt a precautionary approach to development. Implementation of EFlow requirements is necessary to improve the resilience of estuarine ecosystems. Essential elements of climate-related change need to be integrated into estuary management plans and governance policies.

Author Contributions: Conceptualization, J.B.A. and L.V.N.; Data curation, J.B.A. and L.V.N.; Funding acquisition, J.B.A. and L.V.N.; Writing—original draft, J.B.A. and L.V.N.; Writing—review & editing, J.B.A. and L.V.N. All authors have read and agreed to the published version of the manuscript.

Funding: The Water Research Commission, Department of Water and Sanitation, and the National Research Foundation (NRF) have all funded aspects of EFlow studies on estuaries. The National Research Foundation of South Africa through the support of the Department of Science and Technology (DST)/NRF Research Chair in Shallow Water Ecosystems (UID 84375) funded the first author. The second author was supported by a DST/CSIR Parliamentary grant.

Acknowledgments: Stephen Lamberth, Department of Forestry, Fisheries, and Environment is thanked for his input on biotic processes and responses. Carla Dodd is thanked for assistance with the figures and Lesley McGwynne provided editorial input. Teddy Hiscock provided excellent support during completion of this ten pretty principles paper in lockdown. Three anonymous reviewers are thanked for their useful input that improved the final version.

Conflicts of Interest: The authors declare no conflict of interest. The funders had no role in the design of the study; in the collection, analyses, or interpretation of data; in the writing of the manuscript, or in the decision to publish the results.

Ethics Statement: No human or animal samples were involved in this study. We have accurately presented our research findings and objectively discussed the significance of the findings.

References

1. Whitfield, A.K.; Wooldridge, T. Changes in Freshwater Supplies to Southern African Estuaries: Some Theoretical and Practical Considerations. In *Changes in Fluxes in Estuaries: Implications from Science to Management*; Dyer, K., Orth, R., Eds.; Olsen & Olsen: Fredensborg, Denmark, 1994; pp. 41–50.
2. Bate, G.; Adams, J.B. The Effects of a Single Artificial Freshwater Release into the Kromme Estuary. 1: Overview and Interpretation for the Future. *Water SA* **2000**, *26*, 329–332.
3. Alber, M. A Conceptual Model of Estuarine Freshwater Inflow Management. *Estuar. Coasts* **2002**, *25*, 1246–1261. [CrossRef]
4. Estevez, E. Review and Assessment of Biotic Variables and Analytical Methods Used in Estuarine Inflow Studies. *Estuaries* **2002**, *25*, 1291–1303. [CrossRef]
5. Gillanders, B.; Kingsford, M. Impact of Changes in Flow of Freshwater on Estuarine and Open Coastal Habitats and the Associated Organisms. In *Oceanography and Marine Biology: An Annual Review*; Gibson, R., Barnes, M., Atkinson, R., Eds.; Taylor & Francis: New York, NY, USA, 2002; Volume 40, pp. 233–309.
6. Fohrer, N.; Chicharo, L. Interaction of River Basins and Coastal Waters—An Integrated Ecohydrological View. In *Treatise on Estuarine and Coastal Science*; McLusky, D., Wolanski, E., Eds.; Elsevier: Amsterdam, The Netherlands, 2011; pp. 109–150.
7. Van Niekerk, L.; Adams, J.B.; Bate, G.; Forbes, N.; Forbes, A.; Huizinga, P.; Lamberth, S.; MacKay, F.; Petersen, C.; Taljaard, S.; et al. Country-Wide Assessment of Estuary Health: An Approach for Integrating Pressures and Ecosystem Response in a Data Limited Environment. *Estuar. Coast. Shelf Sci.* **2013**, *130*, 239–251. [CrossRef]
8. Van Niekerk, L.; Adams, J.B.; James, N.; Lamberth, S.; MacKay, C.; Turpie, J.K.; Rajkaran, A.; Weerts, S.; Whitfield, A.K. An Estuary Ecosystem Classification That Encompasses Biogeography and a High Diversity of Types in Support of Protection and Management. *Afr. J. Aquat. Sci.* **2020**, *45*, 199–216. [CrossRef]
9. Whitfield, A.K.; Bate, G. *A Review of Information on Temporarily Open/Closed Estuaries in the Warm and Cool Temperate Biogeographic Regions of South Africa, with Particular Emphasis on the Influence of River Flow on These Systems*; WRC Report No. 1581/1/07; Water Research Commission: Pretoria, South Africa, 2007.
10. Whitfield, A.K. A Characterization of Southern African Estuarine Systems. *S. Afr. J. Aquat. Sci.* **1992**, *18*, 89–103.
11. Taljaard, S.; Van Niekerk, L.; Joubert, W. Extension of a Qualitative Model on Nutrient Cycling and Transformation to Include Microtidal Estuaries on Wave-Dominated Coasts: Southern Hemisphere Perspective. *Estuar. Coast. Shelf Sci.* **2009**, *85*, 407–421. [CrossRef]
12. McSweeney, S.; Kennedy, D.; Rutherford, I.; Stout, J. Intermittently Closed/Open Lakes and Lagoons: Their Global Distribution and Boundary Conditions. *Geomorphology* **2017**, *292*, 142–152. [CrossRef]
13. Hadwen, W.L.; Arthington, A. *Ecology, Threats and Management Options for Small Estuaries and ICOLLS*; Technical Report; Sustainable Tourism Cooperative Research Center: Queensland, Australia, 2006.
14. Whitfield, A.K.; Elliot, M. Ecosystem and Biotic Classification of Estuaries and Coasts. In *Treatise on Estuarine and Coastal Science 2*; McLusky, D., Wolanski, E., Eds.; Elsevier: Amsterdam, The Netherlands, 2011.
15. Perissinotto, R.; Stretch, D.; Whitfield, A.K.; Adams, J.B.; Forbes, A.; Demetriades, N. Ecosystem Functioning of Temporarily Open/Closed Estuaries in South Africa. In *Estuaries: Types, Movement Patterns and Climatical Impacts*; Crane, J., Solomon, A., Eds.; Nova Science Publishers: New York, NY, USA, 2010; pp. 1–69.
16. Tweedley, J.R.; Warwick, R.M.; Hallett, C.S.; Potter, I.C. Fish-Based Indicators of Estuarine Condition That Do Not Require Reference Data. *Estuar. Coast. Shelf Sci.* **2017**, *191*, 209–220. [CrossRef]
17. Scanes, E.; Scanes, P. Climate Change Rapidly Warms and Acidifies Australian Estuaries. *Nat. Commun.* **2020**, *11*, 1803. [CrossRef]
18. Chávez, V.; Mendoza, E.; Ramírez, E.; Silva, R. Impact of Inlet Management on the Resilience of a Coastal Lagoon: La Mancha, Veracruz, Mexico. *J. Coast. Res.* **2017**, *77*, 51–61. [CrossRef]

19. Gale, E.; Pattiaratchi, C.; Ranasinghe, R. Processes Driving Circulation, Exchange and Flushing within Intermittently Closing and Opening Lakes and Lagoons. *Mar. Freshw. Res.* **2007**, *58*, 709–719. [CrossRef]

20. Conde, D.; Solari, S.; De Àlava, D.; Rodríguez-Gallego, L.; Verrastro, N.; Chreties, C.; Lagos, X.; Piñeiro, G.; Teixeira, L.; Seijo, L.; et al. Ecological and Social Basis for the Development of a Sand Barrier Breaching Model in Laguna de Rocha, Uruguay. *Estuar. Coast. Shelf Sci.* **2019**, *219*, 300–316. [CrossRef]

21. Behrens, D.K.; Bombardelli, F.A.; Largier, J.L.; Twohy, E. Episodic closure of the tidal inlet at the mouth of the Russian River—A small bar-built estuary in California. *Geomorphology* **2013**, *189*, 66–80. [CrossRef]

22. Jacobs, D.; Stein, E. *Classification of California Estuaries Based on Natural Closure Patterns: Templates for Restoration and Management*; Southern California Coastal Water Research Project: Costa Mesa, CA, USA, 2011; Volume 619.

23. Whitfield, A.K.; Bate, G.C.; Adams, J.B.; Cowley, P.; Froneman, P.; Gama, P.; Strydom, N.A.; Taljaard, S.; Theron, A.; Turpie, J.K.; et al. A Review of the Ecology and Management of Temporarily Open/Closed Estuaries in South Africa, with Particular Emphasis on River Flow and Mouth State as Primary Drivers of These Systems. *Afr. J. Mar. Sci.* **2012**, *34*, 163–180. [CrossRef]

24. Roper, T.; Creese, B.; Scanes, P.; Stephens, K.; Williams, R.; Dela-Cruz, J.; Coade, G.; Coates, B.; Fraser, M. *Assessing the Condition of Estuaries and Coastal Lake Ecosystems in NSW. Monitoring, Evaluation and Reporting Program*; NSW Office of Environment and Heritage: Sydney, Australia, 2011.

25. Behrens, D.K.; Brennan, M.; Battalio, P.E. A quantified conceptual model of inlet morphology and associated lagoon hydrology. *Shore Beach* **2015**, *83*, 33–42.

26. Cooper, J.A.G. Geomorphological Variability among Microtidal Estuaries from the Wave-Dominated South African Coast. *Geomorphology* **2001**, *40*, 99–122. [CrossRef]

27. DWS (Department of Water and Sanitation). Determination of Ecological Water Requirements for Surface Water (River, Estuaries and Wetlands) and Groundwater in the Lower Orange WMA. Buffels, Swartlintjies, Spoeg, Groen and Sout Estuaries Ecological Water Requirement. Available online: http://www.dwa.gov.za/rdm/currentstudies/doc/lowerorange/Buffels,%20Swartlientjies%20Spoeg,%20Groen%20and%20Sout%20EWR%20Report.pdf (accessed on 7 July 2020).

28. Arthington, A.; Bhaduri, A.; Bunn, S.; Jackson, S.; Tharme, R.; Tickner, D.; Young, B.; Acreman, M.; Baker, N.; Capon, S.; et al. The Brisbane Declaration and Global Action Agenda on Environmental Flows. *Front. Environ. Sci.* **2018**, *6*, 45. [CrossRef]

29. Arthington, A.; Bunn, S.; Poff, N.; Naiman, R. The Challenge of Providing Environmental Flow Rules to Sustain River Ecosystems. *Ecol. Appl.* **2006**, *16*, 1311–1318. [CrossRef]

30. Brown, C.; Campher, D.; King, J. Status and Trends in EFlows in Southern Africa. *Nat. Resour. Forum* **2020**, *44*, 66–88. [CrossRef]

31. Van Niekerk, L.; Adams, J.B.; Allen, D.; Taljaard, S.; Weerts, S.; Louw, D.; Talanda, C.; Van Rooyen, P. Assessing and Planning Future Estuarine Resource Use: A Scenario-Based Regional Scale Freshwater Allocation Approach. *Sci. Total Environ.* **2019**, *657*, 1000–1013. [CrossRef] [PubMed]

32. Van Niekerk, L.; Taljaard, S.; Adams, J.B.; Lamberth, S.; Huizinga, P.; Turpie, J.K.; Wooldridge, T. An Environmental Flow Determination Method for Integrating Multiple-Scale Ecohydrological and Complex Ecosystem Processes in Estuaries. *Sci. Total Environ.* **2019**, *656*, 482–494. [CrossRef]

33. Adams, J.B. *Determination and Implementation of Environmental Water Requirements for Estuaries*; Ramsar Technical Report No. 9; CBD Technical Series No. 69; Ramsar Convention Secretariat: Gland, Switzerland; Secretariat of the Convention on Biological Diversity: Montreal, QC, Canada, 2012.

34. Borja, A.; Bricker, S.; Dauer, D.; Demetriades, N.; Ferreira, J.; Forbes, A.; Hutchings, P. Overview of Integrative Tools and Methods in Assessing Ecological Integrity in Estuarine and Coastal Systems Worldwide. *Mar. Pollut. Bull.* **2008**, *56*, 1519–1537. [CrossRef] [PubMed]

35. Ibáñez, C.; Prat, N. The Environmental Impact of the Spanish National Hydrological Plan on the Lower Ebro River and Delta. *Int. J. Water Resour. Dev.* **2003**, *19*, 485–500. [CrossRef]

36. Belmar, O.; Ibáñez, C.; Forner, A.; Caiola, N. The Influence of Flow Regime on Ecological Quality, Bird Diversity, and Shellfish Fisheries in a Lowland Mediterranean River and Its Coastal Area. *Water* **2019**, *11*, 918. [CrossRef]

37. Adams, J.B. A Review of Methods and Frameworks Used to Determine the Environmental Water Requirements of Estuaries. *Hydrol. Sci. J.* **2014**, *59*, 451–465. [CrossRef]

38. Human, L.; Snow, G.; Adams, J.B. Responses in a Temporarily Open/Closed Estuary to Natural and Artificial Mouth Breaching. *S. Afr. J. Bot.* **2016**, *107*, 39–48. [CrossRef]

39. Costanza, R.; de Groot, R.; Braat, L.; Kubisze-Wski, I.; Fioramonti, L.; Sutton, P.; Farber, S.; Grasso, M. Twenty Years of Ecosystem Services: How Far Have We Come and How Far Do We Still Need to Go? *Ecosyst. Serv.* **2017**, *28*, 1–16. [CrossRef]

40. Adams, J.B.; Bate, G.C.; Harrison, T.D.; Huizinga, P.; Taljaard, S.; Van Niekerk, L.; Plumstead, E.; Whitfield, A.K.; Wooldridge, T. A Method to Assess the Freshwater Inflow Requirements of Estuaries and Application to the Mtata Estuary, South Africa. *Estuaries* **2002**, *25*, 1382–1393. [CrossRef]

41. Taljaard, S.; Adams, J.B.; Turpie, J.K.; Van Niekerk, L.; Demetriades, N.; Bate, G.C.; Cyrus, D.; Huizinga, P.; Lamberth, S.; Weston, B. *Water Resource Protection and Assessment Policy Implementation Process. Resource Directed Measures for Protection of Water Resources: Methodology for the Determination of the Ecological Water Requirements for Estuaries*; Version 2; Department of Water Affairs: Pretoria, South Africa, 2004.

42. Turpie, J.K.; Taljaard, S.; Van Niekerk, L.; Adams, J.B.; Wooldridge, T.; Cyrus, D.; Clark, B.; Forbes, N. *The Estuary Health Index: A Standardised Metric for Use in Estuary Management and the Determination of Ecological Water Requirements*; WRC Report No. 1930/1/12; Water Research Commission: Pretoria, South Africa, 2012.

43. Van Niekerk, L.; Adams, J.B.; Lamberth, S.; Mackay, F.; Taljaard, S.; Turpie, J.K.; Weerts, S. *South Afr. National Biodiversity Assessment 2018: Technical Report Volume 3: Estuarine Realm*; Department of Environmental Affairs: Pretoria, South Africa, 2019.

44. DWAF. *Rapid Determination of Resource Directed Measures of the Mdloti Estuary (Including Preliminary Estimates of Capping Flows for Mdloti and Mhlanga Estuaries)*; Specialist Report Prepared by the CSIR for the Department of Water Affairs & Forestry; CSIR: Pretoria, South Africa, 2002.

45. Department of Water Affairs and Forestry (DWAF). *Determination of the Preliminary Ecological Reserve on a Rapid Level for the Siyaya Estuary*; Demetriades, N., Huizinga, P., Van Niekerk, L., Pillay, S., McKay, F., Cyrus, D., Forbes, A.T., Eds.; Department of Department of Water Affairs and Forestry: Pretoria, South Africa, 2006.

46. Whitfield, A.K.; Adams, J.; Bate, G.; Bezuidenhout, K.; Bornman, T.G.; Cowley, P.; Froneman, P.; Gama, P.; James, N.; Mackenzie, B.; et al. A Multidisciplinary Study of a Small, Temporarily Open/Closed South African Estuary, with Particular Emphasis on the Influence of Mouth State on the Ecology of the System. *Afr. J. Mar. Sci.* **2008**, *30*, 453–473. [CrossRef]

47. Snow, G.; Taljaard, S. Water Quality in South African Temporarily Open/Closed Estuaries: A Conceptual Model. *Afr. J. Aquat. Sci.* **2007**, *32*, 99–111. [CrossRef]

48. Terörde, A.; Turpie, J.K. Influence of Habitat Structure and Mouth Dynamics on Avifauna of Intermittently-Open Estuaries: A Study of Four Small South African Estuaries. *Estuar. Coast. Shelf Sci.* **2013**, *125*, 10–19. [CrossRef]

49. Perissinotto, R.; Blair, A.; Connell, A.; Demetriades, N.; Forbes, A.; Harrison, T.D.; Iyer, K.; Joubert, M.; Kibirige, I.; Mundree, S.; et al. *Contributions to Information Requirements for the Implementation of Resource Directed Measures for Estuaries. Responses of the Biological Communities to Flow Variation and Mouth State in Two KwaZulu-Natal Temporarily Open/Closed Estuaries*; WRC Report; Water Research Commission: Pretoria, South Africa, 2004; Volume 2.

50. Kok, H.; Whitfield, A.K. The Influence of Open and Closed Mouth Phases on the Marine Fish Fauna of the Swartvlei Estuary. *S. Afr. J. Zool.* **1986**, *21*, 309–315. [CrossRef]

51. James, N.; Cowley, P.; Whitfield, A.K. Abundance, Recruitment and Residency of Two Sparids in an Intermittently Open South African Estuary. *Afr. J. Mar. Sci.* **2007**, *29*, 527–538. [CrossRef]

52. Vorwerk, P.; Whitfield, A.K.; Cowley, P.; Paterson, A. The Influence of Selected Environmental Variables on Fish Assemblage Structure in a Range of Southeast African Estuaries. *Environ. Biol. Fishes* **2003**, *66*, 237–247. [CrossRef]

53. Bennett, B. The Fish Community of a Moderately Exposed Beach on the South Western Cape Coast of South Africa and an Assessment of This Habitat as a Nursery for Juvenile Fish. *Estuar. Coast. Shelf Sci.* **1989**, *28*, 293–305. [CrossRef]

54. Cowley, P.; Whitfield, A.K. Ichthyofaunal Characteristics of a Typical Temporarily Open/Closed Estuary on the Southeast Coast of South Africa. *Ichthyol. Bull. JLB. Smith Inst. Icthyol.* **2001**, *71*, 1–19.

55. Von der Heyden, S.; Toms, J.; Teske, P.; Lamberth, S.; Holleman, W. Contrasting Signals of Genetic Diversity and Historical Demography between Two Recently Diverged Marine and Estuarine Fish Species. *Mar. Ecol. Prog. Ser.* **2015**, *526*, 157–167. [CrossRef]

56. Lamberth, S.; Drapeau, L.; Branch, G. The Effects of Altered Freshwater Inflows on Catch Rates of Non-Estuarine-Dependent Fish in a Multispecies Nearshore Line-Fishery. *Estuar. Coast. Shelf Sci.* **2009**, *84*, 527–538. [CrossRef]
57. Riddin, T.; Adams, J.B. The Effect of a Storm Surge Event on the Macrophytes of a Temporarily Open/Closed Estuary, South Africa. *Estuar. Coast. Shelf Sci.* **2010**, *89*, 119–123. [CrossRef]
58. Riddin, T.; Adams, J.B. Influence of Mouth Status and Water Level on the Macrophytes in a Small Temporarily Open/Closed Estuary. *Estuar. Coast. Shelf Sci.* **2008**, *79*, 86–92. [CrossRef]
59. Adams, J.B.; Bate, G. The Effect of Salinity and Inundation on the Estuarine Macrophyte Sarcocornia Perennis (Mill.) A.J. Scott. *Aquat. Bot.* **1994**, *47*, 341–348. [CrossRef]
60. Riddin, T.; Adams, J.B. Water Level Fluctuations and Phenological Responses in a Salt Marsh Succulent. *Aquat. Bot.* **2019**, *153*, 58–66. [CrossRef]
61. Riddin, T.; Adams, J.B. The Seed Banks of Two Temporarily Open/Closed Estuaries in South Africa. *Aquat. Bot.* **2009**, *90*, 328–332. [CrossRef]
62. Adams, J.B.; Knoop, W.; Bate, G.C. The Distribution of Estuarine Macrophytes in Relation to Freshwater. *Bot. Mar.* **1992**, *35*, 215–226. [CrossRef]
63. Whitfield, A.K.; Cowley, P. A Mass Mortality of Fishes Caused by Receding Water Levels in the Vegetated Littoral Zone of the West Kleinemonde Estuary, South Africa. *Afr. J. Aquat. Sci.* **2018**, *43*, 179–186. [CrossRef]
64. Van Niekerk, L.; Huizinga, P.; Theron, A. Semi-Closed Mouth States in Estuaries along the South African Coastline. In Proceedings of the Environmental Flows for River Systems, Fourth International Ecohydraulics Symposium, Cape Town, South Africa, 3–8 March 2002.
65. Mbense, S.; Rajkaran, A.; Bolosha, U.; Adams, J.B. Rapid Colonization of Degraded Mangrove Habitat by Succulent Salt Marsh. *S. Afr. J. Bot.* **2016**, *107*, 129–136. [CrossRef]
66. Adams, J.B.; Taljaard, S.; Van Niekerk, L.; Lemley, D.A. Nutrient Enrichment as a Threat to the Ecological Resilience and Health of Microtidal Estuaries. *Afr. J. Aquat. Sci.* **2020**, *45*, 23–40. [CrossRef]
67. Brooker, B.; Scharler, U.M. The Importance of Climatic Variability and Human Influence in Driving Aspects of Temporarily Open-Closed Estuaries. *Ecohydrology* **2020**, e2205. [CrossRef]
68. DEA (Department of Environmental Affairs). *South Afr. Water Quality Guidelines for Coastal and Marine Waters*; Volume 2: Guidelines for Recreational Use; Department of Environmental Affairs: Pretoria, South Africa, 2012.
69. Largier, J.; Hollibaugh, J.T.; Smith, S. Seasonally Hypersaline Estuaries in Mediterranean-Climate Regions. *Estuar. Coast. Shelf Sci.* **1997**, *45*, 789–797. [CrossRef]
70. Wooldridge, T.; Adams, J.B.; Fernandes, M. Biotic Response to Extreme Hypersalinity in an Arid Zone Estuary, South Africa. *S. Afr. J. Bot.* **2016**, *107*, 160–169. [CrossRef]
71. Bennett, B. A Mass Mortality of Fish Associated with Low Salinity Conditions in the Bot River Estuary. *Trans. R. Soc. S.Afr.* **1985**, *45*, 437–447. [CrossRef]
72. an Niekerk, L.; Taljaard, S.; Adams, J.B.; Huizinga, P.; Lamberth, S.; Riddin, T.; Turpie, J.K.; Mallory, S.; Wooldridge, T. *Rapid Assessment of the Ecological Water Requirements for the Bot Estuary*; Report No. CSIR/NRE/CO/ER/2011/0035/B; CSIR: Pretoria, South Africa, 2011.
73. DWA (Department of Water Affairs). *Reserve Determination Studies for Selected Surface Water, Groundwater, Estuaries and Wetlands in the Outeniqua (Groot Brak and Other Water Resources, Excluding Wetlands) Catchment*; Report No. RDM/K10-K30, K40E/00/CON/0307; DWA: Pretoria, South Africa, 2009.
74. Nunes, M.; Adams, J.B. Responses of Primary Producers to Mouth Closure in the Temporarily Open/Closed Great Brak Estuary in the Warm-Temperate Region of South Africa. *Afr. J. Aquat. Sci.* **2014**, *39*, 387–394. [CrossRef]
75. Human, L.; Snow, G.; Adams, J.B.; Bate, G.C.; Yang, S.-C. The Role of Submerged Macrophytes and Macroalgae in Nutrient Cycling in the Great Brak Estuary South Africa: A Budget Approach. *Estuar. Coast. Shelf Sci.* **2015**, *154*, 169–178. [CrossRef]
76. Human, L.; Snow, G.; Adams, J.B.; Bate, G. The Benthic Regeneration of N and P in the Great Brak Estuary. *Water SA* **2015**, *41*, 594–605. [CrossRef]
77. Anchor Environmental Consultants. *Determination of the Ecological Reserve for the Uilkraals Estuary*; Anchor Environmental: Cape Town, South Africa, 2012.
78. Kaselowski, T.; Adams, J.B. Not so Pristine—Characterising the Physico-Chemical Conditions of an Undescribed Temporarily Open/Closed Estuary. *Water* **2013**, *39*, 627–637. [CrossRef]

79. Van Niekerk, L.; Bate, G.C.; Whitfield, A.K. *An Intermediate Ecological Reserve Determination Study of the East Kleinemonde Estuary*; WRC Report No. 1581/2/08; Water Research Commission: Pretoria, South Africa, 2008.

80. Riddin, T.; Adams, J.B. Predicting Macrophyte States in a Small Temporarily Open/Closed Estuary. *Mar. Freshw. Res.* **2012**, *63*, 616–623. [CrossRef]

81. Van Niekerk, L.; Adams, J.B.; Taljaard, S.; Huizinga, P.; Lamberth, S. Advancing Mouth Management Practices in Groot Brak Estuary, South Africa. In *Complex Coastal Systems—Transdisciplinary Learning on International Case Studies*; Slinger, J., Taljaard, S., D'Hont, F., Eds.; Delft Academic Press: Delft, The Netherlands, 2020; pp. 89–104.

82. Dettinger, M.; Diaz, H. Global Characteristics of Stream Flow Seasonality and Variability. *J. Hydrometeorol.* **2000**, *1*, 289–310. [CrossRef]

83. Poff, N.; Ward, J.V. Implications of Streamflow Variability and Predictability for Lotic Community Structure: A Regional Analysis of Streamflow Patterns. *Can. J. Fish. Aquat. Sci.* **1989**, *46*, 1805–1818. [CrossRef]

84. James, N.; Adams, J.B.; Connell, A.; Lamberth, S.; MacKay, C.; Snow, G.; Van Niekerk, L.; Whitfield, A.K. High Flow Variability and Storm Events Shape the Ecology of the Mbhashe Estuary, South Africa. *Afr. J. Aquat. Sci.* **2020**, *45*, 131–151. [CrossRef]

85. Le Roux, J.; Morgenthal, T.; Malherbe, J.; Pretorius, D.; Sumner, P. Water Erosion Prediction at a National Scale for South Africa. *Water* **2008**, *34*, 305–314. [CrossRef]

86. Adams, J.B.; Whitfield, A.K.; Van Niekerk, L. A Socio-Ecological Systems Approach towards Future Research for the Restoration, Conservation and Management of Southern African Estuaries. *Afr. J. Aquat. Sci.* **2020**, *45*, 231–241. [CrossRef]

87. Lemley, D.A.; Adams, J.B.; Rishworth, G.M.; Bouland, C. Phytoplankton Responses to Adaptive Management Interventions in Eutrophic Urban Estuaries. *Sci. Total Environ.* **2019**, *693*, 133601:1–133601:12. [CrossRef]

88. Schallenberg, M.; Larned, S.; Hayward, S.; Arbuckle, C. Contrasting Effects of Managed Opening Regimes on Water Quality in Two Intermittently Closed and Open Coastal Lakes. *Estuar. Coast. Shelf Sci.* **2010**, *86*, 587–597. [CrossRef]

89. Hoeksema, S.; Chuwen, B.; Tweedly, J.; Potter, I.C. Factors Influencing Marked Variations in the Frequency and Timing of Bar Breaching and Salinity and Oxygen Regimes among Normally-Closed Estuaries. *Estuar. Coast. Shelf Sci.* **2018**, *208*, 205–218. [CrossRef]

90. Suari, Y.; Amit, T.; Gilboa, M.; Sade, T.; Krom, M.; Gafny, S.; Topaz, T.; Yahel, G. Sandbar Breaches Control of the Biogeochemistry of a Micro-Estuary. *Front. Mar. Sci.* **2019**, *6*, 224. [CrossRef]

91. Wooldridge, T.; Adams, J.B. The Development and Implementation of a Management Plan for the Seekoei Estuary: A Case Study Guided by Socio-Economic Issues. *Afr. J. Aquat. Sci.* **2020**, *45*, 83–93. [CrossRef]

92. Stretch, D.; Zietsman, I. *The Hydrodynamics of Mhlanga & Mdloti Estuaries: Flows, Residence Times, Water Levels and Mouth Dynamics*; WRC K5/1247 Final Report; Water Research Commission: Pretoria, South Africa, 2004.

93. Whitfield, A.K. A Quantitative Study of the Trophic Relationships within the Fish Community of the Mhlanga Estuary, South Africa. *Estuar. Coast. Mar. Sci.* **1980**, *10*, 417–435. [CrossRef]

94. Holling, C. *Adaptive Environmental Assessment and Management*; John Wiley & Sons: Chichester, UK, 1978.

95. Cilliers, G.; Adams, J.B. Development and Implementation of a Monitoring Programme for South African Estuaries. *Water* **2016**, *42*, 279–290. [CrossRef]

96. Tagliapietra, D.; Povilanskas, R.; Razinkovas-Baziukas, A.; Taminskas, J. Emerald Growth: A New Framework Concept for Managing Ecological Quality and Ecosystem Services of Transitional Waters. *Water* **2020**, *12*, 894. [CrossRef]

97. Taljaard, S.; Van Niekerk, L.; Huizinga, P.; Joubert, W. *Resource Monitoring Procedures for Estuaries for Application in the Ecological Reserve Determination and Implementation Process*; Report No. 1308/1/03; Water Research Commission: Pretoria, South Africa, 2003.

Article

Emerald Growth: A New Framework Concept for Managing Ecological Quality and Ecosystem Services of Transitional Waters

Davide Tagliapietra [1], Ramūnas Povilanskas [2,*], Artūras Razinkovas-Baziukas [3] and Julius Taminskas [4]

[1] Consiglio Nazionale delle Ricerche–Istituto di Scienze Marine (CNR-ISMAR), Arsenale di Venezia, Tesa 104, I-30122 Venice, Italy; davide.tagliapietra@ve.ismar.cnr.it

[2] Department of Natural Sciences, Klaipeda University, Herkaus Manto str. 84, LT-92294 Klaipeda, Lithuania

[3] Institute of Marine Research, Klaipeda University, Universiteto al. 17, LT-92295 Klaipėda, Lithuania; arturas.razinkovas-baziukas@ku.lt

[4] Laboratory of Climate and Water Research, Nature Research Centre, Akademijos str. 2, LT-08412 Vilnius, Lithuania; julius.taminskas@gamtc.lt

* Correspondence: ramunas.povilanskas@ku.lt; Tel.: +370-615-71-711

Received: 17 February 2020; Accepted: 19 March 2020; Published: 22 March 2020

Abstract: The aim of the present paper is to propose and elaborate on the concept of Emerald Growth as a new framework concept for managing ecological quality and ecosystem services of transitional waters. The research approach combines the longstanding experience of the authors of this article in the investigation of transitional waters of Europe with an analysis of relevant European Union directives and a comparative case study of two European coastal lagoons. The concept includes and reassesses traditional knowledge of the environment of lagoons and estuaries as an engine for sustainable development, but also proposes locally tailored approaches for the renewal of these unique areas. The investigation results show that the Emerald Growth concept enables to extricate better specific management aspects of ecosystem services of transitional waters that fill-in the continuum between the terrestrial (Green Growth) and the maritime areas (Blue Growth). It results from adjusting of both Green Growth and Blue Growth concepts, drivers, indicators and planning approaches regarding durable ways of revitalising coastal communities and their prospects for sustainable development. We conclude that the Emerald Growth concept offers a suitable framework for better dealing with complex and complicated issues pertinent to the sustainable management of transitional waters.

Keywords: Adriatic Sea; Baltic Sea; Curonian Lagoon; Emerald Growth; Lesina Lagoon; MSFD; MSPD; transitional waters; WFD

---

1. Introduction

The Water Framework Directive of the European Communities (WFD, 2000/60/EC) was the first official document introducing the term 'transitional waters' in 2000 to describe the aquatic continuum between freshwaters, coastal waters and marine waters. 'Transitional waters' are hence defined by the European Communities as "bodies of surface water in the vicinity of river mouths which are partially saline in character as a result of their proximity to coastal waters but which are substantially influenced by freshwater flows".

Transitional waters are diverse, productive, ecologically important systems of a global scale that are valued for the services they delivered to human societies since the Paleolithic age. Transitional waters provided food, transportation routes, shelter (e.g., Venice). They also served as natural wastewater treatment systems. However, ecosystem goods and services delivered by transitional waters are insufficiently understood, in spite of being essential for a holistic consideration of sustainability conditions of the vast coastal and marine continuum. Therefore, any mismanagement of transitional waters might cost profoundly.

In the context of maritime spatial planning (MSP), the Land-Sea-Interaction (LSI) has been more addressed and studied, and the EU Maritime Spatial Planning Directive (MSPD, 2014/89/EU) refers to an interplay between the Integrated Coastal Zone Management (ICZM) and MSP. Albeit transitional waters play a pivotal role in LSI, and, as a consequence, in the MSP processes, the peculiarities of their management and planning are largely ignored by policymakers. In transitional ecosystems, the dynamic LSI patterns result in peculiar economical uses that need specific management and planning efforts very little discussed in the MSP context to date.

Transitional waters form coherent ecotones between terrestrial, freshwater and marine ecosystems categorized by high temporal variability and spatial heterogeneity [1]. The incidence of the human traces at lagoons and estuaries is recognized since prehistorical times, representing the core of early civilizations and posterior socio-economic formations [2]. Especially in the Mediterranean, we have plenty of documented evidence of multiple uses of coastal lagoons, from transportation to fisheries [3]. In the 1st century CE, Romans used the Tyrrhenian Sea coastal lagoon system as a connection route between Naples and Rome for military and commercial purposes [4].

The notion of 'transitional waters' defines a varied range of ecosystems types, including river estuary ecosystems, coastal lagoons, lakes, fjords and fjards, rias, brackish wetlands and hypersaline bahiras. Due to the hydrological balance between marine and freshwater forces, transitional waters, particularly the lagoons and the rias, are sediment and nutrient sinks, controlled by different variation scales according to the tidal cycles, seasonal and annual rhythms, precipitation cycles and climate variations [5].

Thus described, the "transitional waters" are included in a larger conceptual framework of the Coastal Transitional Ecosystems, which also includes the coastal basins of warm climates, often hyperhaline, not substantially influenced by fresh water but which share many fundamental features with the "transitional waters" [6]. A synthetic scheme of different coastal types denoted by the notion of "transitional waters" is provided in Table 1. The generalized view on the distribution of different coastal transitional ecosystems in Europe is given in Figure 1.

Table 1. Main physiographic forms included under the term transitional waters

Type	Characteristics
Classical estuary	Tidally dominated at the seaward part; salinity reduced by freshwater river inputs; riverine dominance inward
River mouth *	River outlet as a well-defined physiographic coastal feature
Delta *	Low energy, typically shaped, sediment dominated, river mouth area; estuary outflow
Fjord	Land freshwater seepage or seasonal riverine inputs; limited tidal influence; stratified; long narrow, glacially eroded sea inlet, step sided, sill at the mouth
Ria	Drowned river valley, some freshwater inputs; limited exchange
Non-tidal/microtidal lagoon *	Limited exchange with the marine area through a restricted mouth; separated from the sea by sand or shingle banks, bars, etc., shallow area, tidal range < 50 cm
Tidal lagoon	As above but with tidal range > 50 cm
Coastal plume *	Outflow of estuary, or lagoon, notably diluted salinity and hence different biota than in surrounding marine areas

* available in the Baltic Sea.

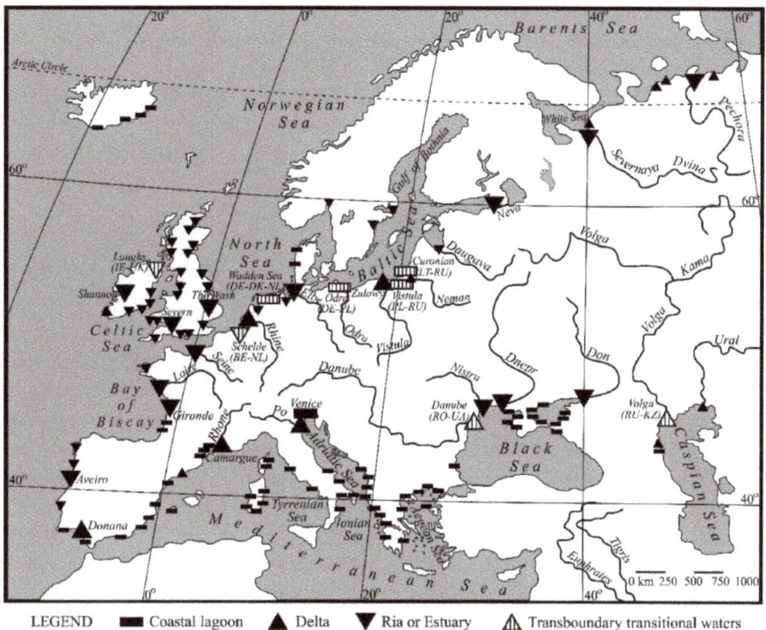

Figure 1. Generalization of major transitional water types in Europe

Transitional waters are under heavy human impact being the locations of major ports or cities. Therefore, these areas are degraded by dredging and contamination from industrial and agricultural activities, fishing, aquaculture as well as from urban sprawl and municipal pollution. These issues do have an enormous impact on human well-being in the regions adjacent to transitional waters since goods and services of these waters, being as diverse as the aquatic ecosystems supporting them, are also adversely affected [7]. Therefore, transitional waters have prompted the need to be categorised into operational types from both the academic and applied points of view [1,2,6].

The notion of 'transitional waters' in the EU WFD interpreation means aquatic areas which are neither fully coastal nor enclosed or flowing freshwater areas [5] whereas, under the more comprehensive term 'coastal transitional ecosystems' could be simply defined as "coastal water bodies with limited seawater supply" [6]. Transitional waters are defined by physiographic features, discontinuities, salinity, or other hydrographic features. Even though the typology provided in the WFD is indispensable for defining environmental descriptors and ensuring environmental integrity, 'transitional waters' is still a complex and not exhaustive term.

The definition is even more challenging if applied to the three largest European transboundary lagoons located on the southern rim of the Baltic Sea [8]. This challenge is complicated because two of these transitional water bodies (Vistula Lagoon and Curonian Lagoon) are shared with the Russian Federation. This country is not an EU member and, therefore, the term 'transitional waters' has no legislative consequences in the Russian parts of both lagoons. Schernewski and Wielgat [9] highlighted that Baltic Sea countries have taken very diverse approaches towards the designation of coastal and transitional waters. Some countries have not designated any transitional waters at all, regardless of the WFD regulations.

In Sweden, there was an attempt to overcome problems with designating transitional waters by suggesting a further category of enclosed, brackish coastal types, whereas Denmark, Finland and Estonia do not have designated transitional waters at all [7]. In Germany, transitional waters were designated for its North Sea estuaries but for none of its Baltic Sea lagoons and estuaries. Lithuania considers the Curonian Lagoon to be a transitional water body [8]. Additionally, the discharge plume from the Klaipeda Strait into the Baltic Sea is also considered as transitional waters [7].

Poland has designated as transitional waters the Polish parts of the Vistula Lagoon and the Odra (Szczecin) Lagoon, a part of the Gulf of Gdansk (the inner Puck Bay) as well as the open parts of the Gulf of Gdansk and the Bay of Pomerania where riverine discharge plumes have an impact [10]. Poland has also designated the coastal areas affected by the riverine plumes discharging into the open Baltic Sea as transitional waters. Latvia treats the Daugava River estuary at Riga and the riverine discharge plume into the Gulf of Riga as a transitional water area [9].

Most of the properties of transitional waters derive from both hydrological balance and land-water interfaces [5]. Strong directional gradients of salinity, organic matter, nutrients and oxygen concentrations featured in these waters can on one hand act as filters for potential coloniser species but on the other host euryoecious invasive species [11]. From a trophic point of view, transitional waters are very productive [12]. The overall hydrological and ecological balance that maintains the ecological status of transitional waters covers scales ranging in time from minutes and hours to years for long-term hydrological balance and large species population dynamics. On the spatial scale, the effects of transitional waters are felt from local to global, regarding the migratory fish and bird species.

In recent decades, the EU has directed many efforts towards new concepts of sustainable growth, first the terrestrial one (Green Growth) and then the maritime one (Blue Growth). According to the Organisation for Economic Cooperation and Development (OECD): "green growth means fostering economic growth and development while ensuring that natural assets continue to provide the resources and environmental services on which our well-being relies" [13]. The governance of marine resource uses ever more focuses on the recently introduced term and concept of Blue Growth [14]. The Blue Growth concept aims to facilitate sustainable economic growth based on the utilisation of marine resources, while at the same time mitigating their degradation, overuse and pollution [15].

The EU´s Blue Growth strategy emphasises the importance of marine areas for innovation and growth in five sectors in addition to increased emphasis on MSP and coastal protection [16]. The Blue Growth concept recognises that diverse maritime sectors, such as fisheries, shipping and tourism, and marine ecosystem services, such as food provisioning, coastal protection and carbon storage, are interconnected. Managing these sectors and services coherently can deliver additional value [16].

In coastal areas, especially in the lagoons and estuaries, the marine and terrestrial domains intertwine intimately, bringing out a unique habitat, with peculiar characteristics, in many aspects, hostile, while others are particularly favourable to human settlement. This environmental mosaic has generated particular socio-ecosystems of transitional waters with required special human skills, adaptive strategies and constant care of the environment. Therefore, in the case of transitional waters, it is challenging to discuss either Green Growth or Blue Growth purely.

In this interim zone, where terrestrial and marine ecosystems interact, the two aspects of sustainable growth go together and have always been intimately connected. In these environments set like emerald gems of the coast, we should instead discuss the Emerald Growth concept, that is the integration of the principles expressed by Green Growth and Blue Growth concepts. It treasures traditional knowledge of the elaborate lagoon and estuarine socio-ecosystems that have developed in these environments over the millennia but also implies new technological and economical solutions. It describes better the management aspects of ecosystem services of transitional waters lying between the terrestrial (Green Growth) and the marine areas (Blue Growth).

'Emerald Growth' is an overarching concept proposed by the authors of this article to address sustainable development and management issues of transitional waters within a broader river basin, coastal and marine interplay framework. The topic of Emerald Growth is especially important for the Baltic Sea due to its marine geography. Eight transboundary transitional waters are located around the Baltic Sea, particularly at its southern rim (Figure 2). They require cross-border cooperation in ecosystem-based MSP between the countries sharing them. Therefore, the task of delivering sustainable use of ecosystem goods and services in such complex environments becomes a real challenge.

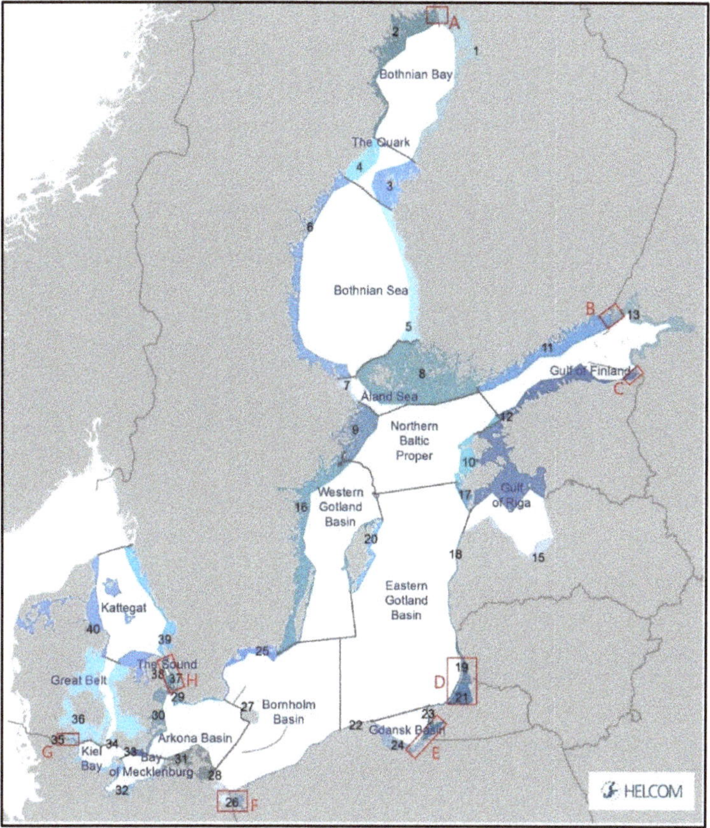

Figure 2. Distribution of coastal and transitional waters in the Baltic Sea Region (Source: Helsinki Commission). Transboundary transitional waters marked in red by the authors.

Considering this task, the main aim of the present paper is to propose and elaborate on the concept of Emerald Growth as a new framework concept for managing ecological quality and ecosystem services of transitional waters. The notion of 'Emerald Growth' is newly coined and presented in this article for the first time. A comprehensive study was undertaken by the authors to delimit all three concepts considering drivers, indicators and planning approaches. While concrete planning guidelines for transitional waters within the Emerald Growth framework pertinent to the EU WFD, MSFD, MSPD and other regulations are still in the conceptual phase, the drivers, indicators and planning approaches, which are proven relevant in spatial planning contexts, are summarised in Table 2.

Table 2. Drivers, indicators and planning approaches relevant in the spatial planning contexts.

Notions	Green Growth	Emerald Growth	Blue Growth
Key drivers	(1) Environmental and climate change and resulting economic policy changes; (2) Circular economy advancement	(1) Depletion of living resources; (2) Eutrophication; and water and sediment pollution; (3) Land reclamation; 4) Growing industrial and recreational use; (5) Sea level rise	(1) Growth of shipping; (2) Marine pollution; (3) Depletion of living resources; (4) Demand for energy and mineral resources; (5) Expanding networks of pipelines and cables
Main indicator groups [13]	(1) Economic growth, productivity and competitiveness; (2) Labor markets, education and income; (3) Carbon and energy productivity; (4) Resource productivity; (5) Multi-factor productivity; (6) Natural asset base; (7) Renewable stocks; (8) Non-renewable stocks		
Main planning approaches	(1) Hierarchy (2) Master-planning (3) Sectorial planning (4) Functional zoning (5) Detailed planning	(1) Hierarchy (2) Master-planning (3) Sectorial planning (4) Functional zoning (5) Trade-offs (6) Ecosystem approach (7) Cross-border links	(1) Master-planning (2) Sectorial planning (3) Functional zoning (4) Trade-offs (5) Ecosystem approach (6) Cross-border links

The Emerald Growth concept also includes and reassesses traditional knowledge of the coastal environment of lagoons and estuaries as an engine for sustainable development, but also proposes locally tailored approaches for the renewal of these unique areas. The Emerald Growth concept can be elicited by combining both Green Growth and Blue Growth concepts considering sustainable ways of enhancing the well-being of coastal communities and their prospects for a prosperous future. It also implies avoiding the adverse effects that may result from coastal population decline or monoculture prevalence (e.g., fisheries or coastal tourism).

On the conceptual level, the Emerald Growth concept is a framework for analysing socio-economic growth and human well-being relying on sustainable use of transitional waters, their resources and ecosystem services. The Emerald Growth concept applies where the complexity of estuarine and lagoon systems needs a goal- and solution-oriented, realistic and practical management approach. Therefore, it is very much coherent with the Blue Growth drivers, principles, indicators and planning approaches [14–16]. On the other hand, the Emerald Growth concept and principles are also quite coherent with the Green Growth concept and principles [13,17,18].

To showcase the usefulness of the Emerald Growth concept and to deliberate its implications, the methods, the results, the discussion and the conclusions are structured in the following way: First, we analyse relevant European Union (EU) directives and regulations regarding their potential to regulate environmental management and spatial planning of transitional waters and hence ensure a proper Emerald Growth framework. Next, ecosystem goods and services of transitional waters are typified and classified. Finally, to demonstrate the practicality of the Emerald Growth concept, in the results section, we apply it for the analysis of two case studies: the Curonian Lagoon (Baltic Sea) and Lesina Lagoon (Adriatic Sea).

The discussion focuses on the essential debatable issues of the perspectives and limitations of the Emerald Growth concept for facilitating sustainable management of various transitional waters. We argue that it is necessary to find a balance between different EU directives (WFD, MSFD, MSPD) and other supranational regulations pertinent to the protection of aquatic environments in each particular case in order to deliver a holistic approach for the transitional waters' management. In addition, the role of the EU Birds and Habitats directives should not be ignored. The conclusions of the paper highlight practical aspects of the Emerald Growth concept.

2. Materials and Methods

2.1. Methods

The in-depth analysis of relevant European Union (EU) directives and regulations concerning their potential to regulate environmental management of transitional waters and, therefore, to set the Emerald Growth framework, has been conducted during a series of six online workshops among the four authors of this article from 2018 to 2020. The discussed question was whether the transitional waters of the EU should be considered within the scope of the Marine Strategy Framework Directive (MSFD, 2008/56/EC) and the MSPD (2014/89/EU). If they do, then which regulatory areas have to be considered when integrating river basin management with the planning and management of transitional waters and marine territories. The answer to this question is pivotal for ecosystem-based planning and sustainable management of aquatic resources, and for delimiting the Green, Blue and Emerald Growth concepts.

The investigations leading to the case study analysis relied on longstanding first-hand experiences of the authors in the investigation of transitional waters in Europe, the Americas, Africa and Australia [3,8,19–26]. Data on ecology, history and economy of Lesina and Curonian Lagoons as the target areas of the comparative case study were obtained from different sources, namely old maps, earlier trade reports and investigations, other archive documents and literature. Data on traditional fisheries, aquaculture and exploitation of living resources in coastal lagoons of the Adriatic Sea and inherent effects on the recent evolution of lagoon ecosystems and landscapes were also derived from [27,28].

The vicissitudes of living resource uses in the Curonian Lagoon after post-Communist reforms and emergence of private fishery and husbandry were investigated first-hand. Since the mid-1990s, we have investigated social and economic transformations and environmental issues in the Baltic lagoon areas and compared them with the Italian lagoon management practice [3,8,22,24,29]. The issues of interest have been explored employing a wide spectrum of quantitative and qualitative methods—from the in situ and remote surveillance and simulation modelling of aquatic ecosystems to an assortment of social field surveys in the broader lagoon regions.

Perceptions and opinions about ecology, environment, ecosystem goods and services, as well as conservation and local development conflicts, were the key issues during the in-depth individual interviews and focus-groups with local interest groups—fishermen, farmers and environmentalists. During the investigation period spanning over two decades, there were over 20 qualitative longitudinal surveys carried out in Lithuanian and Italian lagoon areas as these were the main research target areas of the authors of the paper. These surveys were conducted within the framework of joint projects supported by EU regional (PHARE, TACIS, INTERREG) and academic research (FP6, FP7, Horizons 2020) cooperation programs.

It is confirmed and accepted that individual in-depth interviews deliver coherent, although not identical, ecosystem service information as do focus groups [30]. Therefore, the ratio between the number of those stakeholders with whom the individual in-depth interviews have been conducted, and those who participated in the focus groups was one to four in our surveys. Typically, a focus group comprised five to seven participants representing various local interest groups. In both cases, earnest efforts were taken to ensure the representativeness of the participants in terms of gender, age and trade. As a result, the perceptions and opinions of 180 locals have been sampled in the Curonian Lagoon area and ca. 100 locals in the Adriatic lagoon areas of Italy (mainly, Venice Lagoon and Lesina Lagoon).

2.2. Case Study Area

Lesina Lagoon in Italy (41°52′58″ N, 15°26′21″ E) is a nano-tidal lagoon (mean tidal range 0.3 m) extending parallel to the south coast of the Adriatic Sea for 22.4 km. It has an oblong shape with the width varying between 3.8 and 1.4 km (Figure 3). The mean water depth is 0.8 m with a maximum of about 1.5 m. Three artificially managed channels link the Adriatic Sea with the Lesina Lagoon [27].

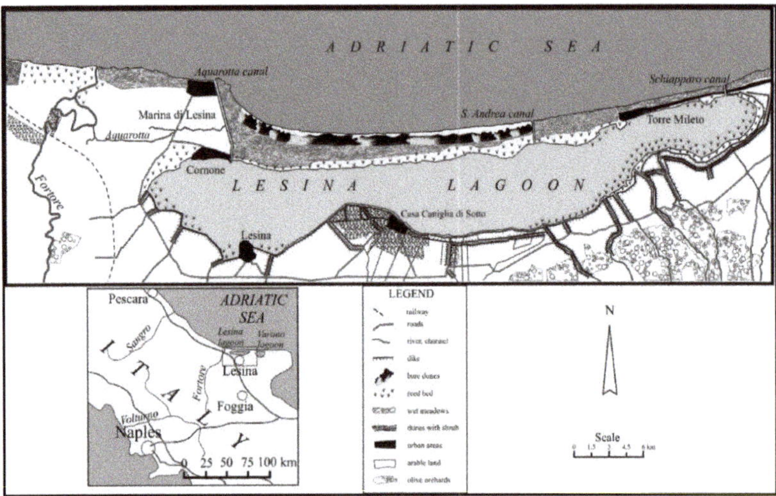

Figure 3. Lesina Lagoon. Source: authors' own plotting.

In the eastern part, fresh groundwater discharges into the lagoon, generating a saline gradient, salinity always remains lower than that of the sea ranging between 11 and 34 psu (average 23.7 psu) [31]. At the mean water level, the surface area of the lagoon is 51.5 km^2 and a volume is 0.04 km^3. The climate of the Lesina Lagoon region is typical Mediterranean one with dry, hot tropical summers and chilly, humid winters. The average annual precipitation level in this area is 427 mm [32].

The non-tidal Curonian Lagoon (55°11′55″ N, 21°03′30″ E) is an enclosed, shallow and almost fresh water body, located in the southeast angle of the Baltic Sea. The Curonian Lagoon is the largest lagoon in all Europe [33]. Its surface area is 1586 km^2, i.e., 30 times larger than that of Lesina Lagoon. The Nemunas River, whose catchment basin is 98,200 km^2, discharges into the Curonian Lagoon on its way to the Baltic Sea. The mean depth of the lagoon is 3.8 m, and the maximum depth is 5.8 m. Salinity varies from just 0.1 psu at the river mouths to 6 psu at the sea entrance [3].

Politically, the Curonian Lagoon is divided into two parts, the southern two-thirds belonging to Kaliningrad Oblast, which is the exclave territory of the Russian Federation, and the northern one-third to Lithuania. The area is in the transition zone from the temperate continental climate to the maritime one. Average annual precipitation level is ca. 750 mm [8]. The Curonian barrier spit separates the lagoon from the Baltic Sea. The east coast of the Curonian Lagoon is or has historically been part of the Nemunas Delta, which stretches about 70 km inland (Figure 4).

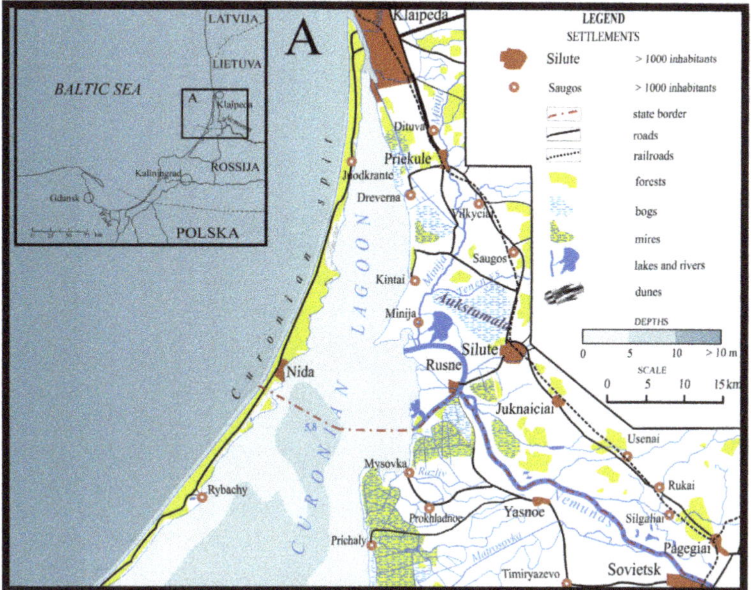

Figure 4. Curonian Lagoon (Lithuanian part). Source: authors' own plotting.

3. Results

3.1. WFD (2000/60/EC), MSFD (2008/56/EC) and MSPD (2014/89/EU): Delimiting Regulation Spheres in Open and Transitional Waters

Although the EU WFD provides an operational definition of transitional waters, still there is some fuzziness resulting from different approaches by the EU Member States in defining transitional waters [5,6,12]. The debates on transitional water definition became relevant within the EU, given the implementation of the WFD. There is also a need to define the limits of scope of the MSFD (2008/56/EC) and, particularly, those of MSPD (2014/89/EU). It states in its Preamble (paragraph 15) that MSP will contribute, among other things, to achieving the aims of the WFD.

The Preamble of the MSPD further states (paragraph 16):"Marine and coastal activities are often closely interrelated. In order to promote the sustainable use of maritime space, maritime spatial planning should take into account land-sea interactions". Article 2 (Scope) of the MSPD explicitly defines the distinction between marine and coastal waters in its very first paragraph: "1. This Directive shall apply to marine waters of Member States, without prejudice to other Union legislation. It shall not apply to coastal waters or parts thereof falling under a Member State's town and country planning, provided that this is communicated in its maritime spatial plans".

Such a definition of the MSPD scope means that the EU Member States should define the boundary between the transitional and coastal waters, which fall within the sphere of regulation of the MSFD and the WFD. The marine waters which are the focus of both the MSFD and the MSPD also are subject to MSP. If the transitional waters or their parts happen to fall under town and country planning, this must be communicated in the descriptive part of the maritime spatial plans. To make matters even more confused, three Baltic Sea countries—Latvia, Lithuania and Poland consider nearshore plumes resulting from the most extensive river discharge as transitional waters. It implies that these nearshore areas fall under the regulation of all three directives—WFD, MSFD and MSPD.

In any case, the essential difference between the WFD, MSFD and the MSPD is that the former two tackle any issues pertinent to the quality management and improvement of the aquatic environment. Meanwhile, the MSPD addresses different issues and its aim is to integrate the ecosystem approach in spatial planning. Therefore, considering the issues of a good water quality status and its indicators, which is at the core of the WFD and the MSFD, the MSPD plays a complementary role. As mentioned above, MSP should contribute, among other things, to achieving the aims of the WFD, i.e., a good status of transitional, coastal and marine waters of the EU Member States.

The WFD process for identifying transitional water body types required the development of new approaches [2]. It further implied the necessity to adopt a standard set of typology factors (tidal range, salinity, vulnerability), and approaches for consistent and comparable typology categorisation across the European regional seas. Although the WFD separated coastal waters from marine ones on the basis of a 1-mile distance according to Article 2–7, it also implied that the estuarine, coastal and marine water body types are not distinct categories that can be identified by a set of factors, but rather a continuum. Therefore, the borderline between the three separate types is often difficult to define [34].

There are still doubts whether transitional waters ought to be excluded from the MSPD focus, if they have a sizeable marine influence, e.g., tidal lagoons and estuaries or where salinity incursion occurs as these are also part of marine systems [5]. According to Tagliapietra et al. [6], in a landscape ecology context, transitional waters as a specific class of habitats should be even approached as 'transitional seascapes' emphasising their marine character. Borja et al. [34] concur that there is a necessity for a seamless and harmonised transition from a watershed through transitional and coastal waters to a marine system.

Cases of transboundary water bodies best illustrate the fuzziness and difficulties with the precise delimitation of different water ecosystem types even within the 'borderless' EU. For example, in the Odra Lagoon, the German and Polish parts belong to different typologies: the German part is designated as coastal waters while the Polish part as transitional waters, which is confusing for both management and research purposes [8]. A similar situation is in Lithuania, where the nearshore freshwater plume discharged from the Curonian Lagoon to the Baltic Sea heading north stops being considered as transitional waters at the Latvian border [9].

3.2. Ecosystem Goods and Services of Transitional Waters

The functioning of transitional water ecosystems, if they are healthy, produces several essential goods and services for humans—biodiversity conservation, biological production, storm and flood protection, river flow purification, transformation and cycling of elements and nutrients, wastewater treatment. However, goods and services of transitional waters are not defined adequately yet [5] because ecological concepts and intangible ecosystem services such as resilience are still meagrely measured for marine and estuarine environments [23]. Yet, these services have to be quantified and related to the management framework to provide a holistic approach to managing these habitats.

Transitional waters provide biological resources commercially exploited since the very Stone Age [35,36]. Due to the geographical position between the firm terrestrial ground and deep sea, these shallow water bodies play an essential role as spawning areas for fish and invertebrates, support rich biodiversity and offer important migration corridors for fish and wetland birds [3]. Despite the high value of the goods and services provided by the transitional waters, the coverage of the provisioning potential of ecosystem goods and services by transitional and marine systems is less documented, if compared with the terrestrial systems [23]. This lack of information hinders the MSP process and the compliance with EU supranational legislation by the Member States [12].

According to the Millennium Ecosystem Assessment, ecosystem goods and services pertinent to transitional waters can be grouped into four broad categories [37]:

1. *Provisioning* of biological and non-biological products such as supplying of food and water. Transitional waters provide fish, shellfish, crustaceans and sea-weeds. They supply building materials such as sand and gravel, and medicinal products from marine plants, microbes and

animals. The definition can also include renewable energies (wind and wave power as well as tidal power systems for estuaries).

2. *Regulating* services are the benefits of regulating ecosystem processes such as climate and disease control. Transitional waters outperform all other ecosystems in terms of regulating services [20]. Transitional waters and their habitats like mangroves, salt marshes and intertidal flats regulate several material flows. They recycle various elements, retain excess nutrients that flow into the sea, protect the hinterland from floods caused by storms or hurricanes, and absorb and process waste materials.

3. *Cultural* services are intangible benefits that people draw from ecosystems, for example through relaxation and aesthetic experiences [38,39]. Cultural heritage is an important trait of cultural services provided by transitional water ecosystems, that is, the development of local cultures with peculiar ethnic connotations. From an inspirational point of view, without referring to the uniqueness of Venice, the Camargue of the Impressionist painters is worth-mentioning.

4. *Supporting* services are those that necessary for the production of all other ecosystem services, like soil formation and nutrient cycling. Primary production is another supporting service as it fuels and maintains the higher trophic levels of the ecosystem and its biodiversity in transitional waters, and in the adjacent sea. For example, coastal lagoons, estuaries and other transitional waters form the main nurseries for juveniles of many commercially harvestable fish species.

More specifically, the services and goods garnered from the transitional water ecosystems can be further categorised into seven groups:

- Harvesting of fish, cray fish, mussels, clams and shrimps [40];
- Growing domestic water fowl, halophytes for fodder and ethnic medicine, spices, fruits and producing traditional local wine;
- Protecting the marine environment from physical disturbances caused by flooding and from chemical disturbances caused by pollution from the watershed [41];
- Conserving aquatic biodiversity, especially the biodiversity of migratory fish and birds [42];
- Providing amenities for water- and nature- tourism and other types of outdoor recreation [19];
- Maintaining coastal cultural and historical heritage values like traditions of combining fisheries and farming, as well as sustainable small-scale aquaculture [3];
- Providing diverse and relatively readily available data for environmental research, education and public awareness illustrating the relationships between ecological, physical and human processes that shape the environment [8].

The increasing use of aquatic resources by all sectors of society and the mismanagement due to many conflicting stakeholder interests are responsible for the deterioration of these ecosystems and the decline of their economic value. The benefits of these ecosystems are threatened by the activities of humans [43]. Transitional waters are under constant pressure, including habitat loss and pollution from their surroundings and catchment areas [26,44]. Examples of this are the decreasing capacity of the transitional waters to supply fishery products or to ensure the circulation of elements [3,27,28].

In the last 60 years, humans have changed the ecosystems of transitional waters faster and more comprehensively than in any comparable period in the past [45]. Although changes in ecosystems have contributed to significant net gains for human well-being and economic development, these gains have had huge costs such as the deterioration of many ecosystem services and the increased risk of adverse changes [46]. The declining provision of ecosystem goods and services from transitional waters worldwide can increase significantly in the future, and adversely influence human well-being.

Adverse changes in ecosystems directly feedback to the socio-economic system that relies on the ecosystem goods and services of transitional waters. An example is the loss of estuarine wetlands as fish nursery areas whereby these juvenile fish develop to become the commercial stocks [46]. For the sustainable management of environmental resources, identifying and quantifying ecosystem goods and services are increasingly required [47]. The decline of the carrying capacity of the transitional waters will have long-term effects. The degradation of the transitional coastal ecosystems also acts as a bottleneck in the movement of wildlife. This bottleneck will hinder the migration of birds from the southern wetlands to the Arctic breeding grounds and the migration of fish from the sea to rivers.

3.3. Case Study: Comparative Analysis of the Curonian Lagoon (Baltic Sea) and Lesina Lagoon (Adriatic Sea)

3.3.1. Emerald Growth Drivers

In this comparative case study, we illustrate the concept and analytical framework of Emerald Growth through the analysis and interpretation of the critical drivers of European transitional waters, using Lesina Lagoon (Italy) and the Curonian Lagoon (Lithuania/Russia) as an example. These areas are nutrient and contaminant traps and valuable fish habitats, and therefore vitally important for the integrity both of terrestrial and marine environment. Due to shallow water conditions and variation in water salinity, coastal lagoons are among the richest biotopes on the Earth with high productivity and rich biodiversity [20].

In our opinion, the usefulness of the Emerald Growth concept for analysing trends and capabilities for sustainable development and management of transitional waters can be best illustrated through the analysis of the Emerald Growth drivers. As in [48], we interpreted drivers as external forces on which the individual actors cannot possibly have an impact with their own means. It is notable, that Emerald Growth drivers and the alternative development scenarios for transitional waters, like the Blue Growth drivers and the alternative development scenarios for marine waters, are closely interrelated.

Different alternative scenarios are not only resulting from the sequence of political reforms or societal developments shaping the potential of the transitional water areas, but are also dependent on the external drivers and forces which have an impact on the future and should be considered seriously in policy formulation, planning and decision-making [48]. The descriptions of drivers and scenarios can be done for each Emerald Growth sector, fisheries, agriculture and outdoor recreation in the case of the Curonian and Lesina Lagoons. For these lagoons, we have identified the main five specific drivers (Table 2): the depletion of living resources, eutrophication, sea level rise, land reclamation, as well as growing industrial and recreational use.

Yet, a comparative analysis of the Curonian and Lesina Lagoon development in the modernity shows that the main driver of Emerald Growth throughout ages was active human interference into natural processes. An untenable interference had caused the depletion of fish resources, reclamation of aquatic-terrestrial ecotones, and shift towards reliance on the subsidised intensive agriculture in the reclaimed areas (Table 3) and the resulting structural decline of the lagoon economy (Figure 5). Similar negative processes had adversely affected many other European lagoons in the modernity as well [3]. Such a nature mismanagement policy had resulted in the declining diversity of migratory fish and bird species and recurring sharp production crises in the lagoon areas.

The societies and environment of the Curonian and Lesina lagoon areas have undergone immense changes during the modernity from the Napoleonic reforms till nowadays. These changes had resulted in economic ups and downs within the course of modern history. There always existed differences between the Emerald Growth drivers in two lagoons, but the similarities prevailed. It is evident from the scheme (Figure 5) what the resulting relative economic share of the yielded local commodities from the lagoons (fish) and their fringes (agricultural products) on the regional scale in various phases of the lagoon ecosystem evolution was. We see that radical changes in the tenure of land and water inevitably led to imminent economic and ecological precarities.

Table 3. Emerald Growth in the European lagoons in the modernity.

Period	Feudal Tenure	Privatisation	Cooperation	Current Stage
Main yield	Fish, cattle forage	Dairy products, vegetables, fish	Dairy products, vegetables, fish	Dairy products, vegetables
Fish stock use	Sustainable	Unsustainable	Quasi-sustainable	Quasi-sustainable
Causes of precarity	Diseases (plague, malaria), wars	Fish stock depletion, wars	Diking, land reclamation	Eutrophication, over-reliance on subsidies
Environmental concerns	Low	Increasing (eutrophication, persistent pollution)	Increasing (eutrophication, persistent pollution)	Stable
Provisional ecosystem services	High	Intermediate	High	Decreasing *
Cultural ecosystem services	Low	Increasing	Increasing	High
Economic role	High	Low	High	Low

* In case of nature protection focus scenario.

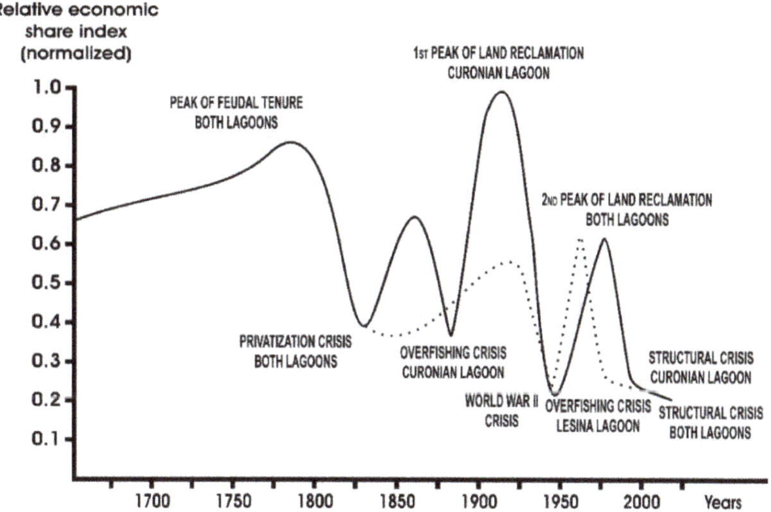

Figure 5. Resulting relative economic share of yielded products from lagoon areas on a regional scale (taking the first peak of land reclamation for 1.0; the dashed line shows the relative share of the products from Lesina Lagoon if it differs from the Curonian Lagoon) (Compiled and plotted by Ramūnas Povilanskas)

Depletion of Living Resources

Availability and eventual depletion of fish resources are the main specific drivers of Emerald Growth. The traditional fishing strategies differ significantly in the Mediterranean, Baltic and Atlantic lagoons, due to different physical processes, which control the water level, water exchange between the sea and the lagoon and water salinity in different seasons. In the Mediterranean lagoons, the fish species, which are the most important for lagoon fisheries, enter the lagoons at the end of winter as newborn fry, while for the sexually mature fish the impulse to migrate back to the sea comes in autumn at the onset of the breeding season. The years between the stage of fry and that of sexual maturity fish spend in the lagoon [27,28].

In the Baltic Sea lagoons, which are controlled by surplus precipitation and an influx of vast volumes of freshwater from large tributaries, the migration pattern of the important commercial fish species—other than eels—is opposite to that of the Mediterranean lagoons. Since the commercial fish like salmon, sea trout or twaite shad prevailed in the catches, the main fishing areas traditionally had been either in the lagoon or in the mouths of the river tributaries and branches or regularly flooded shallow deltaic lakes as well.

While the Curonian Lagoon and its tributaries had always remained a state domain, Lesina Lagoon was a private property till the very 1934. Both lagoons had been famous for centuries for their eels, which were caught during their migration to the sea and sent to the major cities of Italy and Germany where they constituted the traditional Christmas dish. Hence, fishing of eels and other commercially valuable fish made the lagoon resource owners and users wealthier than the farmers in adjacent areas. In both lagoons, however, the mismanagement of fish stock and lack of proper fishery control had led to the depletion of fish resources and the decline of fisheries.

Throughout the modernity, both lagoons experienced disputes among the local communities over the control and use of fishing areas. These disputes have resulted in protracted a period of decadence in the fishery management. In the 1970s, in Lesina Lagoon the annual yield in fish had dropped to less than 40 kg/ha, and the fishermen had dwindled to 40 units, mostly pensioners rounding off their income [27]. The 1970s–1980s also witnessed a dramatic global decline of the eel stock [49], which had a detrimental effect on eel fisheries, and on the welfare of the fishermen in both lagoons.

While entering the 2020s, a slow yet sure decline of commercial and artisanal fishing continues in both lagoons. The economic reforms of the 1990s have radically changed the fishing organisation in the Curonian Lagoon. Many small private fishing enterprises have got the license to fish. It has proved harder to monitor the size of the catches landed, and poaching, which gives a substantial nontaxed profit, has increased drastically [3]. Lack of proper collaboration with Russia on the protection of fish resources also played a negative role. Although fish monitoring in both parts of the lagoon is conducted in a coordinated way [8], yet it does not prevent the decline of the lagoon fishery.

Similarly, Lesina Lagoon experiences a further decline of the artisanal lagoon fishery. The weirs are opened and closed with regard to the fish migrations, and fishing rules are enforced [3]. However, the lagoon is still exploited by small vessels of 40 artisanal fishermen using traditional fishing gears and targeting a broad spectrum of species [50]. The local fishery is still mainly based on the use of "paranza" (Figure 6), a traditional fishing system made up of net walls fixed on stakes and retaining devices (fyke-nets), but also on the use of gillnets and trammel nets. It features low selectivity and produces a considerable amount of by-catch, including juveniles of commercial species [50].

Figure 6. Artisanal stake-nets in Lesina Lagoon (Image: Ramūnas Povilanskas).

Other Specific Drivers of Emerald Growth

With the advent of industrial agriculture, wetland fringes of both lagoons had been reclaimed and turned into agricultural landscapes, whereas the ecotones, which integrated aquatic and terrestrial ecosystems, had been extensively eliminated. Such a situation significantly impaired the capacity of the lagoon areas to fulfil their fundamental ecological functions, whereas, the reclaimed land at the Curonian Lagoon became ever more dependent on the external subsidies in the Soviet period. Central USSR ministries of agriculture and fisheries provided production plans and techniques, financed land reclamation, maintenance of dikes and polders and restocking of fish [29].

After the collapse of the Soviet Union, the main local agro-industrial activities, namely dairy cattle breeding, and milk production have dwindled mostly to the low-productivity farming at smallscale individual farms. Since the maintenance of dikes and water pumping stations needs substantial financial resources, only part of the polder system was maintained properly, due to the lack of proper financial and energy resources, while the rest was left unattended and rapidly declined. Currently, agriculture on the Lithuanian side of the Nemunas Delta depends heavily on the EU agricultural sub-sidies, likewise on the reclaimed fringes of Lesina Lagoon.

In the Lesina Lagoon area, state authorities have drained 1500 ha of freshwater wetlands in the 1950s. These efforts were part of a comprehensive land reclamation program aimed to increase the acreage of land suitable for agricultural purposes, and for maintaining the permanent high-water level within the open lagoon perimeter as well. These areas used to be a prime habitat and feeding ground for the eel. The engineering works had the effect of lowering the salinity at the eastern end of the basin. This intervention has caused the 500 ha, once occupied by eelgrass, to be invaded by reed growth. Thus, the engineering works took away a third of the prime habitat for eel production [3].

The prospect of lucrative recreational fishing services makes the Curonian lagoon communities waver: whether to continue relying on the commercial exploitation of the declining fish stocks or to shift to the provision of the recreational fishing services. The Curonian Lagoon region has been always famous as a major destination for recreational fishermen [3]. Considering the number of tourists visiting the Curonian Lagoon and, particularly, the Nemunas Delta, the fishermen constitute a significant part (ca. 50%). A phenomenon of particular interest is the upstream migration of smelt in winter and early spring. It attracts thousands of recreational fishermen. This 'secondary' wave of tourists gives the major low-season share of income to the local hospitality sector in the region.

Eutrophication is also a pivotal driver of Emerald Growth. Eutrophication itself is usually considered as a result of a vast array of internal and external forcing drivers [31]. Yet, eutrophication related drivers are among the most important factors causing pressure and impacts on the transitional water bodies [26]. Due to their large variety in the turnover rate and, hence, due to considerable exposure to rapid nutrient enrichment, the transitional water ecosystems, in general, and the Curonian and Lesina Lagoons in particular, are vulnerable to eutrophication [31]. Therefore, different Emerald Growth scenarios, from the maintenance of traditional artisanal fisheries to the development of new sectors, e.g., tourism, are directly reliant on the current eutrophication levels and their future trends in both lagoons.

Considering different Emerald Growth scenarios, eutrophication of transitional waters, despite recent positive trends, including re-oligotrophication [51], can be seen as a constraint for sustainable development in transitional waters. The simultaneous increases in nutrient loads and in the rate of sea-level rise may result in adverse synergistic effects. The drowning of coastal marshlands due to sea-level rise and loss of creek-edge marsh [52] due to eutrophication impose unwelcome limits on human activities and may result in coastal areas with a dramatically reduced capacity to provide important ecological and economic services.

Last but not least, the sea level rise can be also considered as a critical driver of Emerald Growth, especially regarding the loss of coastal marshlands, and an ever-accelerating erosion of barrier spits separating coastal lagoons from the ocean or the sea. Lagoons and estuaries are highly adaptable ecosystems. In natural conditions, these systems would have moved along the coast following its development under sea level rise. However, the lack of plasticity of the modern human landscape and the rapidity of climate-related changes put their very existence at risk.

The Emerald Growth concept, envisaging local-scale integrated economy, reuse of materials and innovative widening the use of local materials and provisional services, for instance, the use of salttolerant plants and saltmarsh gardening, may be included in the accommodation measures. State-of-the-art agriculture models must be adopted in the transitional islands with complete water recycling and zero emissions. An opportunity, on the other hand, can be provided by 'depolderization', that is, the planned reconversion of lagoons and wetlands reclaimed during the last two centuries often lying below the main sea level, whose productivity is currently possible only using hydraulic pumping.

3.3.2. Preconditions for Sustainable Emerald Growth Scenarios

Surrounding Lesina Lagoon with an embankment has eliminated the once wide ecotone of shallow water, which was the habitat of wetland birds [3]. With the undersigning of the Ramsar Convention (1971), the Italian government has finally renounced its policy of draining wetlands. Now the surviving wetlands are to be conserved to benefit bird species broadly termed as wetland birds [53]. Therefore, in 1981, the eastern portion of Lesina Lagoon (Sacca Orientale, IT9110031, 927 ha, Figure 7) has been designated as a bird sanctuary. In 1991 it was included within the newly instituted Gargano National Park.

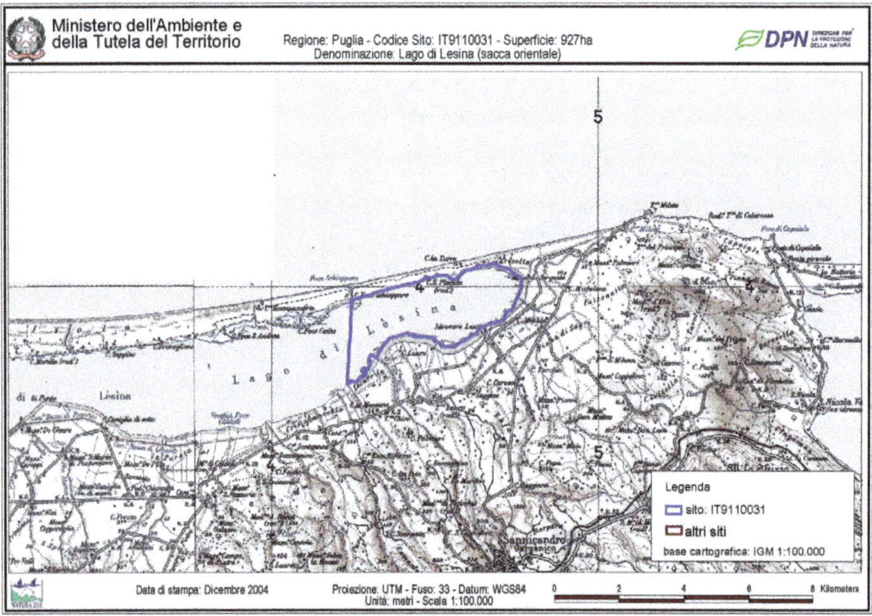

Figure 7. Part of Lesina Lagoon which is a bird sanctuary included into the Gargano National Park (Courtesy: Directorate for Nature Protection).

High concentrations of wetland birds in the Nemunas Delta also classify the area as being of international importance for nature conservation. Nemunas Delta Regional Park established in 1992 was included in the list of wetlands of international importance by the Ramsar Convention [3]. Later, in 2004–2005, the system of protected areas was supplemented with Natura 2000 sites in compliance with the requirements of the EU Birds and Habitats Directives. The whole Curonian Lagoon itself was designated as a biosphere polygon in 2009. Legislative measures create a number of constraints mostly towards the intensive use of natural resources while providing additional incentives for the culural uses of the lagoon area.

This development could be considered typical for the transitional waters featuring important cultural and nature values (e.g., UNESCO heritage sites) and goes well in line with proposed Emerald Growth trends in the European lagoons in the modernity (Table 3). This scenario (nature protection priority) implies an increase in the importance of cultural ecosystem services (including recreational fishery) with a decrease in the value of provisioning ones. However, this does not imply a drop in income or economic value [54]. Indeed, if managed correctly, the nature protection priority could even increase it.

Another Emerald Growth scenario (sustainable development priority) could be proposed for the territories lacking much of natural and cultural values where enviromentally friendly production solutions (e.g., sustainable aquaculture beside sustainable fishery) along with other innovations (alternative energy production, extraction of biologically active chemical compunds from natural sources) could be a future development trend.

4. Discussion

From the presented case study, we see that Emerald Growth can evolve in different directions at different periods in history—from intensive development to extensive development, from sustainable development to decline. We also see that fisheries in transitional waters are ever less important for sustainable Emerald Growth. For the use of transitional waters' ecosystem goods and services to be truly sustainable, the riparian communities using transitional waters need to search for alternative under-utilised ecosystem services to integrate a sustainablefisheries. In the case of transboundary transitional waters, the Emerald Growth deliberations must also take into account the situation and plans of the neighbouring country sharing the transitional waters. These plans may diverge significantly and adversely affect your plans because the aquatic ecosystems do not recognise borders.

The findings of our study support the notion that the WFD process for identifying coastal and transitional water body types requires the development of new typological principles. We need to elicit a standard set of factors and their categories for comparable and consistent categorisation of various transitional waters across the coastal areas of different regional seas. We should also reconsider if estuaries and other transitional waters have to be excluded from MSP as currently, it is the case in the EU. On the contrary, it is necessary to find a balance between different EU directives (WFD, MSFD and MSPD) and other supranational regulations pertinent to the environmental protection of aquatic environments in each particular case in order to deliver a holistic approach for the transitional waters' management [55,56].

The EU nature conservation directives—the Birds Directive (2009/147/EC) and the Habitats Directive (92/43/EEC)—are also highly relevant documents for the Emerald Growth framework since there is a high emphasis on priority coastal habitat types (such as coastal lagoons) in these applications within the EU Natura 2000 scheme (the world's largest network of protected areas). Emerald Growth provides a suitable framework in the need to promote holistic cultural, natural and sustainability management approaches in the Natura 2000 sites [57–59]. It is obvious that protected territories established in both lagoons create new possibilities for sustainable Emerald Growth including eco-oriented tourism (especially birdwatching), recreational fishery and sustainable uses of specific local products.

Interaction between the watersheds of the rivers discharging into the transitional waters and their ecosystems, particularly regarding their productivity and resilience [60,61], is also among the essential issues of the perspectives and limitations of the Emerald Growth concept. This issue is especially pertinent for facilitating sustainable management of the two largest south Baltic coastal lagoons—the Curonian Lagoon and Odra Lagoon being recipient water bodies of large rivers—Nemunas and Odra, respectively. As many environmental problems affecting the transitional waters are generated upstream in the watersheds, like in the case of Blue Growth drivers, the 'sustainability dilemma' strongly relies upon environmental leadership, innovation and a common focus on the circular economy [48].

5. Conclusions

The main practical implication of the study is that on the policy aspect, contrary to a commonly accepted view among policymakers throughout the EU, stronger links should be established between the Emerald Growth and MSP principles. The Emerald Growth concept offers a suitable conceptual framework for an adequate understanding and dealing with complex issues pertinent to environmental protection and sustainable development of the economies of transitional waters, particularly the transboundary ones. The differences in formal designation and planning approaches between the countries sharing the transboundary transitional waters should not be considered as an obstacle to the cross-border co-operation efforts to manage the transitional waters, environmental protection, nurturing of ecosystem goods and services and sharing a joint vision of Emerald Growth.

The investigation results show that the Emerald Growth concept and management framework enables to extricate better specific management aspects of ecosystem services of transitional waters that fill-in the continuum between the terrestrial (Green Growth) and the maritime areas (Blue Growth). The Emerald Growth concept offers a suitable framework for better dealing with complex and complicated issues pertinent to sustainable management of transitional waters. Although each transitional water body is unique in many ways, yet the drivers shaping their future development scenarios are similar and some standard integrated planning and sustainable management procedures are attainable.

Author Contributions: Conceptualization, D.T. and R.P.; methodology, D.T. and A.R.-B.; validation, D.T.; formal analysis, D.T., A.R.-B. and J.T.; investigation, D.T.; data curation, R.P.; writing—original draft preparation, D.T.; writing—review and editing, R.P.; visualization, R.P.; supervision, A.R.-B. and J.T.; project administration, R.P.; funding acquisition, R.P. All authors have read and agreed to the published version of the manuscript.

Funding: This research was funded by the EU Interreg South Baltic Program as part of the work according to the Subsidy Contract No. STHB.04.01.00-22-0111/17 for the ERDF co-financing of SEAPLANSPACE project – Marine spatial planning instruments for sustainable marine governance.

Acknowledgments: Aušrinė Armaitienė has contributed to field studies in the Curonian Lagoon area while Paolo Breber has contributed to field studies in the Lesina Lagoon area. Authors are also grateful to the ano-nymous reviewers of the manuscript whose valuable suggestions helped to substantially improve the article.

References

1. Basset, A.; Sabetta, L.; Fonnesu, A.; Mouillot, D.; Do Chi, T.; Viaroli, P.; Giordani, G.; Reizopoulou, S.; Abbiati, M.; Carrada, G.C. Typology in Mediterranean transitional waters: New challenges and perspectives. *Aquat. Conserv.* **2006**, *16*, 441–455. [CrossRef]
2. Tagliapietra, D.; Volpi Ghirardini, A. Notes on coastal lagoon typology in the light of the EU Water Frame-work Directive: Italy as a case study. *Aquat. Conserv.* **2006**, *16*, 457–467. [CrossRef]
3. Breber, P.; Povilanskas, R.; Armaitienė, A. Recent evolution of fishery and land reclamation in Curonian and Lesina lagoons. *Hydrobiologia* **2008**, *611*, 105–114. [CrossRef]
4. Viaroli, P.; Mistri, M.; Troussellier, M.; Guerzoni, S.; Cardoso, A.C. Structure, functions and ecosystem alterations in Southern European coastal lagoons. Preface. *Hydrobiologia* **2005**, *550*, 7–9.
5. McLusky, D.S.; Elliott, M. Transitional waters: A new approach, semantics or just muddying the waters? *Estuar. Coast. Shelf Sci.* **2007**, *71*, 359–363. [CrossRef]
6. Tagliapietra, D.; Sigovini, M.; Ghirardini, A.V. A review of terms and definitions to categorise estuaries, lagoons and associated environments. *Mar. Freshw. Res.* **2009**, *60*, 497–509. [CrossRef]
7. Razinkovas-Baziukas, A.; Povilanskas, R. Chapter 1. In *troducing transitional waters. In Transboundary Management of Transitional Waters—Code of Conduct and Good Practice Examples*; Nilsson, H., Povilanskas, R., Stybel, N., Eds.; EUCC Germany: Warnemünde, Germany, 2012; pp. 8–14.

8. Povilanskas, R.; Razinkovas-Baziukas, A.; Jurkus, E. Integrated environmental management of trans-boundary transitional waters: Curonian Lagoon case study. *Ocean Coast. Manag.* **2014**, *101*, 14–23. [CrossRef]

9. Schernewski, G.; Wielgat, M. Towards a Typology for the Baltic Sea. In *Managing the Baltic Sea*; Schernewski, G., Löser, N., Eds.; EUCC Germany: Warnemünde, Germany, 2004; pp. 35–52.

10. Krzymiński, W.; Kruk-Dowgiałło, L.; Zawadzka-Kahlau, E.; Dubrawski, R.; Kamińska, M.; Łysiak-Pastuszak, E. Typology of Polish marine waters. In *Baltic Sea Typology*; Schernewski, G., Wielgat, M., Eds.; EUCC Germany: Warnemünde, Germany, 2004; pp. 39–48.

11. Cardeccia, A.; Marchini, A.; Occhipinti-Ambrogi, A.; Galil, B.; Gollasch, S.; Minchin, D.; Narščius, A.; Olenin, S.; Ojaveer, H. Assessing biological invasions in European Seas: Biological traits of the most widespread non-indigenous species. *Estuar. Coast. Shelf Sci.* **2018**, *201*, 17–28. [CrossRef]

12. Elliott, M.; McLusky, D.S. The need for definitions in understanding estuaries. *Estuar. Coast. Shelf Sci.* **2002**, *55*, 815–827. [CrossRef]

13. Aoki-Suzuki, C. Chapter 1. Green Economy and Green Growth in international trends of sustainability indicators. In *The Economics of Green Growth: New indicators for Sustainable Societies*; Managi, S., Ed.; Rout-ledge: London, UK; New York, NY, USA, 2015; pp. 23–46.

14. Eikeset, A.M.; Mazzarella, A.B.; Davíðsdóttir, B.; Klinger, D.H.; Levin, S.A.; Rovenskaya, E.; Stenseth, N.C. What is blue growth? The semantics of 'Sustainable Development' of marine environments. *Mar. Policy* **2018**, *87*, 177–179. [CrossRef]

15. Boonstra, W.J.; Valman, M.; Björkvik, E. A sea of many colours—How relevant is Blue Growth for capture fisheries in the Global North, and vice versa? *Mar. Policy* **2018**, *87*, 340–349. [CrossRef]

16. Burgess, M.G.; Clemence, M.; McDermott, G.R.; Costello, C.; Gaines, S.D. Five rules for pragmatic blue growth. *Mar. Policy* **2018**, *87*, 331–339. [CrossRef]

17. Jänicke, M. Green growth: From a growing eco-industry to economic sustainability. *Energy Policy* **2012**, *48*, 13–21. [CrossRef]

18. Lyytimäki, J.; Antikainen, R.; Hokkanen, J.; Koskela, S.; Kurppa, S.; Känkänen, R.; Seppälä, J. Developing Key Indicators of Green Growth. *Sustain. Dev.* **2018**, *26*, 51–64. [CrossRef]

19. Clara, I.; Dyack, B.; Rolfe, J.; Newton, A.; Borg, D.; Povilanskas, R.; Brito, A.C. The Value of Coastal Lagoons: Case Study of Recreation at the Ria de Aveiro, Portugal in comparison to the Coorong, Australia. *J. Nat. Conserv.* **2018**, *43*, 190–200. [CrossRef]

20. Newton, A.; Brito, A.C.; Icely, J.D.; Valérie, D.; Inês, C.; Stewart, A.; Gerald, S.; Miguel, I.; Ana, I.L.; Ana, I.S.; et al. Assessing, quantifying and valuing the ecosystem services of coastal lagoons. *J. Nat. Conserv.* **2018**, *44*, 50–65. [CrossRef]

21. Newton, A.; Icely, J.; Cristina, S.; Ana, B.; Ana, C.C.; Franciscus, C.; Simona, D.R.; Flemming, G.; Jens, W.H.; Marianne, H.; et al. An overview of ecological status, vulnerability and future perspect-ives of European large shallow, semi-enclosed coastal systems, lagoons and transitional waters. *Estuar. Coast. Shelf Sci.* **2014**, *140*, 95–122. [CrossRef]

22. Povilanskas, R.; Armaitienė, A.; Breber, P.; Razinkovas-Baziukas, A.; Taminskas, J. Integrity of linear littoral habitats of Lesina and Curonian Lagoons. *Hydrobiologia* **2012**, *699*, 99–110. [CrossRef]

23. Razinkovas, A.; Gasiūnaitė, Z.; Viaroli, P.; Zaldívar, J.M. Preface: European lagoons—Need for further comparison across spatial and temporal scales. *Hydrobiologia* **2008**, *611*, 1–4. [CrossRef]

24. Tagliapietra, D.; Magni, P.; Basset, A.; Viaroli, P. Ecosistemi costieri di transizione: Trasformazioni recenti, pressioni antropiche dirette e possibili impatti del cambiamento climatico. *Biol. Ambient.* **2014**, *28*, 101–111.

25. Taminskas, J.; Pileckas, M.; Šimanauskienė, R.; Linkevičienė, R. Wetland classification and inventory in Lithuania. *Baltica* **2012**, *25*, 33–44. [CrossRef]

26. Zaldívar, J.M.; Cardoso, A.C.; Viaroli, P.; Newton, A.; De Wit, R.; Ibañez, C.; Reizopoulou, S.; Somma, F.; Razinkovas, A.; Basset, A.; et al. Eutrophication in transitional waters: An overview. *Transit. Waters Monogr.* **2008**, *2*, 1–78.

27. Breber, P. The situation of lagoons in Italy today. *J. Coast. Conserv.* **1995**, *1*, 173–175. [CrossRef]

28. Breber, P.; Cilenti, L.; Scirocco, T. The vicissitudes of a coastal lagoon from the 19th century to the present day. In *The Changing Coast, Proceedings of Littoral 6th—International Multi-Disciplinary Symposium on Coastal Zone Research, Management and Planning, Porto, Portugal, 5–7 September, 2002*; Veloso Gomes, F., Taveira Pinto, F., das Neves, L., Eds.; EUROCOAST—EUCC: Porto, Portugal, 2002; Volume 1, pp. 375–381.

29. Roepstorff, A.; Povilanskas, R. On the concepts of nature protection and sustainable use of natural resources: A case study from the Curonian lagoon. In *Coastal Conservation and Management in the Baltic Region, Proceedings of the EUCC–WWF Conference, Klaipėda, Lithuania, 2–8 May 1994*; Gudelis, V., Povilanskas, R., Roepstorff, A., Eds.; University Publishers: Klaipėda, Lithuania, 1995; pp. 223–232.

30. Kaplowitz, M.D.; Hoehn, J.P. Do focus groups and individual interviews reveal the same information for natural resource valuation? *Ecol. Econ.* **2001**, *36*, 237–247. [CrossRef]

31. Roselli, L.; Fabbrocini, A.; Manzo, C.; D'Adamo, R. Hydrological heterogeneity, nutrient dynamics and water quality of a non-tidal lentic ecosystem (Lesina Lagoon, Italy). *Estuar. Coast. Shelf Sci.* **2009**, *84*, 539–552. [CrossRef]

32. Manini, E.; Breber, P.; D'Adamo, R.; Spagnoli, F.; Danovaro, R. Lake of Lesina. In *South-Eastern Italian Coastal Systems*; LOICZ: Texel, The Netherlands, 2005; p. 17.

33. Armaitienė, A.; Boldyrev, V.L.; Povilanskas, R.; Taminskas, J. Integrated shoreline management and tourism development on the cross-border World Heritage Site: A case study from the Curonian spit (Lithuania/Russia). *J. Coast. Conserv.* **2007**, *11*, 13–22. [CrossRef]

34. Borja, Á.; Elliott, M.; Carstensen, J.; Heiskanen, A.-S.; van de Bund, W. Marine management—Towards an integrated implementation of the European Marine Strategy Framework and the Water Framework Directives. *Mar. Pollut. Bull.* **2010**, *60*, 2175–2186. [CrossRef]

35. Wilson, L.; Dickinson, P.; Jeandron, J. *Reconstructing Human-Landscape Interactions*; Cambridge Scholars Publishing: Cambridge, UK, 2009; p. 291.

36. Tilley, C. *An Ethnography of the Neolithic: Early Prehistoric Societies in Southern Scandinavia*; Cambridge University Press: Cambridge, UK, 2003; p. 388.

37. Razinkovas–Baziukas, A.; Povilanskas, R.; Armaitienė, A. Chapter 2. Ecosystem goods and services of transitional waters. In *Transboundary Management of Transitional Waters—Code of Conduct and Good Practice Examples*; Nilsson, H., Povilanskas, R., Stybel, N., Eds.; EUCC Germany: Warnemünde, Germany, 2012; pp. 15–22.

38. Urbis, A.; Povilanskas, R.; Newton, A. Valuation of aesthetic ecosystem services of protected coastal dunes and forests. *Ocean Coast. Manag.* **2019**, *179*, 104832. [CrossRef]

39. Urbis, A.; Povilanskas, R.; Šimanauskienė, R.; Taminskas, J. Key aesthetic appeal concepts of coastal dunes and forests on the example of the Curonian Spit (Lithuania). *Water* **2019**, *11*, 1193. [CrossRef]

40. Beaumont, N.J.; Austen, M.C.; Atkins, J.P.; Burdon, D.; Degraer, S.; Dentinho, T.P.; Derous, S.; Holm, P.; Horton, T.; van Ierland, E. Identification, definition and quantification of goods and services provided by marine biodiversity: Implications for the ecosystem approach. *Mar. Pollut. Bull.* **2007**, *54*, 253–265. [CrossRef]

41. Daily, G.C.; Polasky, S.; Goldstein, J.; Kareiva, P.M.; Mooney, H.A.; Pejchar, L.; Ricketts, T.H.; Salzman, J.; Shallenberger, R. Ecosystem services in decision making: Time to deliver. *Front. Ecol. Environ.* **2009**, *7*, 21–28. [CrossRef]

42. Nobre, A.M. An ecological and economic assessment methodology for coastal ecosystem management. *Environ. Manag.* **2009**, *44*, 185–204. [CrossRef] [PubMed]

43. Aubry, A.; Elliott, M. The use of environmental integrative indicators to assess seabed disturbance in estuaries and coasts: Application to the Humber Estuary, UK. *Mar. Pollut. Bull.* **2006**, *53*, 175–185. [CrossRef] [PubMed]

44. Duarte, C.; Conley, D.J.; Carstensen, J.; Sánchez-Camacho, M. Return to Neverland: Shifting baselines affect eutrophication restoration targets. *Estuar. Coasts* **2009**, *32*, 29–36. [CrossRef]

45. Solidoro, C.; Bandelj, V.; Bernardi Aubry, F.; Camatti, E.; Ciavatta, S.; Cossarini, G.; Facca, C.; Franzoi, P.; Libralato, S.; Canu, D.; et al. Response of the Venice Lagoon Ecosystem to Natural and Anthropogenic Pressures over the Last 50 Years. In *Coastal Lagoons: Critical Habitats of Environmental Change*; Kennish, M.J., Pearl, H.W., Eds.; CRC Press: Boca Raton, FL, USA, 2010; pp. 483–511.

46. Hassan, R.M.; Scholes, R.; Ash, N. (Eds.) *Ecosystems and Human Well-Being: Current State and Trends: Findings of the Condition and Trends Working Group of the Millennium Ecosystem Assessment*; Island Press: Washington DC, USA, 2005; p. 917.

47. Troy, A.; Wilson, M.A. Mapping ecosystem services: Practical challenges and opportunities in linking GIS and value transfer. *Ecol. Econ.* **2006**, *60*, 435–449. [CrossRef]

48. Pöntynen, R.; Erkkilä-Välimäki, A. *Blue Growth—Drivers and Alternative Scenarios for the Gulf of Finland and the Archipelago Sea: Qualitative Analysis Based on Expert Opinions*; Publications of the Centre for Maritime Studies—Series A75; Turun yliopiston Brahea-keskus: Turku, Finland, 2018; p. 134.

49. Dekker, W. On the distribution of the European eel (*Anguilla anguilla*) and its fisheries. *Can. J. Fish. Aquat. Sci.* **2003**, *60*, 787–799. [CrossRef]

50. Li Veli, D.; Lillo, A.O.; Lucchetti, A.; Mancinelli, G.; Scirocco, T.; Specchiulli, A.; Cilenti, L. Artisanal fishing in Lesina Lagoon: Critical points and innovative fishing techniques. In Proceedings of the 9th International Conference on Coastal Transitional Environments, Venice, Italy, 21–24 January 2020; Umgiesser, G., Ed.; CNR–ISMAR: Venice, Italy, 2020; p. 107.

51. Ibanez, C.; Peñuelas, J. Changing nutrients, changing rivers. *Science* **2019**, *365*, 637–638. [CrossRef]

52. Deegan, L.A.; Johnson, D.S.; Warren, R.S.; Peterson, B.J.; Fleeger, J.W.; Fagherazzi, S.; Wollheim, W.M. Coastal eutrophication as a driver of salt marsh loss. *Nature* **2012**, *490*, 388–392. [CrossRef]

53. Weller, M.W. *Wetland Birds: Habitat Resources and Conservation Implications*; Cambridge University Press: Cambridge, UK, 1999; p. 271.

54. Pauli, G. *The Blue Economy: 10 Years, 100 Innovations, 100 Million Jobs*; Paradigm Publications: Taos, NM, USA, 2010; p. 308.

55. Boyes, S.J.; Elliott, M. Marine legislation—The ultimate 'horrendogram': International law, European directives & national implementation. *Mar. Pollut. Bull.* **2014**, *86*, 39–47.

56. Da Luz Fernandes, M.; Esteves, T.C.; Oliveira, E.R.; Alves, F.L. How does the cumulative impacts approach support Maritime Spatial Planning? *Ecol. Indic.* **2017**, *73*, 189–202. [CrossRef]

57. Ferranti, F.; Turnhout, E.; Beunen, R.; Behagel, J.H. Shifting nature conservation approaches in Natura 2000 and the implications for the roles of stakeholders. *J. Environ. Plan. Manag.* **2014**, *57*, 1642–1657. [CrossRef]

58. Kati, V.; Hovardas, T.; Dieterich, M.; Ibisch, P.L.; Mihok, B.; Selva, N. The challenge of implementing the European network of protected areas Natura 2000. *Conserv. Biol.* **2015**, *29*, 260–270. [CrossRef] [PubMed]

59. Vlami, V.; Kokkoris, I.P.; Zogaris, S.; Cartalis, C.; Kehayias, G.; Dimopoulos, P. Cultural landscapes and attributes of "culturalness" in protected areas: An exploratory assessment in Greece. *Sci. Total Environ.* **2017**, *595*, 229–243. [CrossRef] [PubMed]

60. Cozzi, S.; Ibáñez, C.; Lazar, L.; Raimbault, P.; Giani, M. Flow regime and nutrient-loading trends from the Largest South European watersheds: Implications for the productivity of Mediterranean and Black Sea's Coastal Areas. *Water* **2019**, *11*, 1. [CrossRef]

61. Day, J.W.; Ibáñez, C.; Pont, D.; Scarton, F. Status and Sustainability of Mediterranean Deltas: The Case of the Ebro, Rhône, and Po Deltas and Venice Lagoon. In *Coasts and Estuaries: The Future*; Wolanski, E., Day, J.W., Elliott, M., Ramachandran, R., Eds.; Elsevier: Amsterdam, The Netherlands; Oxford, UK; Cambridge, MA, USA, 2019; pp. 237–249.

Review

Environmental Flows in the Lower Ebro River and Delta: Current Status and Guidelines for a Holistic Approach

Carles Ibáñez, Nuno Caiola and Oscar Belmar *

IRTA, Program of Marine and Continental Waters, Sant Carles de la Ràpita, 43540 Catalonia, Spain; carles.ibanez@irta.cat (C.I.); nuno.caiola@irta.cat (N.C.)
* Correspondence: oscar.belmar@irta.cat

Received: 9 August 2020; Accepted: 22 September 2020; Published: 24 September 2020

Abstract: Deltas are a particular type of estuarine system in which the dependence on river flow (water, sediments and nutrients) is very strong, especially in river-dominated deltas such as the Mediterranean ones, but environmental flow (e-flow) proposals for deltaic systems are scarce. The Ebro Delta is one of the largest wetland areas in the western Mediterranean and one of the most important estuarine systems in Europe. The aim of this paper is to review the state of the art regarding e-flows and to carry out a critical analysis of the proposals for the lower Ebro River and Delta, in order to highlight the possible environmental and socioeconomic impacts arising from the e-flow regime currently approved. Additionally, based on existing scientific information, methods to establish an e-flow regime that allows the maintenance of the main socio-ecological functions and values are discussed; including those functions and values for which not enough information is available. The study concludes that the currently approved e-flows are not suitable for maintaining most functions and values, as they would not prevent the proliferation of alien fish species and macrophytes in the river, the intrusion of the salt wedge in the estuary, the deficit of sediment/nutrient transport and the degradation of riparian habitats or the decline of coastal fisheries. Socioeconomic consequences on coastal fisheries, river navigation, salt water intrusion, sediment deficit, biodiversity, water quality, aquaculture and hydropower are also considered. Other e-flow proposals such as the proposed by the Catalan government would be more suitable to maintain the main socioecological functions and values of the lower Ebro River and Delta. Nevertheless, additional studies are needed to validate e-flows in some relevant aspects such as the capacity of the river to transport sediments to the delta to avoid coastal regression and mitigate the effects of sea level rise and subsidence, as well as the capacity of floods to control the spread of macrophytes. The lower Ebro River and delta is among the case studies where more quantitative and qualitative criteria to set e-flows with a holistic approach have been established.

Keywords: environmental flows; holistic approach; estuary; Ebro River; delta

---

1. Introduction

Numerous studies have shown the effects of modifying the natural hydrological regime on ecosystems [1]. A reduction in flow alters the width, depth, velocity pattern and shear stress [2,3]. Such variations can modify the distribution and availability of in-stream habitat, which can have detrimental effects on invertebrates and fish populations [4] and produce invasion of non-native species (e.g., [5]). In this context, environmental flows (e-flows) are calculated to define the "quantity, timing, and quality of freshwater flows and levels necessary to sustain aquatic ecosystems which, in turn, support human cultures, economies, sustainable livelihoods, and well-being" [6].

The science and practice of e-flows constitutes an approach to protect and recover aquatic biodiversity, ecosystem integrity and important ecological services in rivers and estuarine systems. The several methodologies that have been developed are organized in three general categories [7]. First, hydrological methods, which are based on the analysis of historic (existing or simulated) streamflow data and the supposition that such e-flow regimes drive ecological processes in streams and rivers, assuming that most of the species can cope with low flows typical of dry periods (i.e., summer flows or drought periods). Second, habitat simulation methods, which consist of physical or hydraulic modelling of the river channel and, then, modelling the corresponding biological response through different parameters (water depth, flow velocity, substrate composition, etc.). These methods are often used as a "biological validation" of the hydrological ones and are based on a single-species approach. Third, holistic methods to overcome limitations of the previous ones. They are sometimes referred to as "expert panels" where a multi-disciplinary team of experts seeks to consider ecological, social, cultural and economic functions and values of e-flows. Holistic approaches result recommendable for complex environments such as large rivers or estuarine systems, where multitude of factors must be taken into consideration.

Holistic e-flow methods have been developed during the past two decades and can be grouped into two main approaches based on either a bottom-up or top-down procedure to establish e-flow regimes [8]. The bottom-up procedures assume that it is possible to identify and quantify the critical components of the e-flow regime that need to be maintained, while top-down procedures assume that the whole natural flow regime is ecologically important but some flow components can be modified or removed without significantly altering ecological functions and values [9]. The first and best-known holistic approach is the Building Block Method (BBM) that was developed in South-Africa in the early 1990s to establish e-flow standards using limited amounts of data, and is designed to set an e-flow regime to maintain a river in a predetermined condition [10]. This method is based on the identification of the most important components, or "building blocks", of the natural low flows and floods essential for river biota. Unlike BBM, the DRIFT methodology (Downstream Response to Imposed Flow Transformation) is a top-down, interactive, scenario-based approach that includes all major abiotic and biotic components that constitute the ecosystem to be managed [11]. The DRIFT method includes four modules: biophysical, sociological, scenario development and economic. Another holistic method, the benchmarking approach, was designed to link data on alterations of natural flow regimes to ecological consequences; the main idea is to evaluate the condition of a range of rivers (or river reaches) that have been subjected to various degrees of flow regulation and water resource use [12]. Finally, the ELOHA framework (Ecological Limits of Hydrologic Alteration) is a more recent consensus approach from a group of internationally recognized environmental flow scientists to determine ecological limits of flow alteration at a regional scale [13]. For this purpose, rivers are classified according to their hydrological regime in order to define the ecological consequences of anthropogenic alteration within each class. Such framework simplifies the task because it avoids working on a river-by-river basis, but special attention must be paid to the types of rivers used and their suitability to discriminate relevant hydrological characteristics [14]. The main weakness of holistic methods is the assumption that the experts have a comprehensive knowledge of what constitutes a critical flow component for the considered functions and values (see [7] for details of the described holistic methods and their limitations). Moreover, some authors argue that classical approaches are not suited to modified and managed rivers where returning to natural conditions is no longer feasible, and their objective is to design e-flows in order to maximize natural capital and support economic growth, recreation or culture [15].

Estuarine systems are paradigmatic coastal landforms depending on river flows where the link between such flows and their socioecological functions and values is typically complex, so holistic methods are even more necessary for establishing e-flows than in rivers. Deltas are a particular type of estuarine system in which the dependence on river flow (water, sediments, nutrients) is very strong, especially in river-dominated deltas such as the Mediterranean ones [16,17], but e-flow proposals for deltaic systems are scarce [18–22]. E-flow requirements of estuaries have been ignored in the past, mostly because of the lack of long-term monitoring data or understanding of the responses to changes in freshwater inflow [23]. Three main countries have developed methods for estuaries (i.e., Australia, South Africa and USA), from practical applications and a learning-by-doing approach [23–28], but some efforts have also been made in other countries like China [22,29]. Recent methods take a holistic and adaptive standpoint and are presented as frameworks that include a number of steps and have elements of risk assessment and adaptive management [23]. Effective e-flows need to be explicit about flow-ecology relationships to adequately determine the amount and timing of water required [30]. This is not an easy task, especially in estuarine systems, so the main challenge is to perform good quality research and monitoring to establish sound quantitative relationships between flow regime features and socioecological functions and values, acknowledging that in many cases it is very difficult and too simplistic looking for single quantitative flow-function/value relationships. In this synthesis paper we use the case of the lower Ebro River and its delta to illustrate the complexity of establishing e-flows in a large river connected to a protected estuarine system, and the need to use a holistic approach in order to consider as many functions and values as possible.

The Ebro Delta is one of the largest wetland areas in the western Mediterranean and one of the most important estuarine zones in Europe [31]. Declared Natural Park in 1983, Special Protection Area (SPA) under the Birds Directive (Directive 2009/147/EC) in 2006 and World Biosphere Reserve in 2013, more than 8000 ha are protected by the Spanish legislation. The delta shows a great diversity of habitats, with endemic faunal (ornithological and ichthyologic) and halophilic floral composition [32], together with important human activities that require flows such as rice agriculture, fisheries, aquaculture and tourism [20]. Consequently, the lower Ebro's hydrology, geomorphology and ecology are strongly impacted by the existence, features and operation of dams [33].

The aim of this paper is to carry out a critical analysis of the e-flow proposals for the lower Ebro River and Delta (Figure 1), in order to highlight the possible environmental and socioeconomic impacts arising from the e-flow regime approved by the PHE. Additionally, based on existing scientific information, guidelines to establish an e-flow regime through a holistic approach that allows the maintenance of the main socio-ecological functions and values are discussed; highlighting those functions and values for which not enough information is available. More specifically, we show an example of determination of e-flows in which existing scientific information can be used (and is partially used by the Governments) to build an e-flow regime by overlapping the river flow needs of different socioecological functions and values. This is a holistic approach in the sense that the main environmental and socioeconomic functions and values are considered and can be progressively integrated into the e-flow regime. From the other hand, we show what goes on in terms of e-flow implementation by the administrations, comparing the e-flow regime in force (Spanish Government) and the e-flow proposed by the Government of Catalonia. We also show other functions and values not considered in these two official proposals that could be considered in the future to advance towards a more holistic e-flow regime.

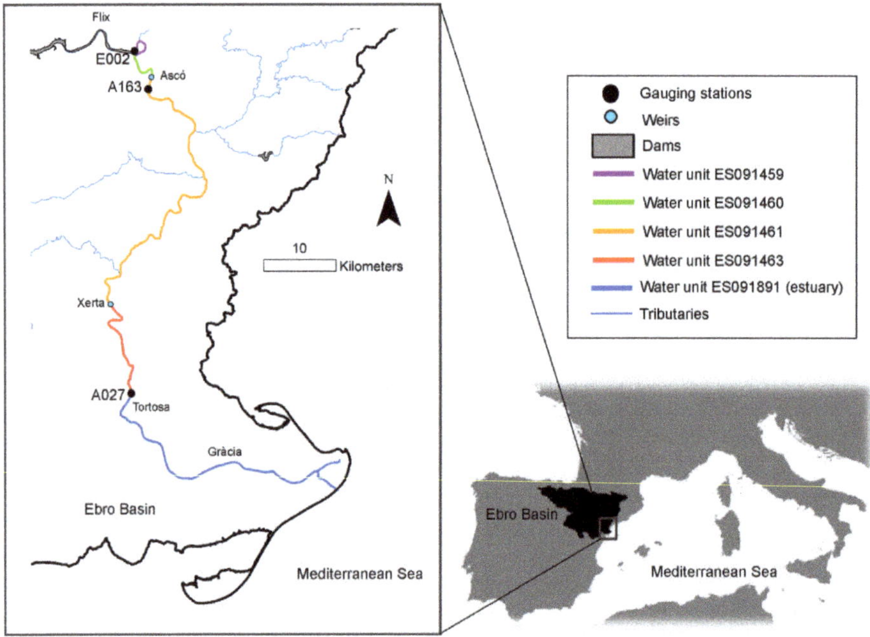

Figure 1. Lower Ebro River and its delta, showing the water units (for hydrological planning) belonging to the river and the estuary. Gauging stations, weirs, dams and tributaries are also represented, as well as the location of the city of Tortosa and the island of Gràcia. Source: modified from Belmar et al. [34].

2. Requirements and Methods to Establish E-Flows in the Lower Ebro River and Delta

The European Commission presented in December 2012 the Blueprint to safeguard Europe's water resources, an action plan to protect Europe's water resources. One of the most important points of the Blueprint is the need to develop a guidance document on how to estimate e-flows. In 2015, the European guide "Ecological flows in the implementation of the Water Framework Directive" was published (European Commission, 2015). The guide calls on member states to make the best possible use of available knowledge in relation to e-flows throughout the implementation process of the Water Framework Directive (WFD), but no specific methods are recommended. In accordance with the WFD, the European guide of e-flows [35] considers that their definition must take into account, (a) the principle of non-deterioration of water bodies, (b) the achievement of good ecological status and, (c) the satisfaction of the specific requirements for protected areas designated to protect habitats and species included in the Natura 2000 network where flows are an important factor for their protection.

The legal and technical requirements for the estimation of environmental flows (e-flows) in the Ebro River basin have been evolving over the last 20 years, in accordance with the changes in legislation in the European Union (EU), Spain and Catalonia. The first regulatory framework to establish e-flows in the lower Ebro River was formulated in 1999 and refers specifically to the mouth area, where "a minimum environmental flow of 100 m^3/s" is set indicatively. Then the Tenth Additional Provision of Law 10/2001, of the National Hydrological Plan (PHN) contemplated the elaboration of the Integral Plan of Protection of the Ebro Delta (PIPDE) with the purpose of "ensuring the maintenance of the special ecological conditions of the Ebro Delta". To achieve this, it was requested to establish a flow regime to allow "the development of the ecological functions of the river, the delta and the nearby marine ecosystem". This specific flow regime for the Ebro Delta has not been established so far, although the Government of Catalonia (through the Catalan Water Agency, ACA) has made several proposals to the authority of the river basin (Ebro Hydrographic Confederation, CHE), which have not been accepted.

The creation by the Government of Catalonia of the Commission for the Sustainability of Terres de l'Ebre (CSTE), legally allowed the elaboration of an e-flow proposal for the lower Ebro River and the delta according to the specific requirements of Law 10/2001 of the PHN (modified by Law 11/2005). Terres de l'Ebre is a region that includes the southern counties of Catalonia located in the lower Ebro River basin. Importantly, CSTE integrates all relevant stakeholders and administrations, including the Government of Spain. Studies to formulate the CSTE's e-flow proposal were conducted by the research institute IRTA at request from ACA [36]. The e-flow proposal was finally approved by CSTE in March 2007 and submitted to CHE. The Government of Spain approved in 2007 the "Regulation of Hydrological Planning" and, later, the "Technical Instruction of Hydrological Planning" (IPH; [37]), which develops regulatory contents in relation to the process of establishing e-flows. In summary, such contents establish that e-flows for rivers must be calculated using hydrological methods validated through fish habitat simulation.

Both the European and the Spanish legislations (WFD, PHN and IPH) establish the need to take into account the e-flow requirements to conserve in good ecological status the transitional and coastal waters (estuaries, coastal lagoons, deltas and marine area influence by rivers). The IPH makes specific reference to the environmental functions and values that e-flows must maintain in the case of transitional waters, and therefore to the main environmental impacts that must be avoided. The three fundamental functions are formulated in this way:

(a) That a prolonged duration of the salt wedge in the estuary does not cause conditions of anoxia, nor a significant change or disappearance of species that are not tolerant to salinity, nor an increase in the frequency and intensity of algal blooms, with detrimental effects on the balance of organisms present in the water body.

(b) That sufficient flows are provided to generate nutrient export rates that maintain the productivity of transitional and coastal waters.

(c) That floods are designed to provide the sediments necessary to maintain the characteristic geomorphological elements (river islands, coastal bars, deltas, etc.) and to contribute positively to the coastal dynamics, as well as to the maintenance of the frequency of washing of accumulated sediments and organic matter.

The IPH states that "to the extent that the protected areas of the Natura 2000 network and the wetlands of the Ramsar Convention may be appreciably affected by e-flow regimes, they must maintain or restore a favorable conservation status of habitats or species, responding to their ecological requirements and maintaining in the long term the ecological functions on which they depend".

In this context, the CSTE took into consideration the following main environmental functions and values in its e-flow proposal, both in 2007 and in 2015, in accordance with the legislation in force:

1. Maintaining habitat diversity and their connectivity (including salinity gradients), to avoid substantial changes in the distribution and presence of flora and fauna.

2. Maintaining appropriate hydrodynamic conditions in order to: (a) ensure the frequency and duration of stratification processes, minimizing the risk of loss of bottom water quality due to anoxia or algal blooms; (b) favor the mechanisms of dispersal of certain species (suspension or transport of eggs and larvae, seeds, etc.); (c) control marine intrusion into adjacent coastal aquifers.

3. Maintaining the spatial and temporal diversity of habitat conditions to meet the needs of different species throughout their life cycles.

4. Synchronizing the seasonal patterns of flow and salinity with other environmental parameters (temperature, light, nutrients, etc.) for the maintenance of certain biological processes (reproduction, migration and dispersal, etc.).

5. Controlling the presence and abundance of the different species of phytoplankton, macrophytes, benthic fauna and fishes, through the control of high flows, to avoid abrasion, erosion and drag in the case of the river and the estuary.

6. Maintaining the physic-chemical conditions of water and sediment in order to: (a) avoid the excessive accumulation of organic matter and anoxia in the river and estuary; (b) favor dilution of point or diffuse pollutants; (c) hinder the conditions conducive to algal blooms; (d) maintain the levels of turbidity that control the light regime in the water column and the primary production of the ecosystems.

7. Improving habitat conditions and availability through geomorphological dynamics in order to: (a) avoid the problems of accumulation of fine particles in the substrate; (b) maintain the distribution of sediment granulometry and mobility; (c) preserve the characteristics of the river channel and its structural elements (river islands, abandoned meanders, etc.); (d) favor the arrival and transport of sediments and nutrients in the estuary, delta and coastal system.

8. Controlling of the hydrological processes that regulate the connection of the transitional waters with the river, the sea and the associated aquifers in order to: (a) control the water flows in the estuary, lagoons and bays; (b) control the frequency, duration and length of the salt wedge; (c) avoid sediment accumulation hindering aquifer-river connectivity.

It should be noted that often there is little information to quantify the effects of flow regime alteration on environmental and socioeconomic functions and values. Further studies are necessary to establish quantitative relationships between flow regime and socioecological functions and values.

In 2015, the CSTE created a technical commission to update the 2007 e-flow proposal, so that the Government of Catalonia could present a new proposal for the next cycle of hydrological planning (Hydrological Plan of the Ebro, PHE). However, the PHE in force (2015–2021) was approved by the Government of Spain without accepting the allegations presented by the Government of Catalonia in relation to the proposed e-flow regime. Thus, the present e-flow regime is very similar to the previous ones, maintaining the indicative 100 m^3/s established for the mouth area, and is considered to be insufficient by many experts to maintain fundamental ecological and socioeconomic functions and values of the lower Ebro River and Delta [20,31]. For the update of the proposal of ecological flows in the lower Ebro River and Delta [19], the following aspects were considered:

- The new regulatory requirements: especially those established in the "Technical Instruction for Hydrological Planning" (IPH; [37]). The methodology for defining the e-flow regime was also adjusted to follow the e-flow European guide published in 2015.
- The selection of the hydrological method: a review of the main hydrological methods for calculating e-flows was carried out, selecting the one that best suits the conditions of the lower Ebro River.
- The time period of the hydrological series: the same series used in the PHCE were selected.
- The simulation model of the natural flow regime: the same simulation method that the CHE was used.

Such aspects were implemented through a protocol for establishing the e-flow regime proposal, as follows:

- An initial proposal was made based on hydrological methods.
- Subsequently, a biological validation of the hydrological proposal was performed (as indicated in the IPH).
- In addition, further requirements of the Ebro estuary to maintain its functionality and "good condition," as established by the IPH were considered.
- The ecological functions of the river, delta and nearby marine ecosystem have also been considered, as required by the tenth additional provision of the PHN Act.
- Finally, the requirements for protected areas dependent on river flow were also considered, as required by Article 4.1.c of the WFD.

These aspects were considered combining the biological validation of the hydrological proposal developed, as stated, with an assessment of the ecological characteristics of the Delta. Then, bibliographical research guided the diagnosis of the flow requirements to maintain such characteristics. Additionally, available biological data allowed determining empirical relationships between flows and ecological properties (e.g., alien species). This provided a more holistic perspective.

3. Current and Proposed E-Flow Regimes in the Lower Ebro River

This section presents the data of the e-flow regime for the lower Ebro River and delta approved by the Government of Spain through the Hydrological Plan of the Ebro [38] and makes a comparison with the proposal elaborated by the Government of Catalonia [19]. The e-flow regime approved by the PHE implies an annual runoff of 3010 hm^3/year that is 17% with respect to the natural regime in the first half of the last century (17,300 hm^3/year), and 34% with respect to the actual flow regime of the last 20 years (8826 hm^3/year). This means that the current runoff in the lower Ebro River is half (51%) the original runoff before dam construction, intensive water use and climate change. Notice that the e-flow regime in force is much lower than the current river flow, while the e-flow proposed by the Government of Catalonia (7732 hm^3/year) is close to the current river flow (see Table 1).

Table 1. E-flow proposals (monthly mean annual flow, m^3/s; total, hm^3/year) for the lower Ebro River and delta elaborated by the Spanish (PHE) and Catalan (CSTE) Governments. The PHE is the proposal presently in force.

Month	PHE	CSTE			
		Dry	Average	Humid	Minimum **
October	80	84	124	192	82
November	80	153	219	326	114
December	91	204	249	396	119
January	95	143	219	321	123
February	150	166	260	316	124
March	150	212	283	410	111
April	91	329 *	410	475	157
May	91	303	410	413	135
June	91	268	310	368	97
July	80	147	180	212	101
August	80	107	132	166	91
September	80	120	151	178	86
Total	3010	5871	7732	9907	3518

* A minimum of 15 days with more than 410 m^3/s must be guaranteed. ** It corresponds to the concept of exceptionality (severe drought) of article 4.6 of the WFD.

This e-flow regime is very similar to that in force in previous river basin plans (PHE), with a mean annual flow close to 100 m^3/s. The monthly flow distribution of the e-flow regime in force (PHE) and the regime proposed by the Government of Catalonia are shown in Table 1, ranging from 80 to 150 m^3/s in the first case and from 124 to 410 m^3/s in the second case (for average years). The last proposal of e-flow regime elaborated by the Government of Catalonia [19] supposes an annual runoff of 5871 hm^3 in dry years, 7732 hm^3 in average years, and 9907 hm^3 in humid years. This e-flow regime corresponds to the Ebro River gauging station in Tortosa (before entering the delta). In order to maintain the hydromorphological conditions of the Ebro River and allow the transport of sediments to the delta, the CSTE proposal also includes a flood regime of an average duration of 27 days during the months of March, April and May. The maximum flow is defined at around 1800 m^3/s and would be applicable in wet years (and in average years if there are sufficient reserves in the reservoirs). These controlled floods overlap with the e-flow regime, so that they represent an additional runoff of 1548 hm^3 (out of 2615 hm^3 flood runoff). The CSTE proposal also establishes a minimum annual runoff of 3518 hm^3 as the minimum flow in exceptional drought conditions (Table 1), which is not considered an e-flow, but corresponds to the concept of temporary deterioration by exceptionality (prolonged drought) in accordance with Article 4.6 of the WFD. This is a measure of "minimum services" to withstand an exceptional period and to be able to recover the good ecological status of the water bodies. In addition to the e-flows, there are irrigation concessions in the delta (370 hm^3/year), some of which (30%) end up in the lagoons of the delta or in the sea (agricultural drainage).

Table 1 shows that the e-flows in force (PHE) are well below the ones proposed by the CSTE, and even below the minimum flow regime of exceptionality (prolonged drought). This implies that, in the event that the planned uses in the basin increase according to water resources allocations granted in the PHE (especially for irrigation), the impacts of the subsequent flow decrease will be progressively higher in the lower Ebro River and Delta, harming both environmental and socioeconomic functions and values [31].

4. Socioecological Functions and Values to Be Considered for the Quantification of the E-Flow Regime in the Lower Ebro River and Delta

Here we carry out the analysis of the potential impacts of the e-flow regime in force (PHE) on the lower Ebro River and delta, as well as the flow requirements that would be necessary for those functions and values for which rigorous and quantifiable technical and scientific information is available. The main functions and values for which a quantitative scientific assessment has been made in relation to a holistic determination of the e-flow regime for the lower Ebro River and Delta are summarized in Table 2. The existing scientific and technical information relating flows and socioecological functions and values is analyzed in the next sub-sections.

Table 2. Relationship between river flows and environmental functions and values that have been quantitatively assessed in the lower Ebro River and delta. Compliance with e-flow proposals is shown.

Functions & Values	E-Flow Requirements	Source	Compliance? CSTE	Compliance? PHE
Species of conservation concern (Twaite shad)	50% of weighed usable area (WUA) for breeding is achieved with a mean spring flow of 252 m^3/s	IRTA [36]	yes	no
Control of alien species (fish) in the river	50% of weighed usable area (WUA) for dominance of native over alien species is achieved with a mean annual flow of 194 m^3/s	Caiola et al. [39] CSTE [19]	yes	no
Good ecological status of the river	A theoretical mean annual flow >>400 m^3/s would be required to achieve a moderate ecological status (WFD fish indicator)	Belmar et al. [34]	no	no
Control of macrophyte spread in the river	Spring flood during 27 days (2615 hm^3) with a peak flow of 1800 m^3/s	Ibáñez et al. [40]	Partially [1]	no
Sediment transport to the delta	Spring flood during 27 days (2615 hm^3) have a sediment transport capacity over 2 Mt/year	IRTA [41]	Yes [2]	no
Control of salt wedge in the estuary	Minimum of 410 m^3/s to wash away salt wedge in spring and 130 m^3/s to stop it in Gracias Island (18 km from river mouth)	Sierra et al. [42] Ibáñez et al. [43]	Yes	no
Coastal fisheries	1 m^3/s of river flow during the anchovy breeding period represents an increase of 0.114 Mt of monthly catch a year later	Salat et al. [44]	n.a. [3]	n.a. [3]

[1] The flood may prevent further macrophyte spread but not remove the current spread. [2] Further research is needed to quantify more precisely the transport capacity under present conditions in case sediment are by-passed trough reservoirs. [3] The relationship between river flow and anchovy production cannot be converted into an e-flow requirement. The question "what is the minimum anchovy landing" is meaningless.

In order to better understand the ecological impacts of human intervention on the lower Ebro River in the last decades and help to interpret the complex links between flow regulation and ecological processes we present a conceptual model (Figure 2). This model shows the main impacts of dam construction and the increase in water use that lead to reduced flow, sediment deficit and eutrophication.

The last 20 years are characterized by a re-oligotrophication due to phosphorus decline as consequence of the implementation of waste water treatment plants and other measure to improve water quality in the Ebro River watershed (see Ibáñez & Peñuelas [45]).

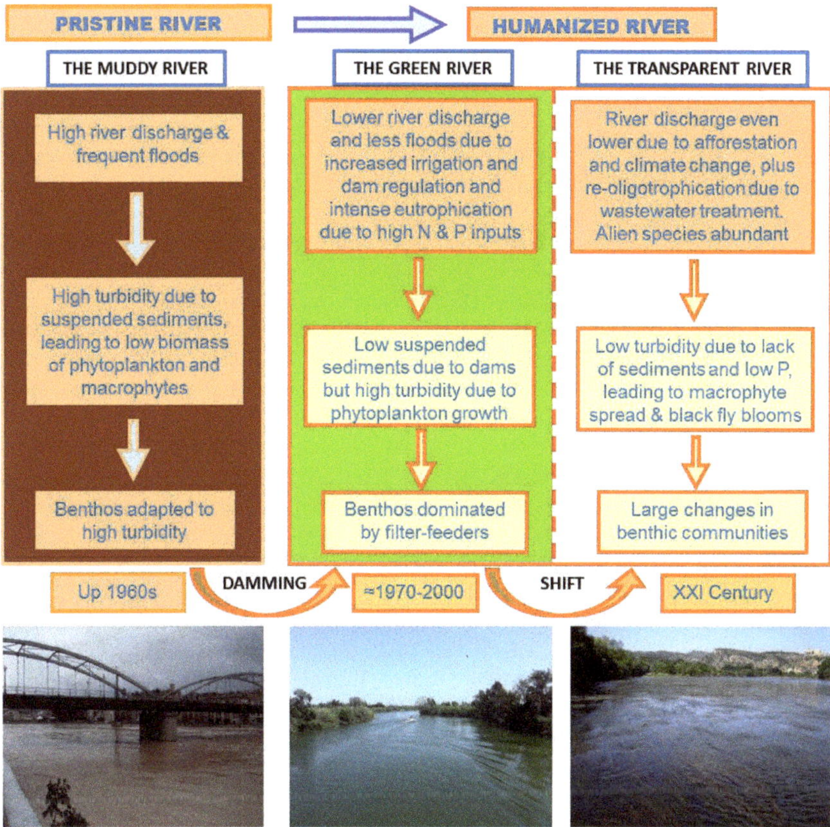

Figure 2. Conceptual model showing the main human impacts and changes in the ecosystem that the lower Ebro River has suffered in the last 60 years.

4.1. Conservation of Protected Species, Habitats and Ecological Quality

Until recently, efforts to quantify the relationship between the flow regime and protected species have focused on the case of the Twaite shad (*Alosa fallax*). This fish species disappeared completely from the river decades ago but has started to recover during the last 20 years. The amount of the Twaite shad reproductive habitat was simulated under different river discharge scenarios in order to determine its relationship with the flow regime [36]. According to this empirical function obtained with field data, a flow of 100 m^3/s (mean e-flow in force) provides only 20% of the potential weighted usable habitat (WUA), while 60% (required by IPH regulations) is achieved with a flow of 252 m^3/s. The e-flows of the CSTE proposal provide potential habitat for Twaite shad in the range of 50–80% of the maximum WUA as required by the IPH. The available information shows that the e-flow in force (PHCE) violates the Spanish regulations (IPH) and does not guarantee the reproduction of this species.

With regard to the species listed in Annex IV (a) of the Habitats Directive 92/43/EEC, as well as the species included in the Catalogue of Threatened Species of Spain, the e-flow regime should ensure adequate conditions to safeguard the ecological functionality of its breeding and resting areas.

The available information and the difficulty to link flow regimes with impacts on specific habitats and species, only allows us to make a qualitative estimate of the impact that the e-flow regime in force may have on the protected areas of the lower Ebro and Delta. The habitats and species of the banks and islands of the Ebro would certainly be affected, as the permanent low flows and the suppression of floods would modify the structure of riparian vegetation and cause its degradation [46,47]. The habitats and species of the Ebro Delta would be affected differently with more or less intensity depending on their characteristics; in general, the habitats and species most dependent on the water supply of the river would be the most affected, especially the estuary and some coastal lagoons and wetlands (Garxal, Buda Island, etc.). Finally, the species in the marine area of influence would be affected by the lower supply of fresh water and nutrients, which would have a negative impact on fish production (see Section 4.6) and seabirds that feed on this production. However, additional studies are needed to better quantify these impacts.

Ensuring sufficient river flows is one of the most important management tools to maintain the conservation status of the Natura 2000 sites. The lower Ebro River and Delta region has more than one million hectares of Natura 2000 surface area (125,000 ha of terrestrial sites, 53,500 ha of coastal sites and 901,000 ha of marine sites) The EU Natura 2000 network includes Special Areas of Conservation (SACs) and Special Protection Areas for Birds (SPAs). In such cases, e-flows should be appropriate to maintain or restore the favorable conservation status of the natural habitats and wild species of fauna and flora of Community interest listed in the Annexes of Directive 92/43/EEC and Directive 2009/147/EC. The three protected areas most affected by river flows are "River banks and Ebro Islands" ZEC, the "Ebro Delta" ZEC and SPA and the "Delta-Columbrets" marine SPA. However, the available information for the Ebro River and Delta is not sufficient to establish quantitative flow criteria, so specific studies are needed to deepen into the relationship between flows and conservation status of habitats in order to carry out the biological validation of the e-flow regime.

Some studies in the Ebro have recently been carried out to evaluated the relationship between the flow regime of the river and some components of biodiversity such as the abundance of protected species of birds in the delta, as well as the ecological quality of the final stretch of the river, measured through fish community [34,48]. Such studies showed that flow regime was significantly related to only a few deltaic bird species, and likely due to an indirect effect on prey availability in rainy years. However, a clearer effect of flow regime on ecological quality (EQ) according to the WFD (measured through fish community) was observed in the lower Ebro River. Mean flow magnitude was directly related to EQ and there were two outcomes relevant for water management. First, the effect on EQ was circumscribed to a specific temporal frame matching the annual cycle (12 previous months). Second, despite the positive relationship found (higher scores of EQ with higher river flow), it would be unfeasible to achieve a good ecological status in the lower Ebro only with an increase in mean flow. In this context, the control of the proliferation of macrophytes (which favor alien fish species) that may be done maintaining floods, naturally occurring in spring [40], would allow to achieve a certain EQ with lower river flows. In recent decades, macrophytes have spread due to reduction in high flows caused by dam management, together with the increase in water transparency due to phosphorus removal [49]. Nevertheless, these results also indicate that further research to assess ecological quality in rivers with great abundance of invasive species is necessary, as such species may condition the determination of ecohydrological relationships and the definition of e-flows.

4.2. Control of Invasive Species in the River

The efforts to quantify the relationship between the flow regime and invasive species have focused on the effects on the ratio between native and alien species, being one of the metrics considered in the elaboration of the ecological status indices of the WFD. Previous works have analyzed the relationship between flows and such ratio using a univariate logistic regression model [39], which was the basis for the biological validation of flows in the CSTE proposal. This was done through the simulation of flow scenarios by applying hydraulic models to test the biological significance of each of the flow proposals, by analyzing the proportion of suitable habitat to maintain the structure and integrity of the fish community in the lower Ebro River (ensuring the dominance of native species over non-native species, listed in Caiola et al. [39]).

Taking as a reference the minimum thresholds of 50 or 60% of the WUA required in the IPH, it can be estimated that a runoff of 6115 hm^3/year would provide 50% and a runoff of 8429 hm^3/year would provide 60% for native species. One can estimate the percentage of WUA that the different proposals for e-flow regimes would represent, using the relationship showed by the Figure 3. The proposed CSTE e-flows, calculated from hydrological methods, always provided a habitat between 49 and 65% of the WUA and therefore follows the criteria established by the IPH.

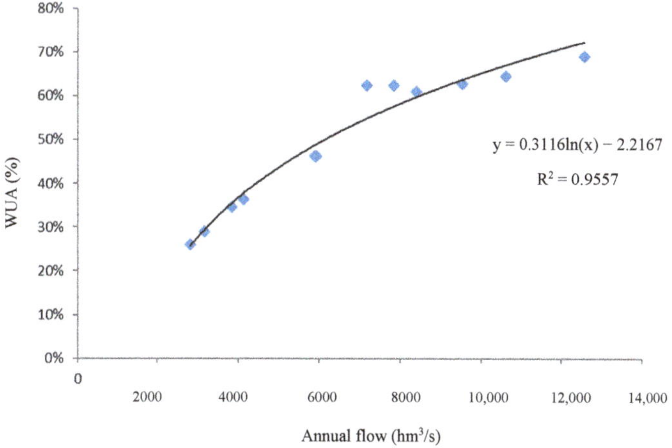

Figure 3. WUA/Flow curve to reach a proportion greater than 50% between native and non-native fish species in the lower Ebro River. Source: modified from CSTE [19].

The available information also allows us to make an estimate of the impact that the application of the e-flow regime in force (PHCE) would have. In this case, the WUA would be only 28%, well below the minimum of 50% required by current Spanish regulation (IPH). This would imply a greater proliferation of invasive species and a worsening of the ecological quality index of the WFD based on fish, violating the principle of non-deterioration of water bodies in the European legislation. In any case, despite the adjustment obtained (Figure 3), additional research is necessary to better understand the relationship between flow regimes and the proportion of native species, given the moderate adjustment of the underlying univariate logistic regression model (R^2 = 0.514; [39]).

4.3. Control of Macrophyte Spread in the River

Current Spanish regulations (IPH) state that the e-flow regime in the case of rivers must include flood flows, in order to control the presence and abundance of different species, maintain the physical and chemical conditions of the water and of the sediment, and improve the conditions and availability of habitat through the maintenance of the geomorphological dynamics. For this purpose, floods play an important role as disruptive elements that control the presence and abundance of species. Since the commissioning of the reservoirs in the lower Ebro River (1960′s), the hydrological regime of the lower Ebro has maintained episodes of floods that have gradually reduced its magnitude and frequency [50]. The environmental monitoring carried out in recent years has allowed establishing threshold values of river flow around 2000 m^3/s (but significant ecological changes have been observed). For the period 2006–2015, three floods were recorded with a maximum average daily flow equal to or greater than 1800 m^3/s. The average runoff of the three floods was 2724 hm^3, with an average duration of 27 days, and an average daily flow of 1170 m^3/s (peak flows were in the range 1800–2100 m^3/s).

In the last 20 years, the proliferation of macrophytes in the lower Ebro River has been great [49] and has had negative effects on the ecological, socioeconomic and hydromorphological system, such as the promotion of invasive species, the proliferation of blackflies, difficulties for navigation and the impact on the cooling system of the Ascó Nuclear Power Plant. The decrease in the floods of the Ebro River due to dam regulation and intensive water use have also led to the deterioration of its hydromorphological status (i.e., riverbed encroachment) and a lower capacity to transport sediments. In 2007, the decrease in macrophyte coverage after a flood could be measured in two sections of about 2 kilometres in Móra d'Ebre and Ginestar [40]. In Móra, the coverage went from 36.6% to 9.80%, and in Ginestar it went from 21.2% to 2.0%. The flood of 2015 presented similar flows but a longer duration (33 days). The inspection of the Miravet-Benifallet section by boat showed a very high effectiveness in removing macrophytes, although the change in coverage could not be measured. Further monitoring and research is needed in order to establish the magnitude and frequency of flood to control macrophyte spread.

The available scientific information allows us to estimate the impact that the e-flow regime in force would have on macrophyte proliferation. A regime of sustained low flows and no floods that is characteristic of the present e-flow regime implies a strong proliferation of macrophytes that will tend to be, with a coverage close to 100% in all areas above 4 m of depth [40]. This would lead to a whole host of negative effects, especially a much stronger proliferation of blackflies, the proliferation of alien species and a decline of native species such as the protected bivalve *Margaritifera auricularia* and other protected species.

4.4. Sediment Transport in the River and Coastal Regression in the Delta

There is a direct and positive relationship between river flow and sediment transport capacity, but in terms of delta growth (and retreat) sand is the critical material, while fine sediment (silt, clay) is either exported offshore or deposited in the delta plain [51]. However, sediment inputs to deltas has been dramatically reduced in the last decades due to dam construction, being the lower Ebro River an extreme case in which sediment transport has been reduced by 99% in relation to pre-dam conditions [33]. Annual sediment transport under pristine conditions was in the order of 30 Mt [52], while at present rounds 0.1 Mt [53]. Despite the transport capacity of the lower Ebro River has been significantly reduced after dam construction [50], the river still keeps a remarkable capacity during spring floods (Figure 4). In the lower Ebro River, flows required to transport sand and avoid coastal delta regression must be greater than 620 m^3/s [54]. The average runoff of spring floods for sediment transport under present conditions has been estimated at 2724 hm^3, with an average duration of 27 days, an average daily flow of 1170 m^3/s and a maximum daily flow of 1894 m^3/s [19]. Based on the characteristics of these floods, both magnitude and duration, it is estimated that they would allow the transport of more than 2 Mt of sediment per year by applying some sediment by-pass system to the Riba-roja reservoir [41], which would substantially slow down the regression of the Ebro Delta coast in the mouth area [55]. It has been estimated that by 2050, considering a sea level rise scenario 4.5 of the IPCC, the annual sediment deficit for the entire delta will be 1.69×10^6 m^3 while by 2100 will be 5.93×10^6 m^3. The minimum deficit to keep the delta above sea level has been estimated at 0.17×10^6 m^3 and 3.30×10^6 m^3 by 2050 and 2100, respectively [56].

The e-flow regime in force does not allow the generation of these floods, which implies a technical impossibility to bring sediments to the delta, and therefore curb the regression of the coast and the subsidence of the deltaic plain. The available information allows concluding that the e-flow regime in force would accentuate the deficit of sediments and the process of regression of the Ebro Delta, which entails a loss of agricultural land and protected natural areas [57]. An option to face this issue would be the transfer of sediments from the Riba-roja reservoir to the lower Ebro River and delta. Specific studies evaluating the by-pass systems, costs and benefits of the sediment transfer are being carried out, while the practice of sediment by-pass in reservoirs is being progressively expanded all around the world, mostly to avoid the loss of water storage capacity of reservoirs [58]. Moreover, further research is needed to quantify more precisely the transport capacity under present conditions in case sediment are by-passed trough reservoirs. The problem relies on the difficulty to calibrate and validate sediment transport models with floods carrying out suspended sediment far below their capacity.

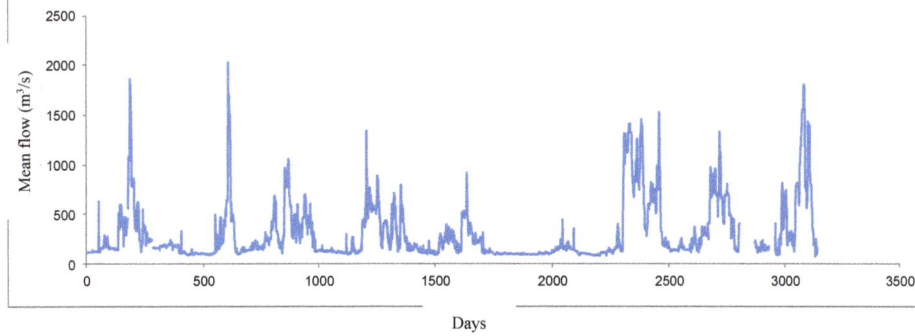

Figure 4. Hydrograph of the lower Ebro River highlighting the main floods for the period 2006–2015. Source: modified from CSTE [19].

4.5. Control of the Salt Wedge in the Estuary

The final stretch of the Ebro River (the last 27 km) behaves like a highly stratified estuary with a salt wedge. The dynamics of this wedge is important from an ecological and economic point of view and is closely linked to the flow of the river, although there are other factors such as bathymetry and sea level that also play an important role. Several studies have been performed through numerical modelling to evaluate the role of flow regime in the dynamics of the salt wedge [42,43,59]. Table 2 summarizes the main results of the studies that have been used in the determination of the e-flow proposal by CSTE, as well as some posterior studies. These studies analyzed the monthly flows under natural conditions based on data from the hydrological model used by CHE, the "Integrated System for Rainfall-Runoff Modelling" (period 1980–2006) and the hydrological series measured at Tortosa gauging station (period 1980–2012). Table 3 compares the flows that, according to different studies, ensure that the river does not have a salt wedge or avoids its pass beyond the island of Gràcia (18 km from the mouth), reaching the town of Amposta (25 km from the mouth). Taking the most recent study as a reference [42], the salt wedge is considered to completely disappear with a flow above 410 m³/s. Based on this requirement, the proposed e-flow regime of the CSTE considered that at least two months a year the flows that wash the salt wedge away are reached (in average years). For wet years, the e-flow is increased to be able to have three months without salt wedge. The aim is to ensure that the presence of salt wedges does not increase significantly with respect to the natural flow regime to avoid hypoxic conditions [43], in order to guarantee the principle of non-deterioration of the ecological status of the estuary.

Table 3. Minimum river flows (m³/s) required for the control of the salt wedge in the Ebro estuary, according to different studies.

Study	To Wash Away Salt Wedge	To Stop Salt Wedge at Gràcia Island *
Sierra et al. [42]	410	130
Ibáñez et al. [43]	342	120
Guillén & Palanques [59]	400	150

* Gràcia Island is located 18 km upstream the river mouth and it constitutes a natural sill to prevent the progression of the salt wedge towards its maximum possible distance (32 km upstream the river mouth).

The available information allows us to estimate the impact that the application of the e-flow regime in force (PCHE) would have on the salt wedge dynamics. In the event that this flow regime becomes the real regime of the river, the salt wedge would be present in the Ebro estuary permanently, and only two months (February and March) would be downstream the island of Gràcia (18 km from the river mouth). This would imply a worsening of the ecological conditions of the estuary, as the stratification of water favors the accumulation of organic matter and the decrease of dissolved oxygen in the water (hypoxia). As a result, it would violate the principle of non-deterioration of the WFD and other legal norms of Spain and Catalonia.

4.6. Coastal Fisheries

Terrestrially enriched river discharge favorably influences biological processes (i.e., growth, survival and recruitment) that affect fisheries production. River effects are most noticeable in oligotrophic seas (e.g., Mississippi River in the Gulf of Mexico, [60]; Rhône River in the Mediterranean, [61]) and in semi-enclosed systems such as the Black Sea [62]. Several studies have shown the influence of river inputs and other environmental factors on coastal fisheries, especially in oligotrophic seas [34,60,62,63]. Some of these studies were carried out in the coastal zone influenced by the Ebro River discharge [34,61,63,64]. In these studies, both pelagic fish (European anchovy, *Engraulis encrasicolus*) and shellfish (caramote prawn, *Penaeus kerathurus* and mantis shrimp, *Squilla mantis*) landings showed to be positively influenced by the Ebro River runoff. The life cycle characteristics of small pelagic fish and larvae (high mobility, plankton-based food chain and short life) make them particularly sensitive to environmental variations. The presence of larvae has been strongly associated with freshwater inputs and, in fact, the main spawning areas of anchovies in the western Mediterranean are near the mouths of the two largest rivers in the region, the Rhône and the Ebro. The European anchovy (*Engraulis encrasicolus*) is distributed throughout the Mediterranean Sea and is one of the main fishing resources, but in the Western Mediterranean, anchovy fishing has shown a declining trend since the 1990s.

An empirical model [44] estimated that an additional contribution of 1 m³/s of Ebro River flow during the anchovy breeding period represents an increase of 0.114 Mt of monthly catch a year later. Moreover, 63% of the variance of the time series of catches is explained by the variance in the flow of the river, with a delay of 12 months (which matches the time span observed in the lower Ebro River, see Belmar et al. [48]).

More recently, Belmar et al. [34] studied the relationship between the outflows of the Ebro River and shellfish species with socioeconomic value such as prawn and mantis shrimp. Like Lloret et al. [44], they found that the temporal scale considered determine the ability of flows or of water quality to explain changes in fish productivity. Nevertheless, such productivity also seemed to depend on the fishing technique. More accurate models would be necessary to improve the relationships found and understand the mechanisms that produce them.

The available information allows us to state that the e-flow regime in force cannot maintain the fisheries of the continental shelf of the Ebro delta. This topic has been studied and partly quantified in the case of anchovies, sardines, prawns and mantis shrimps. However, there will be necessary further scientific studies to determine the quantitative relationship between flows and nutrients of the river Ebro and fisheries production, in order to extend them to other species of high commercial interest where the relationship between flows and fisheries production is more complex.

4.7. River Flow Regime and Socioeconomic Functions and Values

The main socio-economic functions and values that are directly or indirectly dependent on the river flows but for which sufficiently contrasted information is not available to make a quantitative assessment of the impacts derived from the altered flow regime are the following (see summary in Table 4):

- Besides the dependence of coastal fisheries on river flow in terms of maintaining biodiversity and biological productivity, the studied quantitative relationship can be converted into social and economic costs that should be quantified. However, the relationship between river flow and fish landings cannot be converted into an e-flow requirement, as it is the case for many socioeconomic functions and values. Regarding aquaculture, there are evidences that reduced river flows can impact shellfish production (mussels, oysters, etc.; [65]), but no quantification has been done in the delta yet.

- The minimum flow required to guarantee the navigability in the lower Ebro River was set in 120 m^3/s in the Navigability Project (touristic navigation). The e-flow regime in force only exceeds this figure in February and March, while the rest of the year navigation is partially compromised. CSTE e-flow proposal complies with this flow in average and wet years, but not in dry years. The difficulty to maintain this minimum flow has triggered the elaboration of projects to dam the river in order to increase river depth [66], which has created a lot of controversy due to the potential negative environmental impacts and positive socioeconomic impacts.

- Salt water intrusion through the estuary due to low flow conditions may affect soil salinity and productivity of delta rice fields, as suggested by some studies [67]. Further research is needed to quantify these effects and estimate economic losses.

- River sediment deficit due to retention in reservoirs is causing a severe coastal retreat in the mouth area (Buda Island) and hundreds of hectares of wetlands, coastal lagoons and beaches have disappeared [57]; rice production is also threatened in some areas. Besides the loss of biodiversity and ecosystem services there is a socioeconomic loss that has not been quantified yet. Increasing the downstream flow of sediments retained in upstream dams is a crucial measure for dealing with climate change threats (sea-level rise) in the Delta, but public policies do not increase sediment flow. Instead, they implement incremental adaptation at the impacted site (coast), which implies economic costs (small-scale land expropriations) and causes social conflict [68].

- Besides the direct land loss due to coastal regression, biodiversity (habitats and species) is affected by declining flows in many direct and indirect ways, and there are social and economic values associated to biodiversity loss that should be estimated, though often a monetary quantification is limited or not possible.

- Given that there is a direct (but complex) relationship between water quantity and quality in rivers and estuaries [69–72], the decrease in flows would have a potential negative economic impact on aspects such as drinking water, industry, agricultural production and water treatment.

- Decreased flows may also have direct and indirect impacts on tourism: river navigation, blackfly proliferation, sport fishing, habitat degradation, etc., but they have not been quantified.

- The impacts on industrial activities must also be considered, especially on the hydropower production [73] and cooling needs of the Ascó nuclear power plant, but no quantification has been done yet.

- The social impacts of the loss of heritage should also be analyzed, given that the population of Terres de l'Ebre (lower Ebro basin) has a very strong perception that the Ebro River and Delta have important cultural, economic and symbolic value.

Table 4. Relationship between river flows and socio-economic functions and values of the lower Ebro River and delta that have been assessed with existing information, mostly qualitative because of the lack of studies.

Functions & Values	E-Flow Requirements	Source	Comments
Coastal fisheries	Several quantitative relationship between river flow and fisheries of European anchovy, caramote prawn & mantis shrimp	Sant Felíu [64] Salat et al. [44] Belmar et al. [34]	The relationship between river flow and fish landings cannot be converted into an e-flow requirement, but an economic loss can be estimated
River navigation	The minimum flow for touristic river navigation was set to 120 m^3/s	Roset [74]	The e-flow regime in force only exceeds this figure in February and March. CSTE proposal complies with this flow in average and wet years
Salt water intrusion in rice fields	An inverse relationship between river flow in winter and soil salinity in rice fields has been found	Genua-Olmedo et al. [67]	The relationship between river flow and fish landings cannot be converted into an e-flow requirement, but an economic loss can be estimated
Sediment deficit and coastal retreat in the delta	Sediment retention in reservoirs is causing a severe coastal retreat in the river mouth area (Buda Island) and socioeconomic impacts	Valdemoro et al. [57] Zografos [68]	Further research is needed to design river floods with sufficient sediments to stop coastal retreat in the mouth, but an economic loss can be estimated
Biodiversity loss (habitats & species)	Declining river flows are causing biodiversity loss in direct and indirect ways, with significant socioeconomic impacts	Day et al. [31]	Often difficult to quantify the socioeconomic impacts of biodiversity loss, mostly in monetary terms
Water quality	Declining river flows may decrease water quality (nutrients, chemical pollutants, salinity, etc.)	Caruso [70] Zhang et al. [72]	Literature points to ecological and economic impacts of water quality decline, but not yet assessed in the lower Ebro River and delta
Aquaculture production	Declining river flows may negatively affect oyster and mussel production	Powell et al. [65]	Literature points to economic impacts of river flow decline on aquaculture, but not yet assessed in the Ebro delta
Hydropower generation	Declining river flows negatively affect hydropower generation	Kuriqi et al. [73]	Literature points to economic impacts, but not yet quantified in the lower Ebro River

5. Conclusions

The main impacts deriving from the implementation of the e-flow regime in force established in the Hydrological Plan of the Ebro, from which contrasted scientific and technical information (quantitative or qualitative) is available, are summarized below.

5.1. Environmental Impacts

- A mean spring e-flow of 91 m^3/s would imply a potential useful habitat for the Twaite shad (*Alosa fallax*) below 20%, and would not guarantee the reproduction of this fish species of conservation concern.

- The e-flow regime implies a WUA of native species in the lower Ebro River of only 28%, well below the minimum of 50% required by current Spanish regulations. This would imply a greater proliferation of invasive species and a worsening of ecological quality indices.

- The e-flow regime implies that the salt wedge would be present in the Ebro estuary permanently, and only for 2 months below the island of Gràcia (18 km from the mouth). This would imply a worsening of the ecological conditions of the estuary, as the stratification of water favors the accumulation of organic matter and the decrease of dissolved oxygen (hypoxia).

- With regard to macrophytes, an e-flow regime of sustained low flows without floods would imply a strong proliferation, with a coverage close to 100% in all areas above 4 m of depth. This would imply a much stronger proliferation of alien species and blackflies and a decline of native species.

- With regard to sediment transport, the e-flow regime in force does not guarantee the maintenance of spring floods, making impossible in practice the option of transporting sediments to the delta by implementing a sediment by-pass system in the lower Ebro reservoirs.

- With regard to the protected areas, the habitats and species of the banks and islands of the Ebro will be affected, as permanent low flows and the suppression of floods may change the structure of riparian vegetation and cause its degradation.

5.2. Socioeconomic Impacts

- The e-flow regime in force cannot maintain the fishery of the continental marine platform of the Ebro Delta. The link between river flow and fisheries has been studied in the case of anchovies, prawns and mantis shrimps, but more research will be needed to extend them to other species of high commercial interest.
- The available information suggests that the e-flow regime in force would involve a risk of salinization of the rice fields of the Ebro delta, although further studies are needed to quantify the impact.
- The e-flow regime in force will accentuate the sediment deficit and the coastal regression process of the Ebro Delta, which involves a loss of agricultural land and protected natural areas.
- Other socioeconomic activities such as river navigation, tourism, hydropower production or aquaculture will likely be impacted by the e-flow regime in force, although some studies are needed to make a quantitative assessment in socio-economic terms.

The e-flow regime proposed by the Government of Catalonia [19] would be more suitable to maintain the main socioecological functions and values of the lower Ebro River and delta that have been analyzed in the present work. The lower Ebro River and delta is among the case studies where more quantitative and qualitative criteria to set e-flows with a holistic approach have been established. The existing information clearly suggests that the e-flows in force established by the Government of Spain should be increased to guarantee the maintenance of the main socioecological functions and values. However, additional studies are needed to validate the proposal of the Government of Catalonia in some relevant aspects such as the capacity of the river to transport sediments to the delta to avoid coastal regression and mitigate the effects of sea level rise and subsidence, as well as the capacity of floods to controls the spread of macrophytes.

Author Contributions: Conceptualization, C.I. and N.C.; investigation, N.C. and O.B.; writing—original draft preparation, C.I.; writing—review and editing, O.B. All authors have read and agreed to the published version of the manuscript.

Funding: This study was developed within the framework of an agreement between IRTA ("Institut de Recerca i Tecnologia Agroalimentàries") and ACA ("Agència Catalana de l'Aigua"), which acted as financial supporter.

Acknowledgments: The authors acknowledge support from the CERCA Programme/Generalitat de Catalunya.

Conflicts of Interest: The authors declare no conflict of interest. The funders had no role in the design of the study; in the collection, analyses, or interpretation of data; in the writing of the manuscript, or in the decision to publish the results.

References

1. Poff, N.L.; Zimmerman, J.K.H. Ecological responses to altered flow regimes: A literature review to inform the science and management of environmental flows. *Freshw. Biol.* **2010**, *55*, 194–205. [CrossRef]
2. Armitage, P.D.; Petts, G.E. Biotic score and prediction to assess the effects of water abstractions on river macroinvertebrates for conservation purposes. *Aquat. Conserv. Mar. Freshw. Ecosyst.* **1992**, *2*, 1–17. [CrossRef]
3. Statzner, B.; Higler, B. Stream hydraulics as a major determinant of benthic invertebrate zonation patterns. *Freshw. Biol.* **1986**, *16*, 127–139. [CrossRef]
4. Wood, P.J.; Armitage, P.D.; Cannan, C.E.; Petts, G.E. Instream mesohabitat biodiversity in three groundwater streams under base-flow conditions. *Aquat. Conserv. Mar. Freshw. Ecosyst.* **1999**, *9*, 265–278. [CrossRef]
5. Brown, L.R.; Ford, T. Effects of flow on the fish communities of a regulated California river: Implications for managing native fishes. *River Res. Appl.* **2002**, *18*, 331–342. [CrossRef]

6. Arthington, A.H.; Bhaduri, A.; Bunn, S.E.; Jackson, S.E.; Tharme, R.E.; Tickner, D.; Young, B.; Acreman, M.; Baker, N.; Capon, S.; et al. The Brisbane Declaration and Global Action Agenda on Environmental Flows. *Front. Environ. Sci.* **2018**, *6*, 45. [CrossRef]

7. Linnansaari, T.; Monk, W.; Baird, D.; Curry, R. Review of approaches and methods to assess Environmental Flows across Canada and internationally. *DFO Can. Sci. Advis. Secr. Res. Doc.* **2012**, *39*, 1–74.

8. Tharme, R.E. A global perspective on environmental flow assessment: Emerging trends in the development and application of environmental flow methodologies for rivers. *River Res. Appl.* **2003**, *19*, 397–441. [CrossRef]

9. Arthington, A. *Comparative Evaluation of Environmental Flow Assessment Techniques: Review of Holistic Methodologies. Land Water Resour. Res. Dev. Corp*; Occasional Paper; Land and Water Resources Research and Development Corporation: Canberra, Australia, 1998.

10. Tharme, R.E.; King, J.M. *Development of the Building Block Methodology for Instream Flow Assessments, and Supporting Research on the Effects of Different Magnitude Flows on Riverine Ecosystems*; Report No. 576/1/98; Water Research Commission: Pretoria, South Africa, 1998.

11. King, J.; Brown, C.; Sabet, H. A scenario-based holistic approach to environmental flow assessments for rivers. *River Res. Appl.* **2003**, *19*, 619–639. [CrossRef]

12. Arthington, A.H.; Pusey, B.J. Flow restoration and protection in Australian rivers. *River Res. Appl.* **2003**, *19*, 377–395. [CrossRef]

13. Poff, N.L.; Richter, B.D.; Arthington, A.H.; Bunn, S.E.; Naiman, R.J.; Kendy, E.; Acreman, M.; Apse, C.; Bledsoe, B.P.; Freeman, M.C.; et al. The ecological limits of hydrologic alteration (ELOHA): A new framework for developing regional environmental flow standards. *Freshw. Biol.* **2010**, *55*, 147–170. [CrossRef]

14. Belmar, O.; Velasco, J.; Martínez-Capel, F.; Peredo-Parada, M.; Snelder, T. Do Environmental Stream Classifications Support Flow Assessments in Mediterranean Basins? *Water Resour. Manag.* **2012**, *26*, 3803–3817. [CrossRef]

15. Acreman, M.; Arthington, A.H.; Colloff, M.J.; Couch, C.; Crossman, N.D.; Dyer, F.; Overton, I.; Pollino, C.A.; Stewardson, M.J.; Young, W. Environmental flows for natural, hybrid, and novel riverine ecosystems in a changing world. *Front. Ecol. Environ.* **2014**, *12*, 466–473. [CrossRef]

16. Day, J.; Ibáñez, C.; Scarton, F.; Pont, D.; Hensel, P.; Day, J.; Lane, R. Sustainability of Mediterranean Deltaic and Lagoon Wetlands with Sea-Level Rise: The Importance of River Input. *Estuaries Coasts* **2011**, *34*, 483–493. [CrossRef]

17. Day, J.W.; Ibáñez, C.; Pont, D.; Scarton, F. Status and Sustainability of Mediterranean Deltas: The Case of the Ebro, Rhône, and Po Deltas and Venice Lagoon. In *Coasts and Estuaries*; Wolanski, E., Day, J.W., Elliott, M., Ramachandran, R., Eds.; Elsevier: Amsterdam, The Netherlands, 2019; pp. 237–249.

18. Beilfuss, R.; Brown, C. Assessing environmental flow requirements and trade-offs for the Lower Zambezi River and Delta, Mozambique. *Int. J. River Basin Manag.* **2010**, *8*, 127–138. [CrossRef]

19. CSTE. *Revisió i Actualització de la Proposta de Règim de Cabals Ecològics al Tram Final del Riu Ebre, Delta i Estuari*; Departament de Medi Ambient i Habitatge, Generalitat de Catalunya: Barcelona, Spain, 2015.

20. Ibáñez, C.; Prat, N. The environmental impact of the Spanish national hydrological plan on the lower Ebro river and delta. *Int. J. Water Resour. Dev.* **2003**, *19*, 485–500. [CrossRef]

21. Ramírez-Hernández, J.; Rodríguez-Burgueño, J.E.; Kendy, E.; Salcedo-Peredia, A.; Lomeli, M.A. Hydrological response to an environmental flood: Pulse flow 2014 on the Colorado River Delta. *Ecol. Eng.* **2017**, *106*, 633–644. [CrossRef]

22. Sun, T.; Yang, Z.F.; Cui, B.S. Critical Environmental Flows to Support Integrated Ecological Objectives for the Yellow River Estuary, China. *Water Resour. Manag.* **2008**, *22*, 973–989. [CrossRef]

23. Adams, J.B.; Van Niekerk, L. Ten Principles to Determine Environmental Flow Requirements for Temporarily Closed Estuaries. *Water* **2020**, *12*, 1944. [CrossRef]

24. Adams, J.B.; Bate, G.C.; Harrison, T.D.; Huizinga, P.; Taljaard, S.; van Niekerk, L.; Plumstead, E.E.; Whitfield, A.K.; Wooldridge, T.H. A method to assess the freshwater inflow requirements of estuaries and application to the Mtata estuary, South Africa. *Estuaries* **2002**, *25*, 1382. [CrossRef]

25. Flannery, M.S.; Peebles, E.B.; Montgomery, R.T. A percent-of-flow approach for managing reductions of freshwater inflows from unimpounded rivers to Southwest Florida estuaries. *Estuaries* **2002**, *25*, 1318–1332. [CrossRef]

26. Pelrson, W.L.; Nittim, R.; Chadwick, M.J.; Bishop, K.A.; Horton, P.R. Assessment of changes to saltwater/freshwater habitat from reductions in flow to the Richmond River estuary, Australia. *Water Sci. Technol. A J. Int. Assoc. Water Pollut. Res.* **2001**, *43*, 89–97.

27. Richter, B.D.; Mathews, R.; Harrison, D.L.; Wigington, R. Ecologically sustainable water management: Managing river flows for ecological integrity. *Ecol. Appl.* **2003**, *13*, 206–224. [CrossRef]

28. Robins, J.B.; Halliday, I.A.; Staunton-Smith, J.; Mayer, D.G.; Sellin, M.J. Freshwater-flow requirements of estuarine fisheries in tropical Australia: A review of the state of knowledge and application of a suggested approach. *Mar. Freshw. Res.* **2005**, *56*, 343–360. [CrossRef]

29. Sun, T.; Zhang, H.; Yang, Z.; Yang, W. Environmental flow assessments for transformed estuaries. *J. Hydrol.* **2015**, *520*, 75–84. [CrossRef]

30. Davies, P.M.; Naiman, R.J.; Warfe, D.M.; Pettit, N.E.; Arthington, A.H.; Bunn, S.E. Flow–ecology relationships: Closing the loop on effective environmental flows. *Mar. Freshw. Res.* **2014**, *65*, 133–141. [CrossRef]

31. Day, J.W.; Maltby, E.; Ibáñez, C. River basin management and delta sustainability: A commentary on the Ebro Delta and the Spanish National Hydrological Plan. *Ecol. Eng.* **2006**, *26*, 85–99. [CrossRef]

32. Ibáñez, C.; Pont, D.; Prat, N. Characterization of the Ebre and Rhone estuaries: A basis for defining and classifying salt-wedge estuaries. *Limnol. Oceanogr.* **1997**, *42*, 89–101. [CrossRef]

33. Ibáñez, C.; Prat, N.; Canicio, A. Changes in the hydrology and sediment transport produced by large dams on the lower Ebro river and its estuary. *Regul. Rivers Res. Manag.* **1996**, *12*, 51–62. [CrossRef]

34. Belmar, O.; Ibáñez, C.; Forner, A.; Caiola, N. The Influence of Flow Regime on Ecological Quality, Bird Diversity, and Shellfish Fisheries in a Lowland Mediterranean River and Its Coastal Area. *Water* **2019**, *11*, 918. [CrossRef]

35. European Comission. *Ecological FLows in the Implementation of the Water Framework Directive*; Guidance Document No. 31; European Comission: Brussels, Belgium, 2015.

36. IRTA. *Proposta de Règim de Cabals Ambientals del Tram final del Riu Ebre i Validació Biològica Preliminar*; IRTA, Department of Agriculture, Generalitat de Catalunya: Barcelona, Spain, 2008.

37. MARM. *Orden ARM/2656/2008, de 10 de Septiembre, por la que se Aprueba la Instrucción de Planificación Hidrológica*; Ministerio de Medio Ambiente Rural y. Marino: Madrid, Spain, 2008.

38. CHE. *Plan Hidrológico del Ebro 2015-2021*; Confederación Hidrográfica del Ebro: Zaragoza, Spain, 2015.

39. Caiola, N.; Ibáñez, C.; Verdú, J.; Munné, A. Effects of flow regulation on the establishment of alien fish species: A community structure approach to biological validation of environmental flows. *Ecol. Indic.* **2014**, *45*, 598–604. [CrossRef]

40. Ibáñez, C.; Caiola, N.; Rovira, A.; Real, M. Monitoring the effects of floods on submerged macrophytes in a large river. *Sci. Total Environ.* **2012**, *440*, 132–139. [CrossRef] [PubMed]

41. IRTA. *Tècniques per a la Gestió del Sediment als Embassaments: Els Cabals D'arrossegament i els Corrents de Densitat*; IRTA, Department of Agriculture, Generalitat de Catalunya: Barcelona, Spain, 2009.

42. Sierra, J.P.; Sánchez-Arcilla, A.; Figueras, P.A.; González Del Río, J.; Rassmussen, E.K.; Mösso, C. Effects of discharge reductions on salt wedge dynamics of the Ebro River. *River Res. Appl.* **2004**, *20*, 61–77. [CrossRef]

43. Ibàñez, C.; Canicio, A.; Day, J.W.; Curcó, A. Morphologic development, relative sea level rise and sustainable management of water and sediment in the Ebre Delta, Spain. *J. Coast. Conserv.* **1997**, *3*, 191–202. [CrossRef]

44. Lloret, J.; Palomar, I.; Salat, J.; Solé, I. Impacto de los Aportes Fluviales sobre la Productividad de la Población de Anchoa del Sur de Cataluña. 2011. Available online: https://digital.csic.es/handle/10261/100871 (accessed on 23 September 2020).

45. Ibáñez, C.; Peñuelas, J. Changing nutrients, changing rivers. *Science* **2019**, *365*, 637–638. [CrossRef] [PubMed]

46. Arheimer, B.; Hjerdt, N.; Lindström, G. Artificially Induced Floods to Manage Forest Habitats Under Climate Change. *Front. Environ. Sci.* **2018**, *6*. [CrossRef]

47. Staentzel, C.; Beisel, J.-N.; Gallet, S.; Hardion, L.; Barillier, A.; Combroux, I. A multiscale assessment protocol to quantify effects of restoration works on alluvial vegetation communities. *Ecol. Indic.* **2018**, *90*, 643–652. [CrossRef]

48. Belmar, O.; Vila-Martínez, N.; Ibáñez, C.; Caiola, N. Linking fish-based biological indicators with hydrological dynamics in a Mediterranean river: Relevance for environmental flow regimes. *Ecol. Indic.* **2018**, *95*, 492–501. [CrossRef]

49. Ibáñez, C.; Alcaraz, C.; Caiola, N.; Rovira, A.; Trobajo, R.; Alonso, M.; Duran, C.; Jiménez, P.J.; Munné, A.; Prat, N. Regime shift from phytoplankton to macrophyte dominance in a large river: Top-down versus bottom-up effects. *Sci. Total Environ.* **2012**, *416*, 314–322. [CrossRef]

50. Batalla, R.J.; Gómez, C.M.; Kondolf, G.M. Reservoir-induced hydrological changes in the Ebro River basin (NE Spain). *J. Hydrol.* **2004**, *290*, 117–136. [CrossRef]

51. Ibáñez, C.; Alcaraz, C.; Caiola, N.; Prado, P.; Trobajo, R.; Benito, X.; Day, J.W.; Reyes, E.; Syvitski, J.P.M. Basin-scale land use impacts on world deltas: Human vs natural forcings. *Glob. Planet. Chang.* **2019**, *173*, 24–32. [CrossRef]

52. Xing, F.; Kettner, A.J.; Ashton, A.; Giosan, L.; Ibáñez, C.; Kaplan, J.O. Fluvial response to climate variations and anthropogenic perturbations for the Ebro River, Spain in the last 4000years. *Sci. Total Environ.* **2014**, *473–474*, 20–31. [CrossRef] [PubMed]

53. Rovira, A.; Ibáñez, C.; Martín-Vide, J.P. Suspended sediment load at the lowermost Ebro River (Catalonia, Spain). *Quat. Int.* **2015**, *388*, 188–198. [CrossRef]

54. Núñez-González, F.; Rovira, A.; Ibáñez, C. Bed load transport and incipient motion below a large gravel bed river bend. *Adv. Water Resour.* **2018**, *120*, 83–97. [CrossRef]

55. Rovira, A.; Ibáñez, C. Sediment management options for the lower Ebro River and its delta. *J. Soils Sediments* **2007**, *7*, 285–295. [CrossRef]

56. Genua-Olmedo, A. *Modelling Sea Level Rise Impacts and the Management Options for Rice Production: The Ebro Delta as an Example*; Universitat Rovira i Virgili: Tortosa, Spain, 2017.

57. Valdemoro, H.I.; Sánchez-Arcilla, A.; Jiménez, J.A. Coastal dynamics and wetlands stability. The Ebro delta case. *Hydrobiologia* **2007**, *577*, 17–29. [CrossRef]

58. Kondolf, G.M.; Gao, Y.; Annandale, G.W.; Morris, G.L.; Jiang, E.; Zhang, J.; Cao, Y.; Carling, P.; Fu, K.; Guo, Q.; et al. Sustainable sediment management in reservoirs and regulated rivers: Experiences from five continents. *Earth's Future* **2014**, *2*, 256–280. [CrossRef]

59. Guillén, J.; Palanques, A. Sediment dynamics and hydrodynamics in the lower course of a river highly regulated by dams: The Ebro River. *Sedimentology* **1992**, *39*, 567–579. [CrossRef]

60. Grimes, C.B. Fishery Production and the Mississippi River Discharge. *Fisheries* **2001**, *26*, 17–26. [CrossRef]

61. Lloret, J.; Lleonart, J.; Solé, I.; Fromentin, J.-M. Fluctuations of landings and environmental conditions in the north-western Mediterranean Sea. *Fish. Oceanogr.* **2001**, *10*, 33–50. [CrossRef]

62. Daskalov, G. Relating fish recruitment to stock biomass and physical environment in the Black Sea using generalized additive models. *Fish. Res.* **1999**, *41*, 1–23. [CrossRef]

63. Lloret, J.; Palomera, I.; Salat, J.; Sole, I. Impact of freshwater input and wind on landings of anchovy (Engraulis encrasicolus) and sardine (Sardina pilchardus) in shelf waters surrounding the Ebre (Ebro) River delta (north-western Mediterranean). *Fish. Oceanogr.* **2004**, *13*, 102–110. [CrossRef]

64. San Feliu, J.M. *Influencia de los Aportes del río Ebro sobre la Producción Pesquera de la Zona*; Dirección General de Pesca Marítima: Gijon, Spain, 1975.

65. Powell, E.N.; Klinck, J.M.; Hofmann, E.E.; McManus, M.A. Influence of Water Allocation and Freshwater Inflow on Oyster Production: A Hydrodynamic–Oyster Population Model for Galveston Bay, Texas, USA. *Environ. Manag.* **2003**, *31*, 0100–0121. [CrossRef] [PubMed]

66. COCINT. *Estudi de Navegabilitat del Riu Ebre*; COCINT: Tortosa, Spain, 2007; p. 30.

67. Genua-Olmedo, A.; Alcaraz, C.; Caiola, N.; Ibáñez, C. Sea level rise impacts on rice production: The Ebro Delta as an example. *Sci. Total Environ.* **2016**, *571*, 1200–1210. [CrossRef]

68. Zografos, C. Flows of sediment, flows of insecurity: Climate change adaptation and the social contract in the Ebro Delta, Catalonia. *Geoforum* **2017**, *80*, 49–60. [CrossRef]

69. Alizadeh, M.J.; Kavianpour, M.R.; Danesh, M.; Adolf, J.; Shamshirband, S.; Chau, K.-W. Effect of river flow on the quality of estuarine and coastal waters using machine learning models. *Eng. Appl. Comput. Fluid Mech.* **2018**, *12*, 810–823. [CrossRef]

70. Caruso, B.S. Regional river flow, water quality, aquatic ecological impacts and recovery from drought. *Hydrol. Sci. J.* **2001**, *46*, 677–699. [CrossRef]

71. Duan, W.; He, B.; Chen, Y.; Zou, S.; Wang, Y.; Nover, D.; Chen, W.; Yang, G. Identification of long-term trends and seasonality in high-frequency water quality data from the Yangtze River basin, China. *PLoS ONE* **2018**, *13*, e0188889. [CrossRef]

72. Zhang, Y.; Xia, J.; Liang, T.; Shao, Q. Impact of Water Projects on River Flow Regimes and Water Quality in Huai River Basin. *Water Resour. Manag.* **2010**, *24*, 889–908. [CrossRef]

73. Kuriqi, A.; Pinheiro, A.N.; Sordo-Ward, A.; Garrote, L. Flow regime aspects in determining environmental flows and maximising energy production at run-of-river hydropower plants. *Appl. Energy* **2019**, *256*, 113980. [CrossRef]

74. Roset, J. *Estudi del Comportament del Riu Ebre, en el Tram Móra D'Ebre—Tortosa, per a Analitzar la Navegabilitat en Aquest Tram*; Universitat Politècnica de Catalunya: Barcelona, Spain, 2004.

 *water*

Article

Flow Regime and Nutrient-Loading Trends from the Largest South European Watersheds: Implications for the Productivity of Mediterranean and Black Sea's Coastal Areas

Stefano Cozzi [1,*], Carles Ibáñez [2], Luminita Lazar [3], Patrick Raimbault [4] and Michele Giani [5]

[1] CNR—ISMAR Marine Science Institute, 34149 Trieste, Italy
[2] IRTA—Aquatic Ecosystems Program, Sant Carles de la Ràpita, 43540 Catalunya, Spain; carles.ibanez@irta.cat
[3] NIMRD—National Institute for Marine Research and Development, 900581 Constanța, Romania; llazar@alpha.rmri.ro
[4] Aix Marseille Université, CNRS/INSU, Université de Toulon, IRD, Mediterranean Institute of Oceanography (MIO), 13288 Marseille, France; patrick.raimbault@mio.osupytheas.fr
[5] OGS—Istituto Nazionale di Oceanografia e di Geofisica Sperimentale, 34151 Trieste, Italy; mgiani@inogs.it
* Correspondence: stefano.cozzi@ts.ismar.cnr.it; Tel.: +39-040-375-6874

Received: 14 November 2018; Accepted: 12 December 2018; Published: 20 December 2018

Abstract: In the last century, large watersheds in Southern Europe have been impacted by a combination of anthropogenic and climatic pressures, which have rapidly evolved to change the ecological status of freshwater and coastal systems. A comparative analysis was performed for Ebro, Rhône, Po and Danube rivers, to investigate if they exhibited differential dynamics in hydrology and water quality that can be linked to specific human and natural forces acting at sub-continental scales. Flow regime series were analyzed from daily to multi-decadal scales, considering frequency distributions, trends (Mann–Kendall and Sen tests) and discontinuities (SRSD Method). River loads of suspended matter, nutrients and organic matter and the eutrophication potential of river nutrients were estimated to assess the impact of river loads on adjacent coastal areas. The decline of freshwater resources largely impacted the Ebro watershed on annual (-0.139 km^3 yr^{-1}) and seasonal (-0.4% yr^{-1}) scales. In the other rivers, only spring–summer showed significant decreases of the runoff coupled to an exacerbated flow variability (0.1–0.3% yr^{-1}), which suggested the presence of an enhanced regional climatic instability. Discontinuities in annual runoff series (every 20–30 years) indicated a similar long-term evolution of Rhône and Po rivers, differently from Ebro and Danube. Higher nutrient concentrations in the Ebro and Po (+50%) compared to Rhône and Danube and distinct stoichiometric nutrient ratios may exert specific impacts on the growth of plankton biomass in coastal areas. The overall decline of inorganic phosphorus in the Rhône and Po (since the 1980s) and the Ebro and Danube (since the 1990s) mitigated the eutrophication in coastal ecosystems inducing, however, a phase in which the role of organic phosphorus loads (Po > Danube > Rhône > Ebro) on coastal productivity could be more relevant. Overall, the study showed that the largest South European watersheds are differently impacted by anthropogenic and climatic forces and that this will influence their vulnerability to future changes of flow regime and water quality.

Keywords: river runoff; anthropogenic pressures; climate change; water quality; coastal productivity; time series; Mediterranean; Black Sea

---

1. Introduction

Over the past half of the century, anthropogenic activities have largely impacted freshwater and coastal ecosystems, changing river loads of suspended matter, nutrients, organic matter and causing the mobilization of a variety of human-derived chemicals from the land to receiving water bodies. The increase of river nutrient loads, besides sustaining the productivity of large marine regions, have caused widespread eutrophication and changes in the structure of fluvial, estuarine and marine food webs [1–5]. High freshwater and nutrient loads enhance the stratification of the water column and the accumulation of algal biomass in estuaries and coastal areas, frequently leading to hypoxic conditions that cause the degradation of pelagic and benthic habitats [6–12]. Harmful algal blooms can be promoted by high nutrient loads or by altered nutrient ratios, with negative consequences for marine fauna and human health [6,7]. The delivery of terrestrial organic matter to the coastal zones, in combination to the enrichment of nutrients, can affect the growth of marine bacteria [13,14]. Pulsed or reduced inputs of suspended particulate modify the morphology of deltas, coasts and bottom sediments [15–17]. All these alterations have been linked to changes in abundance and structure of pelagic and benthic communities, mass mortality events and jellyfish proliferation.

The Mediterranean and Black Sea are the largest semi-enclosed marine basins on the Earth, in which the effects of river discharges can be assessed [6,18–20]. The current river inputs of total nitrogen, phosphorus and dissolved silica to the Mediterranean (1077 kt-N yr^{-1}, 49 kt-P yr^{-1}, 1028 kt-Si yr^{-1}) are similar to those to the Black Sea (1116 kt-N yr^{-1}, 55 kt-P yr^{-1}, 861 kt-Si yr^{-1}), which is a basin $\approx$ 5 times smaller than the Mediterranean one, but surrounded by a large drainage region [21]. For comparison, the Mediterranean Sea exports through the Gibraltar Strait nitrate and phosphate respectively at 1947 kt-N yr^{-1} and 147 kt-P yr^{-1} [22] and it receives them from the atmosphere as a bulk deposition at 1246 kt-N yr^{-1} and 34 kt-P yr^{-1} [18].

In Southern Europe, river nutrient discharge has increased during the last century, because of the growth of industrial and urban settlements and the introduction of intensive practices in livestock farming and agriculture, causing problems of eutrophication in the receiving coastal zones. This process has occurred mostly in 1960–1980, being it afterwards partially mitigated by stricter environmental policies [23–25]. For this reason, several environmental studies were focused on specific drainage basins e.g., [4,26–29] and on their impacts in the adjacent coastal zones e.g., [6,9–11,30,31].

However, the differences at regional scale in river water and nutrient loads and in the responses of fluvial, estuarine and coastal ecosystems are still poorly investigated in Southern Europe, due to the inhomogeneous availability of long-term datasets covering multiple watersheds. Comparative studies are of basic importance as drainage and coastal systems can experience divergent evolutions, even in the presence of common anthropogenic impacts, if they are subjected to climatic conditions with distinct meso-scale features [32]. The combination of anthropogenic and natural pressures can induce in these ecosystems specific hydrological and biogeochemical transformations at highly variable spatial and temporal scales [30,33–35]. For example, comparative studies showed a more pronounced shift toward dry climatic conditions in the Ebro basin with respect to Adige and Sava basins [36]. A recent common trend towards pluvial-torrential regimes was observed for the Rhône, Po and Danube, with a higher runoff in early spring originated by snowmelt in the mountainous areas [37]. Differences in the extension of drainage basins were found to affect annual and seasonal loads of water and nutrient in some rivers in the W Mediterranean [25].

In this study, a comparative analysis of the discharges of freshwater, suspended matter, nutrients and organic matter to the sea of the four most important South European watersheds (Ebro, Rhône, Po and Danube), from daily to decadal scales, is performed through the compilation of an extended dataset and a harmonized data analysis (Figure 1). The impacts of the changes in river loadings on fluvial and coastal ecosystems is also discussed through the estimate of the eutrophication potential of river nutrients in the coastal zones and through a meta-analysis of the available literature. The aim is to compare the evolution of flow regime in South European watersheds, since existing literature either includes specific studies focused on single drainage basins or global analysis, in order to investigate how the hydrology and water quality (nutrients) can be linked to specific anthropogenic pressures and climatic conditions acting at a regional scale, and to infer the current role that flow regime exerts on the productivity of the coastal ecosystems.

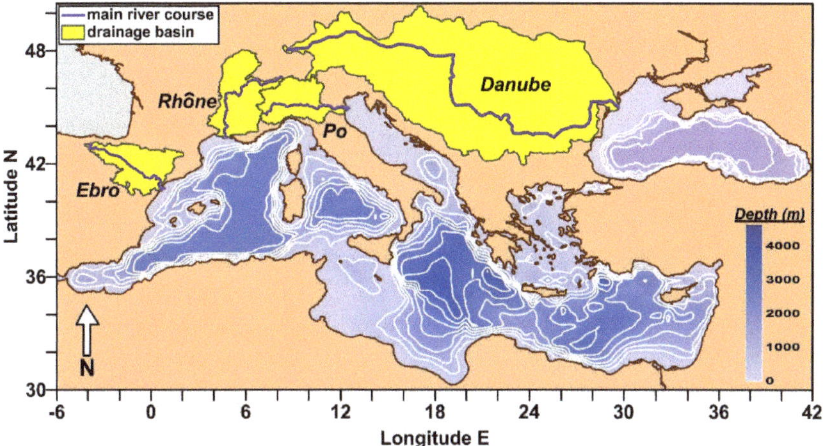

Figure 1. Drainage basins and main river courses of the Ebro, Rhône, Po and Danube.

2. Materials and Methods

2.1. Geographical Settings of River Basins

The Ebro is the largest river in the Iberian Peninsula draining to the Mediterranean Sea, and it has a watershed surrounded by the mountainous areas of Cantabrian Range, Pyrenees, Iberian Massif and parts of the Castillian "meseta" (Table 1). The river and the tributaries are subjected to a controlled flow, due to the presence of several dams and irrigation canals mostly built in 1920–1980 for farming and hydropower generation purposes. The last dam along the main river course in Ribarroja (100 km upstream the river mouth) controls the flow of the lower Ebro [38]. Agriculture covers ≈ 50% of the Ebro basin and it is the main source for nitrate inputs, whereas urban and industrial activities are the most important sources for phosphate inputs [36,39,40]. A long-term decrease of the runoff has characterized this river since the 1960s, whereas a decline of phosphate and ammonium was detected since the 1990s. This process caused a re-oligotrophication of the fluvial ecosystem with the collapse of phytoplankton and the spread of macrophytes [4,27]. Changes in freshwater phytoplankton community were reported due to the regulation of the flow caused by dams and water reservoirs [41]. The decrease in TSM due to dam construction has been also dramatic, as much as 99%, with severe consequences on the erosion of the Ebro Delta [38].

The Rhône has its source in Switzerland and it has most of the drainage basin in mountainous areas of Alps, Jura, Cevennes and Vosges. This feature causes a large runoff compared to the river length, because of the contributions of snowmelt and rainfall [28]. Fifty kilometers upstream the mouth, the Rhône splits in two branches of "Grand Rhône" and "Petit Rhône", which carry about 90% and 10% of the total flow, respectively. After the damming of the Nile in 1968, the Rhône has become the major source of freshwater in the Mediterranean [42]. This river is characterized by large loads of nitrogen and phosphorus, which has increased in the period 1970–1990 [43,44]. The environment of the lower Rhône has been largely modified by human activities, mostly since the 1970s, leading to a stronger regulation of the flow and an increase of nutrient inputs [45].

Table 1. General characteristics of drainage basins and river runoff (km^3 yr^{-1}) of the Ebro, Rhône, Po and Danube.

River (Gauging Station)	Drainage Basin [†] (km^2)	River Length [‡] (km)	Inhabitants ($n \cdot 10^6$)	Water Flow Data Series (yr)	Annual Water Flow (km^3 yr^{-1})		
					Median	1th–3rd	Quartile Range
Ebro (Tortosa)	86,800	930	3	1914–1931, 1951–2012	13.0	8.7–17.1	3.8–34.1
Rhône (Beaucaire)	157,950	813	18	1920–2012	53.9	45.8–61.7	22.8–78.0
Po (Pontelagoscuro)	74,000	682	16	1914–2012	45.5	38.8–55.7	26.2–82.7
Danube (Ceatal Izmail)	801,500	2857	82	1931–2012	201.2	178.6–221.0	134.2–300.2
Total	1,120,250	5282	120	1951–2012	316.8	297.9–344.9	208.6–424.0

[†] included tributaries; [‡] main river course.

The Po is the largest river in Italy, with a drainage basin that hosts urban and industrial settlements and large areas devoted to intensive cropping and livestock activities [46]. As a consequence, nutrient transport by Po River is mainly of anthropogenic origin and it is due to rain-driven diffuse sources (20% for TN, 20% for TP), point sources (40% for TN, 80% for TP) and groundwater, springs and tributaries (40% for TN) [29,47]. Freshwater loads of the Po have shown a complex and partially unresolved long-term variability, which included strong multi-year oscillations [34] and a shift towards early spring peaks of the runoff [37]. A decoupling between the transport of dissolved and suspended elements was also observed during the floods. This phenomenon was due to the contrasting effects of water flow on erosion, groundwater, dilution and biological processes in the river environment [48].

The Danube is the largest river in Mediterranean and Black Seas and it is about three times longer than the other rivers considered in this study (Table 1). Its catchment covers 33% of the whole Black Sea drainage basin; it is shared by 19 highly industrialized countries and it hosts 82 million inhabitants [26]. The river has a large Delta ($\approx$7000 km^2) characterized by a moderate continental climate [16]. Since the early 1970s, the regime of the Danube has changed with respect to a previous more pristine condition, because of the construction of water reservoirs, dams and hydropower plants, which significantly regulated its flow [49]. In the period 1960–1990, nitrogen discharge from Danube basin has increased about five times whereas phosphate doubled due to the increase of anthropogenic inputs. During the most recent years, a reduction of nutrient loads was observed because of political and economic changes occurred in several eastern European countries [26].

2.2. River Flow Data

Daily flow rates (m^3 s^{-1}) were obtained from publications and databases maintained by river and government authorities. Gauging stations were selected on the basis of the availability of long-term data series and of their representativeness of total water discharge from the drainage basins, upstream of the partition of flows among the arms that can be encountered in the deltas.

Ebro River flows in 1913–1931 and in 1952–2012 (Tortosa station; 40.82° N, 0.51° E; 25 m AMSL) were published in the "Anuarios de Aforos" by Ministry of Agriculture, Food and Environment of Spain (URL: http://hercules.cedex.es/anuarioaforos/afo/). Rhône River flows in 1920–2012 (Beaucaire station; 43.79° N 4.65° E; 6 m AMSL) were provided by HYDRO data center (URL: http://www.hydro.eaufrance.fr/) of Ministry of Ecology, Sustainable Development and Energy of France. Po River flows in 1917–2012 (Pontelagoscuro station; 44.88° N, 11.60° E; 8 m AMSL) were published in the "Annali Idrologici" (URL: http://www.acq.isprambiente.it/annalipdf/) by Hydrographic and Mareographic National Service of Italy (SIMN) and by the Regional Environmental Protection Agency of Emilia Romagna (ARPA; URL: http://www.arpa.emr.it/). Danube River flows in 1931–2012 (Ceatal Izmail station; 45.18° N, 28.80° E; 1 m AMSL) were provided by the Global Runoff Data Center (GRDC; URL: http://www.bafg.de/GRDC/EN/) of the German Federal Institute of Hydrology.

2.3. Chemical Parameters

The concentrations of total suspended matter (TSM), nitrate (NO_3^-), nitrite (NO_2^-), ammonium (NH_4^+), reactive phosphorus (PO_4^{3-}), reactive silicate (SiO_2), total nitrogen (TN), total phosphorus (TP) and total organic carbon (TOC) in river waters were obtained by monitoring programs, research projects, water treatment companies and published datasets. These chemical data were mostly collected monthly or seasonally, but monthly average concentrations were calculated when a higher sampling resolution was available.

For the Ebro, chemical data at Tortosa station (1980–2011) were provided by Confederacion Hidrografica del Ebro (CHE; URL: http://www.chebro.es/) and by the Water Consortium of Tarragona (CAT). For the Rhône, chemical data (1980–2012) were provided by MOOSE Program (Mediterranean Oceanic Observing System for the Environment) at Arles station (43.67° N, 4.62° E, 4 m AMSL). This station is located in the main course of the Rhône, downstream the diffluence of the Petit Rhône. For the Po, chemical data (1969–2012) were obtained from the literature [34], from monitoring programs of environmental institutions (ARPA Emilia Romagna, ICRAM and IRSA) and from past research projects focused on Adriatic ecosystem (PRISMA and MAT). Po river data referred to three sampling stations (Pontelagoscuro, Polesella and Serravalle; 44.88–44.97° N, 11.60–12.04° E; 6–8 m AMSL) located on the main river course at the enclosure of the drainage basin, 45–90 km upstream the river mouth. For the Danube, nutrient data at Reni station (45.46° N, 28.25° E, 4 m AMSL) were provided by the International Commission for the Protection of the Danube River (ICPDR; URL: www.icpdr.org), except for SiO_2 whose time series was obtained by NIMRD at Sulina station (45.15° N, 29.66° E, 2 m AMSL).

TSM was collected on membrane filters and determined by gravimetry. Nutrient concentrations were determined by standard colorimetric methods. TN and TP were determined in unfiltered samples by persulfate oxidation method followed by the colorimetric determinations of $NO_3^- + NO_2^-$ and PO_4^{3-}. TOC was determined by High Temperature Catalytic Oxidation method in unfiltered water samples. The concentration of dissolved inorganic nitrogen (DIN) was calculated as $NO_3^- + NO_2^- + NH_4^+$, whereas the concentrations of organic nitrogen (ON) and phosphorus (OP) in the dissolved and particulate pools were calculated as $TN - DIN$ and $TP - PO_4^{3-}$, respectively. The analytical methods used for the determination of these chemicals are outlined at the national level by different environmental regulations and they could have been partially changed during the period of study. However, chemical data analyzed here can be considered comparable to those applied in a wide range of environmental [50] and oceanographic [51] studies due to the high concentrations usually encountered in the river waters. The detection limits of these techniques applied to the analysis in river waters can be assumed as 1 mg L^{-1} for TSM, 0.1 µmol L^{-1} for NO_3^-, 0.5 µmol L^{-1} for NH_4^+, 0.05 µmol L^{-1} for NO_2^-, 0.03 µmol L^{-1} for PO_4^{3-}, 1 µmol L^{-1} for SiO_2, 1 µmol L^{-1} for TN, 0.3 µmol L^{-1} for TP.

2.4. Data Analysis

Variability and trends of river runoff were analyzed from daily to multi-decadal scales, in order to explore the complex dynamics that characterizes these watersheds. Daily flow distributions ($m^3 s^{-1}$) were analyzed in each river through the calculation their symmetry properties: mode (M_0), moment coefficient of skewness (γ) and Kurtosis (β). These parameters represent respectively the most frequent value of the flow, a measure of the asymmetry of tailed distributions and an index of higher ($\beta > 0$) or lower ($\beta < 0$) weights of the outliers in the population data compared to a normal distribution ($\beta = 0$). The presence of annual cycles in river water discharges was evaluated through the nonparametric statistical analysis of monthly data series [52].

Significance and magnitude of long-term trends in the series of water discharges were assessed by Mann–Kendall test (MKT) and Sen's test (ST). Mann–Kendall test is used to verify the significance of a monotonic change of a parameter over a selected period. The Sen's test quantifies this change through the calculation of the slope of the linear interpolation model [53]. These tests were applied to the time series of annual water loads ($km^3 yr^{-1}$), monthly water loads ($km^3 month^{-1}$) and maximum flow variability in each month. Maximum flow variability was calculated as difference between the highest and lowest daily flows in each month and normalized as percentage ($[Q_{max} - Q_{min}] \times 100/Q_{max}$; %).

The analysis of the trends was completed with that of the discontinuities in the series of annual water discharge, using SRSD method (Sequential Regime Shift Detector) [54]. This algorithm consists in a sequential application of the t-test to identify abrupt changes in the series and it allows their analysis avoiding a lower performance towards the end of the dataset.

Annual integrated river transport of TSM, nutrients, TN, TP and TOC (F; expressed in kt yr^{-1} of C, N, P and Si) was estimated using the equation based on discharge weighted means of daily loads:

$$F = [\sum_{i=1}^{n}(C_i \cdot Q_i) / \sum_{i=1}^{n} Q_i]Q_{yr} \cdot m_A \cdot 10^{-9} \tag{1}$$

where C_i and Q_i are nutrient concentration (mol m^{-3}) and flow (m^3 s^{-1}) for each day of sampling during the year ($n \geq 4$), Q_{yr} is the annual water discharge (m^3 yr^{-1}) and m_A is the atomic mass of the element. This method allows the best estimate of river transport through the compensation of the biases originated by (i) the variability of flows encountered in concomitance of the samplings and by (ii) the different resolutions of chemical and flow data [23]. The statistical analysis of river nutrient loads was performed using the same nonparametric methods previously mentioned for freshwater loads.

The eutrophication potential of river nutrient loads in the coastal zones (EP; kt-C yr^{-1}) was also estimated for all considered watersheds. It was calculated by annual river transports of N, P and Si, expressed as carbon biomass by Redfield's model, using the ratios proposed for the Mediterranean Sea: C:N:P = 169:23.3:1 [55] and Si:N = 0.84 [56]. EP represents the fraction of new production of algal biomass potentially sustained, in the receiving coastal water bodies, by the delivery of nitrogen, phosphorus and silicon of riverine origin [2]. Despite it does not include other processes that supply new nutrients in the coastal zones (e.g., upwelling of deeper waters, wastewater loads, atmospheric deposition), as well as the regenerated production, this calculation highlights the role of the rivers to induce nitrogen or phosphorus limitations of the new production in the coastal zones and to favor (depress) the growth of non-siliceous algal species delivering large (scarce) quantities of both these nutrients compared to silicon. Moreover, considering the ability of several autotrophic plankton species to use organic phosphorus for their growth after enzymatic hydrolysis [57,58], the eutrophication potential of DIN was compared to both those of PO_4^{3-} and TP.

3. Results

3.1. Daily to Multi-Decadal Variability of River Flows

The comparison of flow series indicated that the largest South European watersheds have distinct characteristics, which have to be analyzed on a multi-scale resolution. The most frequent values of daily flow (M_0) varied in the order: Danube > Rhône > Po > Ebro (4,214, 1,036, 929, 113 m^3 s^{-1}, respectively). The moment coefficient of skewness indicated a heavy-tailed distribution of the flow for Ebro ($\gamma = 2.59$), a river characterized by infrequent but extreme freshets that generate discharges 40-times higher than the regular flow. This asymmetry progressively reduced for Po ($\gamma = 2.30$), Rhône ($\gamma = 1.88$) and Danube ($\gamma = 0.56$), which is a river showing persistent high flows (3000–10,000 m^3 s^{-1}) and infrequent freshets in the range 10,000–15,900 m^3 s^{-1}. The values of Kurtosis confirmed that the weight of the outliers was consistent to the asymmetry and in the order: Ebro > Po > Rhône > Danube ($\beta = 10.6, 7.5, 5.6, -0.2$, respectively) and that only the distribution of the Danube was characterized by a degree of peakedness lower than the normal distribution (Figure A1; Appendix A).

On monthly scale, the analysis of integrated water loads indicated that these rivers also have distinct annual cycles of the regime (Figure 2). The Ebro has a single period of low runoff in July–October, with the lowest median in August (0.34 km^3 month^{-1}). During the other months, it is relatively high with a maximum in March (1.53 km^3 month^{-1}). The regime of the Rhône is similar to that of the Ebro, although its higher water discharge (2.54 km^3 month^{-1} in September,

5.13 km^3 month^{-1} in January). The annual cycle of the Po is characterized by two dry periods in January–February (>1.35 km^3 month^{-1}) and July–September (>0.34 km^3 month^{-1}), alternated by two periods of high runoff that reach the highest median in May (5.00 km^3 month^{-1}). The Danube shows only one main period of high discharge in March–June (20.38–23.82 km^3 month^{-1}) and the lowest runoff in October (11.41 km^3 month^{-1}). Ebro, Rhône and Po rivers also show the largest variability of the flows during the months of high discharge, whereas the Danube shows a similar dispersion of monthly data regardless of the level of the flow.

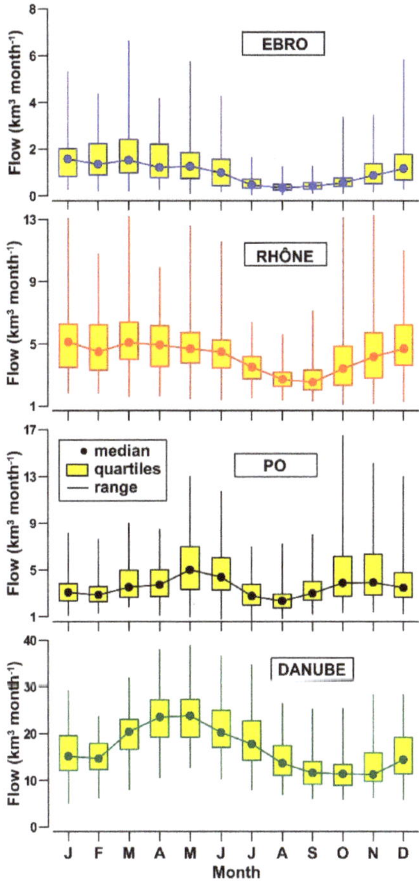

Figure 2. Box Whisker Plot (median, quartiles, data range) of monthly integrated river water discharge (km^3 month^{-1}).

On a decadal scale, the analysis of the presence of flow trends indicated for the Ebro a highly significant impoverishment of freshwater resources in all the months (from -0.002 to -0.020 km^3 month^{-1} yr^{-1}), with special regards to the periods of January–June and November–December (Table A1; Appendix A). For the other three rivers, flow reduction during the last century was significant only in late spring and summer months: May–August for the Rhône (-0.07 to -0.010 km^3 month^{-1} yr^{-1}), June–August for the Po (-0.08 to -0.015 km^3 month^{-1} yr^{-1}) and only June for the Danube (-0.052 km^3 month^{-1} yr^{-1}). For these rivers, flow trends were almost always not significant during the other periods of the year.

The range of variability of river flow can show long-term increase (decreases) in each month that, independently from the total water loads, suggests a possible exacerbation (mitigation) of the alternation between freshets and droughts. For the Ebro, the decrease of monthly water discharge previously mentioned was concomitant to a decrease of flow instability during almost all the months (Table A2; Appendix A). This reduction was pronounced in summer (−0.48% per year) and in autumn–winter (−0.25% per year). For the Rhône and Po, flow variability significantly increased in June–August (up to +0.28 and +0.13% per year, respectively), in contrast to the decrease of water discharges. In the case of the Danube, date series showed a persistent period of enhanced flow irregularity (April–August; up to +0.20% per year), although the decline of water discharge was statistically significant only in June.

The time series of annual water load indicated the presence of further changes of river regime from interannual to multi-decadal scales, which go beyond to the seasonal component of flow variability (Figure 3). MKT indicated that Ebro was the only river showing a highly significant decrease of the annual runoff during the whole of the last century ($\alpha = 0.001$), with a mean slope estimated by ST of -0.139 km^3 yr^{-1}. This trend lead to a persistent scarce water discharge since the early 1980s. The other rivers were characterized by a strong interannual variability with, however, little (0.031–0.041 km^3 yr^{-1}) and not significant ($\alpha > 0.1$) overall increments of annual runoff during the century.

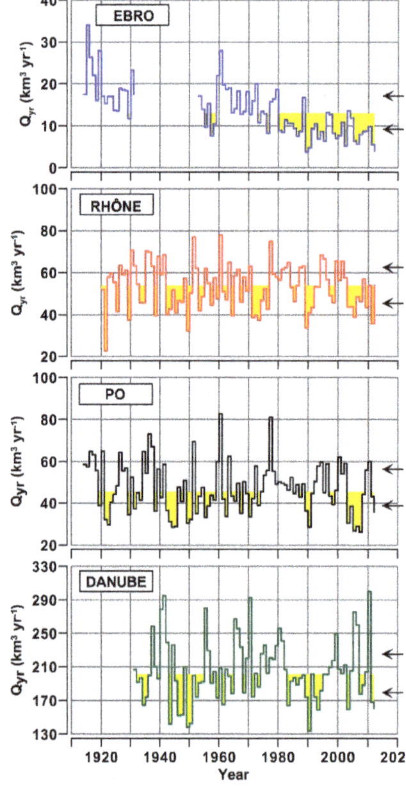

Figure 3. Time series of annual discharge (Q_{yr}; km^3 yr^{-1}) of the Ebro, Rhône, Po and Danube during the last century. Yellow areas indicate the median of water discharge and the arrows indicate 1st and 3rd quartile (data in Table 1).

Several abrupt changes were also detected in these time series (Table 2). For the Ebro, an important reduction of the runoff was observed in 1980 (-7.4 km^3 yr^{-1}), which differentiated an early long period of high flow in 1914–1979 from a period of low flow in 1980–2010. A further decrease was observed in 2011, whose persistence will have to be evaluated in the future. The Rhône was characterized by two periods of high flow (1920–1941, 1977–2002) alternated by two periods of low flow (1942–1976, 2003–2012), which implied the occurrence of three discontinuities (-10.1 to $+5.8$ km^3 yr^{-1}). The behavior of the Po was similar with the alternation of three periods of high (1914–1941), low (1942–1974) and high (1975–2002) runoff, in which the mean annual runoff changed about 7 km^3 yr^{-1}. The last two changes in 2003 and 2009 indicated respectively the beginning and the end of the strongest drought experienced by this river during the last century (-17.9 km^3 yr^{-1}). For the Danube, five alternated periods of high and low runoff were detected since the 1930s. The flows were high in 1931–1945 and 1965–1982 and low in 1946–1964 and 1983–2010, with differences of 23–36 km^3 yr^{-1}. After the highest discharge of the time series in 2010 (300 km^3 yr^{-1}), the Danube was again characterized by a low flow.

Table 2. Analysis of the discontinuities in the series of annual river water discharge by SRSD method (cut-off length = 10 yr, Huber's weight = 1): periods of homogenous flow, year of the shift, change of annual flow (Δ-flow) and confidence level (α).

River	Period	Mean Flow km^3 yr^{-1} (m^3 s^{-1})	Duration yr	Year of Change yr	Δ-Flow km^3 yr^{-1}	Confidence α
Ebro	1914–1979	16.5 (539)	66			
	1980–2010	9.1 (292)	31	1980	-7.4	0.0001
	2011–2012	4.9 (154)	2	2011	-4.2	0.0209
Rhône	1920–1941	55.9 (1772)	22			
	1942–1976	50.9 (1614)	35	1942	-5.0	0.1292
	1977–2002	56.7 (1797)	26	1977	$+5.8$	0.0284
	2003–2012	46.6 (1478)	10	2003	-10.1	0.0018
Po	1914–1941	51.3 (1593)	28			
	1942–1974	43.7 (1390)	33	1942	-7.6	0.0163
	1975–2002	50.4 (1626)	28	1975	$+6.7$	0.0183
	2003–2008	32.6 (1058)	6	2003	-17.8	0.0006
	2009–2012	48.7 (1551)	4	2009	$+16.1$	0.0548
Danube	1931–1945	212.4 (6729)	15			
	1946–1964	189.0 (5990)	19	1946	-23.4	0.0910
	1965–1982	225.3 (7141)	18	1965	$+36.3$	0.0016
	1983–2010	201.5 (6381)	28	1983	-23.8	0.0208
	2011–2012	164.6 (5213)	2	2011	-36.9	0.0003

3.2. Nutrients and Organic Matter in River Waters

Since the early 1990s, water loads of these rivers have been characterized by a high median concentration of DIN (102–171 µmol L^{-1}), which was due to the large abundance of NO$_3^-$ in the pool of dissolved inorganic nitrogen (Table 3). NO$_2^-$ and NH$_4^+$ were secondary nitrogen forms, although the concentration of NH$_4^+$ was slightly higher in Danube waters (15.0 µmol L^{-1}) than in the other rivers (2.8–5.3 µmol L^{-1}). The concentration of PO$_4^{3-}$ was in the range 1.3–2.3 µmol L^{-1}, whereas that of SiO$_2$ was 61–130 µmol L^{-1}. The concentrations of inorganic N, P and Si were about 50% higher in the Ebro and Po compared to Rhône and Danube, suggesting that a larger nutrient accumulation occurs in river basins with smaller flows. Nutrient ratios indicated a large excess (sensu Redfield's model) of inorganic nitrogen with respect to inorganic phosphorus (DIN/PO$_4^{3-}$ = 72–94) and to reactive silicate (Si/DIN = 0.5–0.7) in all the rivers.

Table 3. Concentration (median, 1st–3rd quartile) of TSM (mg L^{-1}), nutrients and organic matter (μmol L^{-1}) in river waters and nutrient molar ratios, in 1990–2012.

River (Station)	Ebro (Tortosa)	Rhône (Arles)	Po (Polesella, Serravalle, Pontelagoscuro)	Danube (Reni, Sulina)
TSM (mg L^{-1})	7.5 (5.0–11.0)	12.3 (7.4–29)	46.8 (28.4–110)	26.5 (15.0–44)
NO$_3$$^-$ (μmol L^{-1})	165 (142–194)	97 (76–119)	161 (124–202)	111 (82–141)
NH$_4$$^+$ (μmol L^{-1})	2.8 (2.2–5.5)	3.3 (1.5–5.6)	5.3 (2.6–10.2)	15.0 (9.1–25)
NO$_2$$^-$ (μmol L^{-1})	1.5 (1.1–2.0)	1.6 (1.1–2.0)	2.0 (1.3–3.0)	2.1 (1.4–3.4)
DIN (μmol L^{-1})	171 (146–201)	102 (84–125)	168 (302–216)	132 (105–163)
PO$_4$$^{3-}$ (μmol L^{-1})	2.3 (1.5–4.5)	1.5 (1.1–1.9)	1.9 (1.6–2.6)	1.3 (0.6–2.0)
SiO$_2$ (μmol L^{-1})	130 (93–160) [†]	64 (51–78)	113 (85–134)	61 (37–82)
TN (μmol L^{-1})	211 (180–234) [†]	123 (101–148) [*]	249 (197–304) [+]	147 (121–187) [°°]
TP (μmol L^{-1})	2.9 (2.6–3.6)	2.6 (2.2–3.2) [*]	4.8 (4.2–6.1)	2.8 (1.9–4.0)
TOC (μmol L^{-1})	270 (208–341) [†]	241 (199–304)	340 (268–460) [°]	398 (294–569) [**]
ON (μmol L^{-1})	28.1 (22–46) [†]	18.2 (12.7–29) [*]	76 (41–145)	25.3 (19.2–33) [°°]
OP (μmol L^{-1})	1.0 (0.4–1.7)	1.2 (0.8–1.8) [*]	2.9 (1.9–4.1)	1.2 (0.6–2.1)
DIN/PO$_4$$^{3-}$ (molar)	72 (36–118)	73 (52–100)	85 (61–107)	94 (62–228)
Si/DIN (molar)	0.7 (0.5–1.0) [†]	0.6 (0.5–0.7)	0.6 (0.5–0.8)	0.5 (0.4–0.8) [°°]
TN/TP (molar)	58 (44–70) [†]	45 (37–59) [*]	49 (37–64)	55 (38–82) [°°]
ON/OP (molar)	20 (11–31)	15 (10–22) [*]	25 (16–46)	17 (9–41) [°°]
ON/TN (%)	15 (11–21) [†]	16 (10–22) [*]	34 (20–49)	16 (13–22) [°°]
OP/TP (%)	32 (13–57)	49 (35–63) [*]	60 (47–72)	49 (30–72)

[†] 1990–2005; [*] 1996–2012; [**] 2006–2007; [°] 1995–2002; [°°] 1996–2012.

Median concentration of TN was in the range 123–249 μmol L^{-1} and that of TP was in the range 2.6–4.8 μmol L^{-1}, with the highest levels encountered in both cases in the Po (Table 3). The pool of nitrogen was mainly constituted by the inorganic forms since ON was 15–34% of TN. The partitioning of phosphorus was more balanced, with OP corresponding to 32–60% of TP. The excess of nitrogen compared to phosphorus was less pronounced in the pool of organic matter (ON/OP = 15–25) compared to that of dissolved nutrients. The concentration of TOC was high in river waters (241–398 μmol L^{-1}) and followed the order Danube > Po > Ebro > Rhône, but monitoring data covered only two years for the Danube.

TSM followed a distinct pattern from those of nutrients, as it is mainly regulated by the geomorphological characteristics of drainage basins and by the presence of dams. TSM concentration was relatively high for the Po and Danube (46.8 and 26.5 mg L^{-1}, respectively) and low for the Rhône and Ebro (12.3 and 7.5 mg L^{-1}, respectively).

Compiled data indicate that all chemical parameters in river waters were highly variable a long time. For some nutrients, like NH$_4$$^+$ and PO$_4$$^{3-}$, most of the highest concentrations occurred in early 1990s, suggesting that an improved management of nutrients in river basins has mitigated high anthropogenic loads during the last two decades. The relationships between nutrient concentration and daily flow were always complex for these rivers. However, there was a common tendency towards low concentrations of PO$_4$$^{3-}$ and high concentrations of SiO$_2$ with the increase of the flow in the cases of Ebro, Rhône and Po.

The analysis of monthly distributions of nutrient concentrations indicated the presence of pronounced annual cycles of DIN and SiO$_2$ for all the rivers (Figure A2, Appendix A). The concentration of DIN was lower in summer compared to winter and autumn, due to periodic decreases of NO$_3$$^-$ (60–120 μmol L^{-1}) and of NO$_2$$^-$ (1–2 μmol L^{-1}). SiO$_2$ also showed large annual variations (40–90 μmol L^{-1}) due to a strong decrease in April–September. For PO$_4$$^{3-}$, seasonal oscillations were observed for Po and Danube, whereas they appeared in the Rhône waters only in the most recent years (2007–2012). The seasonal oscillations of ON and OP were scarce and, therefore, those of TN and TP were mainly due to the changes of inorganic nutrients. For TOC, available data did not allow a robust analysis of seasonal variability as the monitoring of this parameter in river waters was adopted by national regulations later than that of inorganic nutrients and often with an incomplete temporal coverage.

Stoichiometric ratios of the nutrients also changed seasonally in Ebro, Rhône and Po waters. The highest excess of nitrogen, both in inorganic and total pools, were found in winter and early spring (DIN/PO_4^{3-} = 71–108, TN/TP = 47–76). In summer, DIN/PO_4^{3-} and TN/TP ratios were less unbalanced (30–88 and 36–55, respectively), due to a larger decrease of nitrogen concentration compared to phosphorus concentration. The early annual decrease of SiO_2 concentration in the waters of these three rivers caused lower Si/DIN ratios in March–April (0.42–0.60) with respect to summer (0.60–1.14). However, silicate was seldom in excess with respect to inorganic nitrogen. For the Danube, seasonal oscillations of the concentration of DIN and PO_4^{3-} were similar in amplitude and in phase, determining rather constant values of DIN/PO_4^{3-} ratio. By contrast, the strong decrease of SiO_2 concentration in summer reduced the values of Si/DIN ratio in this season with respect to winter and autumn. TN/TP ratio varied from 41 to 82 and it was maximum in February–June (Figure A3, Appendix A).

3.3. River Discharges of Biogenic Elements and Eutrophication Potential in Coastal Zones

Since the 1990s, median discharges of TSM were 71 kt yr^{-1} for the Ebro, 2031 kt yr^{-1} for the Rhône, 5547 kt yr^{-1} for the Po and 6565 kt yr^{-1} for the Danube, indicating a relatively high discharge of suspended matter in comparison to the flow for the Po. The values of DIN load (20–412 kt-N yr^{-1}) indicated that the discharge of inorganic nitrogen by the Danube was 4–20 times higher than in other rivers. PO_4^{3-} transport varied from 0.5 kt-P yr^{-1} for the Ebro to 8.3 kt-P yr^{-1} for the Danube, whereas the transport of SiO_2 (30–360 kt-Si yr^{-1}) was similar to that of DIN. The transport of TN (26–446 kt-N yr^{-1}) and TP (0.8–19.0 kt-P yr^{-1}) and TOC (29–1140 kt-C yr^{-1}) again indicated that these rivers directly deliver large quantities of particulate and dissolved organic matter to the receiving coastal zones (Table A3, Appendix A).

Annual loads of river nutrients showed long-term trends and a high interannual variability that can have important effects in fluvial and coastal ecosystems (Figure 4). In the Po case, DIN and PO_4^{3-} loads increased until mid-1980s and, in the following three decades, their trends differed with a significant decrease of PO_4^{3-} load and variable, but not decreasing, loads of DIN. The decline of PO_4^{3-} transport was common to the other three rivers, but it occurred for the Ebro and Danube over one decade later than for the Rhone and Po. Annual loads of SiO_2 were mainly characterized by a strong variability with, however, a tendency toward low values in the most recent years.

The analysis of the trends by MKT and ST, performed for all chemical parameters over the available time series, indicated significant decreases ($\alpha \leq 0.1$) of the transport of TSM, NO_2^-, DIN, PO_4^{3-}, TOC and TP for the Ebro (Table A4, Appendix A). For the Rhône, the most important change was a long-term reduction of PO_4^{3-} load (-0.22 kt yr^{-2}), as NO_2^- decrease (-0.04 kt yr^{-2}) is only a secondary nitrogen form and the time series of ON, even if characterized by a decrease (-1.13 kt yr^{-2}), also showed a strong variability. For the Po, significant decreases of NH_4^+, NO_2^-, PO_4^{3-} and TP occurred after mid 1980s. Since 1997, a tendency towards reduced loads of inorganic and organic nitrogen was observed in the case of the Danube but, in this more recent period, the decrease was statistically significant only for NH_4^+ (-3.58 kt yr^{-2}).

The contribution of the discharge of river nutrients to the growth of algal biomass in the receiving coastal zones was estimated by the eutrophication potential [2]. For all the rivers, EP based on the river loads of DIN was higher than that of PO_4^{3-}, suggesting that the supply of land-borne inorganic phosphorus was fundamentally the factor limiting the new production in the coastal marine environment (Figure 5). However, the scarcity of phosphorus could be mitigated by the fraction of the organic phosphorus pool potentially bioavailable for algal growth [57,58]. During the last two decades, this compensation was potentially greater for the Rhône and Po (TP-based production respectively equal to 61% and 79% of DIN-based production) compared to the Ebro and Danube (TP-based production equal to 45% and 54% of DIN-based production).

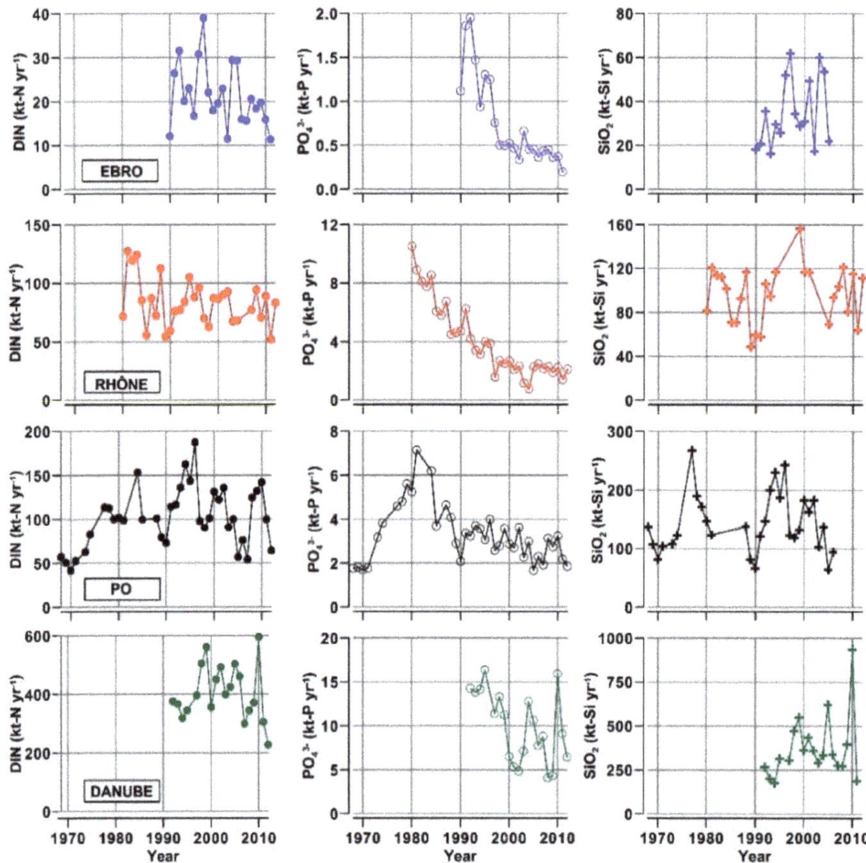

Figure 4. Time series of annual transport (kt yr^{-1}) of the main nutrients (DIN, PO$_4$$^{3-}$ and SiO$_2$) in the four considered rivers.

EP estimated with SiO$_2$ loads was often similar to that of PO$_4$$^{3-}$ in the period from 1980 to the early 1990s (Figure 5). Afterwards, the decline of phosphorus transport induced a persistent condition of excess of SiO$_2$ compared to PO$_4$$^{3-}$ for algal growth (EP-SiO$_2$ 2–6 folds higher than EP-PO$_4$$^{3-}$, in 2000–2012).

Despite TOC series are not so long like those of nutrients, a comparison between the new carbon production potentially generated by river nutrients in the coastal zones and the direct discharge of riverine TOC can be done. In the current post-eutrophic phase, available data indicated a median TOC transport of 29 kt-C yr^{-1} for the Ebro, 155 kt-C yr^{-1} for the Rhone, 246 kt-C yr^{-1} for the Po and 1140 kt C yr^{-1} for the Danube (Table A3). These values suggest that the supply of river organic carbon is not negligible compared to the fraction of new production currently sustained by riverine nutrients, when PO$_4$$^{3-}$ is considered the nutrient limiting the growth of the biomass (median values of 46, 141, 194 and 889 kt-C yr^{-1}, respectively). The importance of riverine TOC decreases for all the rivers ($\approx$50%) when the potential new production is based on TP loads (median values of 69, 317, 587 and 1484 kt-C yr^{-1}, respectively).

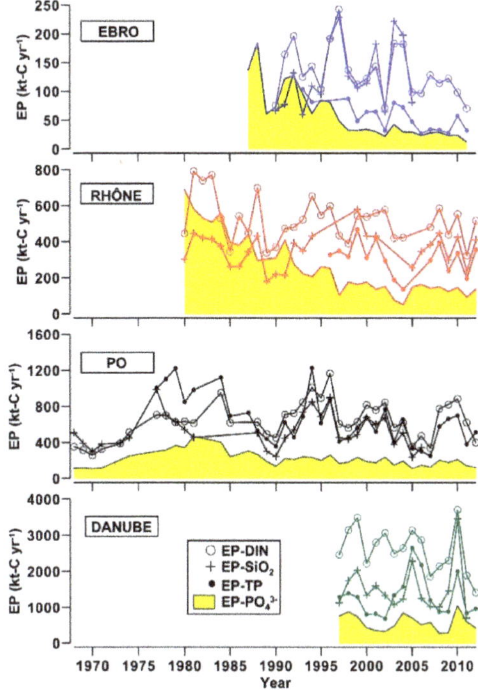

Figure 5. Eutrophication potential of river nutrient loads (EP; kt-C yr^{-1}) estimated by Redfield's model and annual river discharges of DIN, SiO$_2$, TP and PO$_4$$^{3-}$. For each year, the lowest value of EP identifies the nutrient that potentially limits the new production sustained by river loads.

4. Discussion

The watersheds considered in this study cover wide ranges of longitude (4° Long. W–30° Long. E), surface (74,000–801,500 km^3), inhabitants (3–82 × 10^6) and flows (9–15,900 m^3 s^{-1}). Moreover, they constitute the main contribution to the runoff in the Mediterranean and Black Seas. The comparative analysis of their characteristics has highlighted a multi-scale variability at sub-continental scales that cannot be ignored as it reflects in distinct hydrological and ecological impacts across freshwater and coastal environments in Southern Europe (Table 4).

Table 4. Summary of the descriptors analyzed in this study and of the potential impacts in the adjacent coastal zones discussed in the Sections 4.1 and 4.2.

Descriptor	Ebro NE Shelf of Spain	Rhône Gulf of Lion	Po NW Adriatic Sea	Danube NW Black Sea	Environmental Impacts
Annual freshwater discharge	Long-term decrease.	Interannual to multi-decadal oscillations increased recently.	Interannual to multi-decadal oscillations increased recently.	Interannual to multi-decadal oscillations.	Marked regional differences of river discharges. Increased droughts in summer more marked in SW Europe than in SE Europe. Increased oscillations of summer runoff in SE Europe.
Long-term trend of monthly flows	Decrease in all months.	Decrease in late spring and summer.	Decrease in summer.	Small decrease in summer.	
Long-term trend of flow oscillations	Decrease in all months.	Increase in summer.	Increase in summer.	Increase in spring and summer.	
Dry seasons	Summer.	Summer.	Winter and summer.	Winter and summer.	
Flow regime	Low with a high incidence of freshets.	High with low incidence of freshets.	Intermediate with a high incidence of freshets.	Very high with low incidence of freshets.	
Concentrations of TSM and nutrients	Low TSM, high nutrients.	Low TSM, low nutrients.	High TSM, high nutrients.	Medium TSM, medium nutrients.	Impact of the decrease of TSM transport on estuarine and coastal areas. Distinct impacts of seasonal changes of nutrients on riverine, estuarine and coastal ecosystems.
Variability of nutrient concentrations	Seasonal cycle (except PO_4^{3-}).	Seasonal cycle (for PO_4^{3-} only since 2007).	Seasonal cycle.	Seasonal cycle	
Variability of DIN/PO_4^{3-}, TN/TP and Si/DIN ratios	Seasonal oscillations (low N/P in summer, low Si/DIN in spring).	Seasonal oscillations (low N/P in summer, low Si/DIN in spring).	Seasonal oscillations (low N/P in summer, low Si/DIN in spring).	Rather constant DIN/PO_4^{3-} through the year, high TN/TP in spring.	
Incidence of organic nitrogen and phosphorus on TN and TP pools	ON: low OP: low	ON: low OP: medium	ON: high OP: high	ON: low OP: high	Potential growth of marine plankton species able to utilize riverine ON and OP.
Recent trends of annual loads of TSM, nutrients and OM	Decreases of TSM, N, P and TOC transport since the 1990s. Oscillations of SiO_2.	Decrease of PO_4^{3-} and ON transport since the 1980s. Oscillations of SiO_2 and DIN.	Decreases of NH_4^+, NO_2^- and PO_4^{3-} and TP transport since the 1980s. Oscillations of SiO_2.	Decreased transport of NH_4^+ since the 1990s. Oscillating N, P transports since the 2000s.	Reduction of PO_4^{3-} transport leading to phytoplankton biomass reduction. Oscillation of nutrient loads linked to runoff variability.
Marine region of freshwater influence	Small, limited by continental shelf orography.	Medium, limited by continental shelf orography.	Large, enhanced by continental shelf orography.	Very large, enhanced by continental shelf orography.	Larger impacts in the coastal zones of Po and Danube, even if river loads are reduced.
Eutrophication potential of river nutrient loads in the receiving coastal zones	Until 1995, high and balanced nutrient loads. Afterwards, excesses of DIN and SiO_2 over PO_4^{3-}. Low weight of OP in TP.	Until 1990, high and balanced nutrient loads. Afterwards, excesses of DIN and SiO_2, over PO_4^{3-}. Medium weight of OP in TP.	Until 1990, high and balanced nutrient loads, with a surplus of OP. Afterwards, excesses of DIN and SiO_2, over PO_4^{3-}. Persistent high weight of OP in TP.	Since 2000s, excesses of DIN and SiO_2, over PO_4^{3-}. Medium weight of OP in TP.	Shift form eutrophic conditions to oligotrophic, but still degraded, conditions around the 1990s. PO_4^{3-} scarcity sometime potentially compensated by OP bioavailability.

4.1. Seasonal to Decadal Trends of River Flows and Coastal Hydrology

Despite flow regulation systems are nowadays present in all drainage basins [16,38,45,46] flow dynamics is still distinct for these rivers, suggesting that their effects on the receiving coastal zones differ depending both on the quantity and on the regime of freshwater discharges. Smaller drainage basins are primarily the cause of a low mean runoff, on which are superimposed infrequent peaks of high discharge generated by rainfall and snowmelt (Table 1; Figure A1). The Ebro watershed has a Mediterranean climate with continental characteristics, with a current precipitation regime (annual mean of 600 mm) characterized by the occurrence of persistent droughts in the middle and lower basin alternated with heavy rainfall events [41,59]. Annual precipitation in the Rhône watershed (1030 mm) depends on a seasonal contribution of snowfall and by the rainfall, which exhibits a high variability mainly due to localized storms occurring on the lower river basin [28]. A high interannual variability of the precipitation (1200 mm) is also typical of mild continental climatic conditions in the Po Basin [17,37]. Mean Annual precipitation in the Danube watershed varies from 500 mm in the central plain and delta regions to 2000 mm in the western mountainous regions [60]. For the Rhône, Po and Danube, the contribution of glacier-melt water to river runoff, although not of primary importance, is a term that should be better quantified [28,37].

The presence of distinct annual cycles of river runoff has consequences on the hydrology of the coastal zones as the advection of river waters interacts with stratified (mixed) seawater in summer (winter). Summer is the driest season for the Ebro, Rhône and Po (Figure 2). In the case of the Danube, the driest period occurs in late summer early autumn, due to a larger contribution of rainfall and snowmelt in April–July [16]. For the Po, it was shown that scarce summer discharges contribute to the weakening of the circulation in NW Adriatic [61], enhancing the retention of coastal waters and the eutrophication problems even in the presence of a low nutrient supply [31]. By contrast, low runoff reduces the impact of Ebro and Rhône on the mesoscale circulation on their continental shelves as they have a greater depth and large open boundaries with the sea [9,15].

The Po and Danube are also characterized by a dry season in winter, contrary to Ebro and Rhône. The timing of the freshets in late winter and early spring is important as it causes the earliest stratification of the water column and the earliest allochthonous nutrient enrichment during the year, mostly favoring diatom blooms. In the other seasons, river discharges can favor non-siliceous species that often lead to undesirable effects in the coastal waters [2]. For this reason, the possible alteration of the timing of river freshets, due to dam regulation and climate changes, is a factor able to modify the succession of plankton communities.

The characteristics of continental shelf regions further concur to differentiate the impacts of these rivers. For the Ebro, the long-term decline of the runoff, coupled to a rather narrow continental shelf and to a relatively strong coastal current, limits the extension of the river plume [15,30]. Meteorological conditions and large-scale circulation in NW Mediterranean regulate the extension of the coastal front generated by the Rhône in the Gulf of Lion, usually preventing eutrophication problems [60]. The fine haline structure of Rhône plume, which includes the presence of persistent low-salinity surface water lenses, can favor a high production sustained by river nutrients, but these processes are usually temporarily [33].

The cases of Po and Danube are opposite. The wide, shallow and semi-enclosed continental shelf of N Adriatic strongly enhances the impact of Po discharges. Particularly during summer, a weak circulation and stable meteorological conditions cause the retention of low-salinity waters on the shelf and the formation of coastal fronts with meanders and gyres that favor alternated eutrophic and oligotrophic conditions [20], as well as seasonal hypoxia [11,31]. For the Danube, a persistent large water discharge (3000–10,000 m^3 s^{-1}) with a scarce incidence of the freshets (Figure A1), coupled to the orography of the shallow continental shelf of NW Black Sea, maintains a permanently low-salinity upper layer as far as 20 km from river mouths, which isolates the deeper layer favoring bottom hypoxia [16].

The analysis of long-term trends indicated a significant decrease of the runoff all along the year in the Ebro basin (Table A1, Figure 3), concomitant to a decrease of the variability of river discharge (Table A2). An important part of this trend has anthropogenic causes like an increased usage of freshwater for cropping in the middle river basin and a stronger regulation of river flow after dam building. However, the afforestation in mountain areas of the basin following the progressive abandonment of farming has also contributed to the decline of the runoff in the last decades [38]. At the same time, climate change effects such as large droughts observed in Iberian Peninsula in the 2000s, increased evapotranspiration and recurrent heat waves have exacerbated the scarcity of freshwater in this region [32,36]. These events had significant socio-economic impacts, in particular on crop and livestock productions, hydropower generation, tourism and population health [32,38].

In the other three river basins, significant trends were detected essentially in summer, a season where the decline of the runoff was counterbalanced by an increased flow irregularity. An overall reduction of the runoff is expected in Southern Europe in a scenario of climatic warming, due to the increase of evapotranspiration and to the reduction of snow cover and precipitation [32], but present results support the importance of a higher resolution assessment of the related environmental and socio-economic impacts on river drainage basins [47]. Similarly, reliable streamflow simulations at regional scales are of basic importance to study these differences, but they have to be obtained applying appropriate downscaling techniques to global precipitation models [61,62]. The shortage of freshwater in the Iberian Peninsula along most of the year that gradually reduces eastward to only summer in the Balkans indicates that the anthropogenic usage of freshwater and regional climatic conditions are yet differently affecting each of these drainage and coastal systems. For the Danube, current data show only a large variability, but model results suggest an oncoming decrease of the runoff in the whole basin in summer and in the middle and lower basins in autumn in the next 30 years. By contrast, winter water loads should be maintained by the presence of a heavy seasonal rainfall and by an early annual snowmelt [62].

Long-term series of annual water discharge also showed several discontinuities, which indicate alternate positive or negative changes of the runoff in Southern Europe, often separated by intervals of 20–30 years (Table 2). Adjacent drainage basins like those of Rhône and Po showed concomitant changes, different from those of the Ebro and Danube. This feature points out the importance of Mediterranean climate patterns, which are characterized by a latitudinal transition from a maritime coastal climate to a subtropical desert climate and by a longitudinal variability due to the influence of Atlantic circulation, South Asian monsoons in summer and Siberian high-pressure systems in winter. Mesoscale climatic features are significantly modulated by land orography, in particular in mountainous regions like the Alps [63]. This variability easily induces distinct regional regimes of snowmelt, rainfall and evapotranspiration and a distinct variability of river discharges from daily to multi-decadal scales [37,61,64]. Moreover, human interventions for the regulation of river flows and the demographic and economic growth in South European countries during the last century can have contributed to differentiate long-term trends of the runoff in these watersheds.

Multi-year phases of anomalous freshwater discharge can modify the biogeochemistry of marine systems at basin scales as river plumes affect the pelagic and benthic compartments in extended portions of the continental margins [19]. This process was observed in N Adriatic basin in 2003–2007, when a prolonged period of extremely low runoff caused the increase of salinity and the decreases of nutrient concentration and phytoplankton biomass [20,34].

4.2. Loading of Biogenic Elements and Impacts on Fluvial and Coastal Systems

Currently, most of nitrogen and phosphorus loads in these drainage basins are due to human activities, with an overload of NO_3^- mainly originated by agriculture and atmospheric deposition and PO_4^{3-} inputs mainly linked to urban and industrial wastewaters [26,29,36,40,44]. The dissolution of SiO_2 is a natural process caused by the weathering of rocks and sediments. However, the regulation of river flow caused by dam building was shown to reduce silicate transport of the Danube [49].

The concentration of nutrients and organic matter is further modified by the presence of the reservoirs along the rivers [41] and by the hydrological characteristics of deltas [16] as they affect the residence time of waters, the sedimentation of particulate matter and the growth of plankton communities. The combination of these complex processes causes distinct levels of biogenic elements in these four rivers and a variability of nutrient ratios different from that of the concentrations, with excesses of DIN compared to PO_4^{3-} in the order: Danube > Po > Rhône > Ebro, with more similar Si/DIN ratios in all these rivers and with rather high ON/OP ratios in the Po (Table 3).

Another factor that have contributed to the differentiation of nutrient trends in these rivers was the timing of adoption of the measures for the reduction of eutrophication problems by National and Environmental Authorities in each drainage basin. They have included reduction of polyphosphate in the detergents, reduction of the use of fertilizers, improvement of wastewater treatment methodologies and the reconversion of animal farms and industries. However, it should be noticed that changes in nutrient management can have delayed effects in freshwater and coastal ecosystems due to the different turnover times of these elements in groundwater and soils, which have to be monitored for years after their adoption [26,27,29,46].

The current annual cycle of river nutrients generates the largest overloads of nitrogen in winter and autumn and a scarcer and more balanced nutrient supply in summer (Figures A2 and A3). Phytoplankton blooms are favored by large river discharges in winter-early spring and, with a minor intensity, in autumn in the coastal zones of the Rhône [65–67], Po [68,69] and Danube [8]. However, periods of scarce river nutrient discharge not always prevent the eutrophication in the coastal waters due to possible opposite effects of meteorological conditions and circulation, which are particularly important in the shallow, semi-enclosed and more extended continental shelves of N Adriatic and NW Black Sea. Coastal hypoxia is enhanced in summer by a long retention of Po River nutrients in warm and highly stratified shallow waters, even in the presence of a scarce runoff [20,31]. The upwelling of hypoxic near-bottom waters in the coastal zone of the Danube can occur in all the seasons under the influence of western winds [16]. In the coastal zones of the Ebro and Rhône, river inputs can also induce a high biological productivity in early spring and autumn, however, the orography of the continental margins favors the dispersion of nutrient loads limiting eutrophication problems [9,30,33,70].

The transport of TSM is regulated by the morphological characteristics of drainage basins and by the frequency of freshets, which determine the erosion of sediments, but the presence of dams limits again the delivering of suspended sediments to the sea. The present study indicates that the Po is a river with a particularly high transport of TSM, because of the large release of sediments from the mountainous area of Apennines in the southern part of its drainage basin and the absence of dams along its lower course. The decrease of TSM transport due to human activities in the drainage basins can enhance the erosion of deltas and sandy coasts, in particular if it is coupled to changes of storm frequency, circulation and sea level [16,17,38].

The analysis of annual nutrient budgets indicated that the decline of PO_4^{3-} transport has been the most important long-term change characterizing river loadings. This reduction occurred earlier for Rhône and Po compared to Ebro and Danube (Figure 4). Annual discharges of DIN and SiO_2 have shown lower values during the last decade, but the interannual variability of their budgets, mainly originated by the changes in water discharge [34], have been the prominent characteristics of their behavior. The oligotrophication due to phosphorus decline is more relevant in estuaries and semi-enclosed coastal zones with strong riverine influence, where the runoff sustains the largest fraction of the production [4,30,71]. Trophic levels in coastal zones differ at regional scales, due to distinct environmental and climatic conditions. However, this significant decrease of continental inputs of PO_4^{3-} has certainly strengthened a recent P-limitation in the whole Mediterranean and Black Sea [10,19–21,35]. For this reason, future research is needed to better assess the impact of such changes on a variety of ecosystem services [5,67].

Primary production sustained by rivers represents only less than 2% of the production in the Mediterranean and less than 5% in the Black Sea, but it can be ten times higher in the regions of freshwater influence [21]. During the last century, all considered rivers have passed through three phases characterized by "pristine", eutrophic and post-eutrophic conditions [4,26,27,34,72]. In the eutrophic periods (c.a. 1970–1990), the increase of the discharges of inorganic nutrients with rather balanced N:Si:P ratios caused severe eutrophication in the lower river environments, estuaries and coastal zones of N Adriatic and NW Black Sea, which were not previously observed [4,6,10,73]. In the current phase, eutrophication potential of river nutrients based on PO_4^{3-} uptake is always lower compared to that of DIN and SiO_2 (Figure 5), indicating that the impact of these rivers on the marine environment is mitigated by the scarcity of inorganic phosphorus.

For Ebro River, this trend of oligotrophication caused a quick decrease in chlorophyll and a large increase transparency in river waters, triggering the subsequent colonization of macrophytes [4]. The Ebro estuary shifted from a highly eutrophic condition, with summer anoxia in the salt wedge and nearly absence of macroinvertebrates to meso- and oligo-trophic conditions with a total recovery of oxygen levels and a rich macroinvertabrate community [73]. In the Gulf of Lions, the increase of nutrient loadings by the Rhône in the period 1970–1990 had an important impact on the productivity of coastal waters [33,44]. Afterwards, phosphorus limitation of algal growth was proved in coastal waters affected by river inputs [70].

In N Adriatic, an increase of offshore and coastal hypoxia was observed until the 1980s [11,31], coupled to the degradation of the benthic compartment [12] and to recurrent dinoflagellate blooms [74]. Afterwards, the more recent nutrient imbalance in river waters of this region was related to a shift towards oscillating trophic conditions that can have favored the appearance of mucilaginous aggregates [20,74,75].

In the shallow coastal shelf of W Black Sea, river nutrient loads have increased since the 1970s, specifically during the 1980s, causing a long-term increase of eutrophication, which included chronic harmful algal blooms, reduced transparence of the water column, persistent hypoxia, jellyfish proliferation and mass mortality events of pelagic and benthic organisms [35,76]. Since the early 1990s, PO_4^{3-} decline had an important role in the transition of this coastal system from a eutrophic state to a low-energy and degraded food web dominated by non-siliceous plankton communities, which is not comparable to a pristine condition [10,35,77]. For both Po and Danube coastal zones, the regeneration of organic nitrogen and phosphorus entrapped in marine sediments became an additional source of nutrients for plankton communities able to maintain eutrophic conditions independently of the runoff [8,31].

Present budgets further indicate that the imbalance in the delivery of river nitrogen and phosphorus in the coastal zones might be significantly reduced, if particular and dissolved organic phosphorus is considered at least partially available for marine plankton growth (Figure 5). The bioavailability of riverine particulate phosphorus is limited in the estuarine zones, when it is in a mineral form or if the sedimentation of particles is fast [72]. Nevertheless, a high concentration of bioavailable particulate phosphorus was found in the Danube Delta [78]. Moreover, many freshwater and marine algae species are able to synthesize alkaline phosphatase to obtain PO_4^{3-} by the enzymatic hydrolysis of phosphormonoesters, which are an important component of the natural pool of dissolved organic phosphorus [57]. This enzymatic catalysis can significantly sustain phytoplankton productivity in the marine environments characterized by a persistent deprivation of PO_4^{3-}, like the Northern Adriatic [58]. For these reasons, the real eutrophication potential of river nutrient loads would be better assessed in term of DIN/TP ratio rather than DIN/PO_4^{3-} ratio [79] and the role of river inputs of organic phosphorus to sustain autotrophic production should be further investigated in these drainage systems.

Finally, it can be noticed that in the current post-eutrophic phase the discharge of TOC by these rivers is not negligible compared to the new carbon production potentially sustained by river nutrients in the receiving coastal zones, assuming that is limited by phosphorus. Despite the overall production on continental margins also depends by nutrient remineralization, upwelling of nutrient-rich deep waters, wastewater discharges and atmospheric deposition, these data indicates that the consequences of the loads of riverine TOC in the coastal environments should be better investigated. The particulate organic carbon discharged by these rivers in the adjacent continental shelves can be suspended, buried or transferred to the deep sea depending on the river flows and marine circulation [15,42,48,80]. This organic material, together to terrestrial dissolved organic carbon, can be at least partially made available for marine food webs through bacterial degradation [14]. The discordance between the large delivery of terrestrial TOC to the global ocean and its limited quantity usually detected in coastal marine environments suggests that river TOC is used by the marine food webs, both at regional and at global scales [13].

5. Conclusions

Results show that the largest watersheds in Southern Europe have distinct features (Table 4), which originate by the combination of regional climatic conditions, anthropogenic usage of freshwater resource, adopted policies for the mitigation of eutrophication and oceanographic features of the receiving coastal water bodies. Their flow regimes show a complex dynamics that have to be analyzed from daily to multi-decadal scales. Notwithstanding the inhomogeneous coverage of time series of TSM, nutrients and organic matter in river waters, these data are fundamental to assess the possible combined impacts of anthropogenic and climatic pressures on coastal environments. Comparative analysis of these aquatic ecosystems are not common, but they should be intensified to better understand their distinct responses to environmental changes at sub-continental scales. The following main results are highlighted by the present study:

- Flow dynamics of Ebro, Rhône, Po and Danube rivers exhibits a different incidence of freshets and droughts and distinct annual cycles. This feature suggests the importance of regional climatic factors in these drainage basins, despite the widespread presence of flow regulation systems.

- Annual water discharges of the Ebro significantly decreased during the last century, whereas those Rhône, Po and Danube showed multi-decadal oscillations. For the Ebro, this difference is consistent with the rise of anthropogenic usage of freshwater in the drainage basins and with regional climate changes. For the other rivers, interannual variability of water discharge is still prevailing on long-term trends.

- The decrease of water discharge of the Ebro was concomitant to a reduction of flow variability in all the seasons. For the Rhône, Po and Danube, the decrease of discharges occurred mostly in summer, with a concomitant increase of flow variability that suggests a greater instability of climatic conditions in their regions.

- The concentrations of inorganic nutrients, TN and TP in the waters of Ebro and Po are about 50% higher than in those of Rhône and Danube. This finding suggests that the former two watersheds might be the most impacted ones by nutrient pollution, in a future scenario of reduced runoff, even in the presence of constant inputs due to agricultural, urban and industrial activities.

- The concentrations of DIN and SiO_2 show a clear annual cycle in these rivers, with the lowest levels (-50%) being reached in spring and summer. This cycle changes seasonally quantity and composition of the nutrient pool delivered into the receiving coastal water bodies.

- The analysis of nutrient budgets indicated that these rivers have changed from a past condition characterized by large discharges of nutrients, with a rather balanced N:Si:P ratio, to overloads of DIN and SiO_2 with respect to PO_4^{3-}. This process has reduced the eutrophication of rivers, estuaries and coastal marine environments inducing, however, changes of ecological conditions that have to be further assessed.

- Phosphorus scarcity is a common feature of these river and coastal ecosystems, but its potential ability to limit primary production significantly reduces if organic phosphorus is considered at least partially available for the growth of phytoplankton. For this reason, the real bioavailability of riverine organic phosphorus for auto- and hetero-trophs should be better investigated as it could play a key role in the regulation of the productivity and structure of plankton communities.

- In the current post-eutrophic phase, the discharge of riverine TOC to the coastal zone is not negligible with respect to the eutrophication potential of river nutrients.

Author Contributions: Conception and design of the research, S.C. and M.G.; Data curation, S.C., C.I., L.L., P.R. and M.G.; Methodology, S.C.; Writing—original draft, S.C.; Writing—review & editing, S.C., C.I., L.L., P.R. and M.G.

Funding: This research was funded by European Project PERSEUS, "Policy-oriented marine Environmental Research in the Southern EUropean Seas" (FP 7, n. 287600).

Acknowledgments: The authors thank national institutions and environmental agencies of Spain (MAGRAMA, CHE, CAT), France (MEDDE, MIO), Italy (SIMN, ARPA Emilia Romagna, ICRAM, IRSA) and Romania (ABA Dobrogea-Litoral), as well as the European Environment Agency (EEA) for supplying information, flow and chemical data in river waters. For the Rhône, monitoring data were provided by the MOOSE Program with the support of the "Agence de l'eau Rhône–Méditérranée–Corse". For the Danube, flow data were provided by the Global Runoff Data Center (GRDC - BFG) and monitoring data were provided by the International Commission for the Protection of the Danube River (ICPDR).

Conflicts of Interest: The authors declare no conflict of interest. The funders had no role in the design of the study; in the collection, analyses, or interpretation of data; in the writing of the manuscript, or in the decision to publish the results.

Appendix A

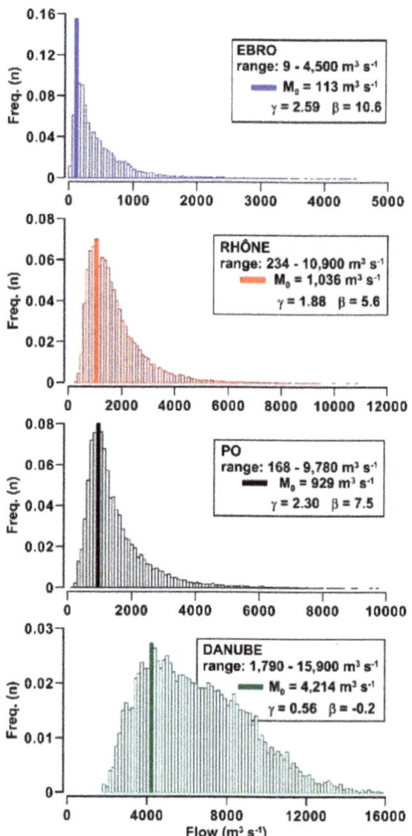

Figure A1. Frequency distributions of daily flows ($m^3\ s^{-1}$) of the rivers and symmetry properties of population data (M_0 = mode, γ = moment coefficient of skewness, β = Kurtosis).

Table A1. Slope (km^3 month^{-1} yr^{-1}) of monthly integrated river discharge (km^3 month^{-1}) vs. time (yr) in flow series of Table 1 estimated by MKT and ST. The levels of significance of these trends are highlighted in gray (*** α = 0.001, ** α = 0.01, * α = 0.05, + α = 0.1, no symbols α > 0.1).

Month	Ebro	Rhône	Po	Danube
		km^3 month^{-1} yr^{-1}		
Jan.	−0.013 ***	+0.011	+0.001	+0.045
Feb.	−0.014 ***	+0.008	+0.001	+0.041 *
Mar.	−0.020 ***	−0.001	−0.009 +	−0.005
Apr.	−0.019 ***	−0.002	−0.007	+0.025
May	−0.015 ***	−0.010 +	−0.001	−0.015
Jun.	−0.014 ***	−0.010 +	−0.015 *	−0.052 +
Jul.	−0.005 ***	−0.009 *	−0.016 **	−0.040
Aug.	−0.002 *	−0.007 **	−0.008 *	−0.014
Sep.	−0.003 ***	−0.004	+0.001	+0.011
Oct.	−0.005 ***	−0.001	−0.004	+0.020
Nov.	−0.010 ***	−0.001	−0.005	+0.006
Dec.	−0.013 ***	+0.012 +	0.000	+0.007

Table A2. Slope (% yr^{-1}) of maximum monthly flow variability ($[Q_{max} - Q_{min}] \times 100/Q_{max}$; %) vs. time (yr) in flow series of Table 1 estimated by MKT and ST. The levels of significance of these trends are highlighted in gray (*** α = 0.001, ** α = 0.01, * α = 0.05, + α = 0.1, no symbols α > 0.1).

Month	Ebro	Rhône	Po	Danube
		% yr^{-1}		
Jan.	−0.116 +	+0.065	−0.065	−0.123 *
Feb.	−0.211 ***	−0.002	−0.054	−0.135
Mar.	−0.060	+0.019	−0.106	+0.028
Apr.	−0.019	+0.082	−0.021	+0.074 +
May	+0.063	+0.047	+0.007	+0.201 ***
Jun.	−0.040	+0.105 +	+0.119 *	+0.123 *
Jul.	−0.420 ***	+0.279 ***	+0.134 **	+0.127 *
Aug.	−0.417 ***	+0.128 *	−0.012	+0.091 +
Sep.	−0.481 ***	+0.020	−0.086	+0.072
Oct.	−0.251 **	+0.006	−0.042	+0.004
Nov.	−0.222 ***	+0.062	−0.009	−0.012
Dec.	−0.165 *	+0.017	+0.038	+0.034

Table A3. Statistics of the dataset of annual transport (F; kt yr^{-1}) of TSM, nutrients and total nitrogen and phosphorus at the river mouths of the Ebro, Rhône, Po and Danube, in 1990–2012.

Year	F-TSM kt yr^{-1}	F-NO$_3^-$ kt-N yr^{-1}	F-NH$_4^+$ kt-N yr^{-1}	F-NO$_2^-$ kt-N yr^{-1}	F-PO$_4^{3-}$ kt-P yr^{-1}	F-SiO$_2$ kt-Si yr^{-1}	F-TOC kt-C yr^{-1}	F-TN kt-N yr^{-1}	F-TP kt-P yr^{-1}
				Ebro					
Median	70.91	19.17	0.50	0.19	0.50	30.37	29.28	26.12	0.81
1th Quartile	51.70	15.74	0.27	0.15	0.44	21.64	25.70	21.85	0.51
3rd Quartile	98.91	24.64	0.69	0.28	1.08	50.20	43.21	34.11	1.14
Min.	21.03	21.03	0.18	0.09	0.20	16.16	12.83	13.68	0.42
Max.	224.10	38.15	1.48	0.43	1.95	62.14	49.33	47.24	1.59
				Rhône					
Median	2030.56	79.98	2.91	1.27	2.33	105.05	155.20	101.91	4.85
1th Quartile	1262.63	65.71	2.39	0.98	2.11	78.11	143.92	89.27	4.08
3rd Quartile	3972.30	83.93	3.43	1.53	3.28	116.62	229.02	109.66	5.39
Min.	405.14	49.67	1.16	0.44	0.79	58.02	124.25	62.79	2.09
Max.	7998.60	99.54	4.59	2.52	6.24	156.88	279.43	115.80	7.13
				Po					
Median	5546.90	102.11	3.99	1.44	2.88	137.49	246.42	156.80	8.53
1th Quartile	3434.51	85.43	3.05	1.16	2.28	118.75	229.18	111.37	6.27
3rd Quartile	7771.78	126.30	6.08	1.85	3.30	183.28	299.69	227.11	10.25
Min.	1030.47	51.42	1.60	0.54	1.67	64.39	225.14	94.19	3.83
Max.	16,292.57	179.19	9.69	2.53	4.00	243.46	366.53	295.16	18.71
				Danube					
Median	6565.15	338.77	62.72	7.95	8.28	360.13	1139.75	446.30	18.96
1th Quartile	4820.07	297.08	36.09	5.67	6.19	298.29	968.46	374.27	13.17
3rd Quartile	10,478.27	406.26	75.93	8.92	11.27	452.54	1311.04	587.20	21.95
Min.	2645.97	188.40	26.08	2.14	4.13	189.11	797.17	270.00	10.39
Max.	19,077.89	535.22	89.79	12.05	15.96	935.85	1482.33	714.05	40.47

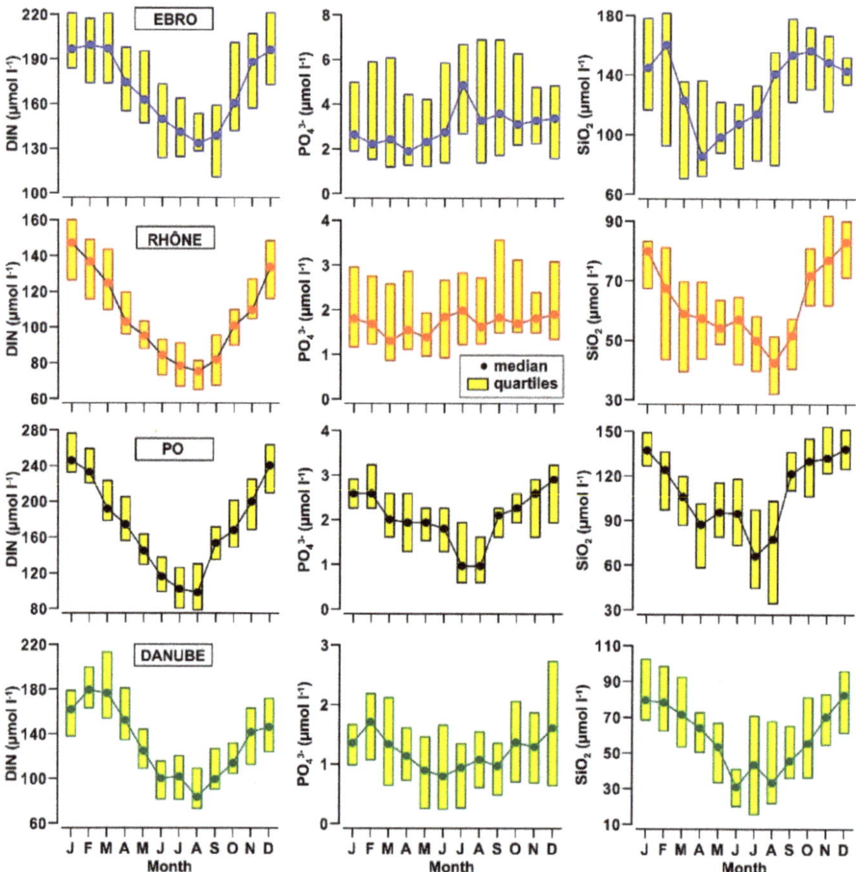

Figure A2. Box Whisker Plot (median and quartiles) of monthly distribution of nutrient concentration (μmol L^{-1}) in the Ebro, Rhône, Po and Danube waters, in 1990–2012.

Table A4. Slope (kt yr^{-2}) of annual river transport (F; kt yr^{-1}) of TSM, nutrients and organic matter vs time (yr) estimated by MKT and ST since the 1980s for Ebro, Rhône and Po and since 1997 for the Danube. The levels of significance (Mann-Kendall test) of these trends are highlighted in gray (*** $\alpha = 0.001$, ** $\alpha = 0.01$, * $\alpha = 0.05$, + $\alpha = 0.1$, no symbols $\alpha > 0.1$).

Parameter	Ebro	Rhône	Po	Danube
		kt yr^{-2}		
F-TSM	−1.2 $^+$	−10.6	−171	+190
F-NO$_3^-$	−0.23	−0.11	−0.35	−8.31
F-NH$_4^+$	+0.01	−0.06	−0.28 ***	−3.58 **
F-NO$_2^-$	−0.01 *	−0.04 ***	−0.06 ***	−0.18
F-DIN	−0.36 $^+$	−0.35	−0.81	−11.1
F-PO$_4^{3-}$	−0.06 ***	−0.22 ***	−0.06 **	−0.18
F-SiO$_2$	+1.3	+0.04	−1.06	−7.04
F-TOC	−1.36 $^+$	-	-	-
F-TN	−0.22	−0.52	−1.07	−32.5
F-TP	−0.05 **	−0.09	−0.12 $^+$	0.00
F-ON	+0.01	−1.13 *	+0.09	−4.30
F-OP	−0.01	−0.02	−0.04	+0.08

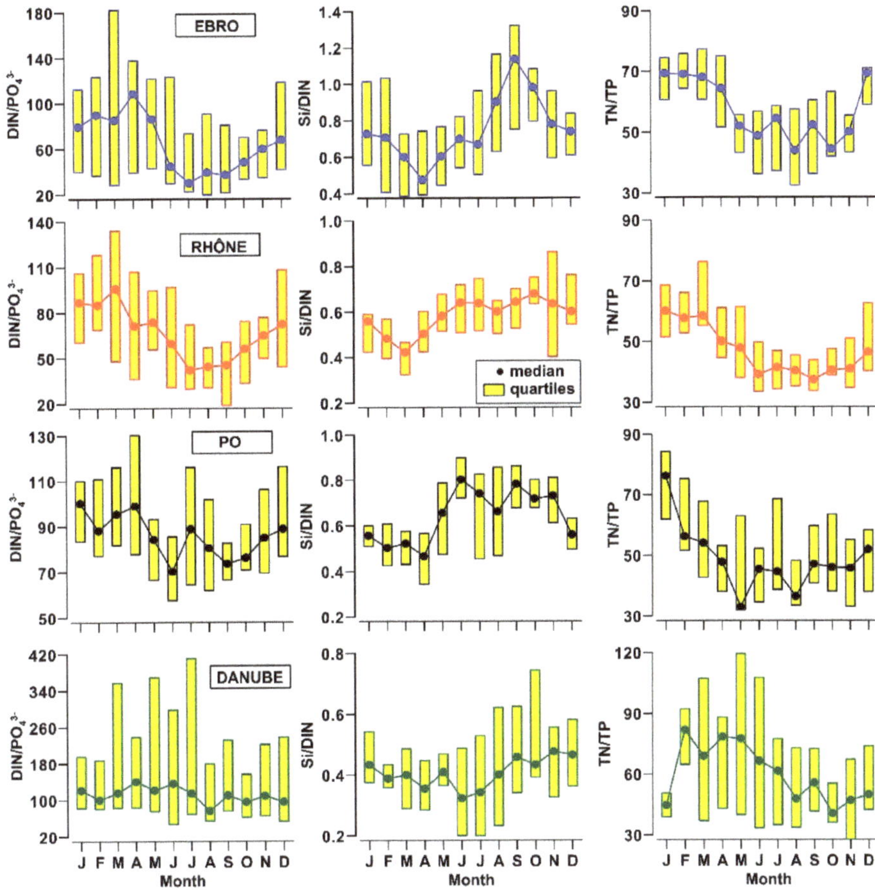

Figure A3. Box Whisker Plot (median and quartiles) of monthly distribution of molar ratios of nutrients in the Ebro, Rhône, Po and Danube waters, in 1990–2012.

References

1. Smith, V.H. Eutrophication of freshwater and coastal marine ecosystems. A global problem. *Environ. Sci. Pollut. Res.* **2003**, *10*, 1–14. [CrossRef]
2. Garnier, J.; Beusen, A.; Thieu, V.; Billen, G.; Bouwman, L. N:P:Si nutrient export ratios and ecological consequences in coastal seas evaluated by the ICEP approach. *Glob. Biogeochem. Cycles* **2010**, *24*, GB0A05. [CrossRef]
3. Howarth, R.; Chan, F.; Conley, D.J.; Garnier, J.; Doney, S.C.; Marino, R.; Billen, G. Coupled biogeochemical cycles: Eutrophication and hypoxia in temperate estuaries and coastal marine ecosystems. *Front. Ecol. Environ.* **2011**, *9*, 18–26. [CrossRef]
4. Ibáñez, C.; Alcaraz, C.; Caiola, N.; Rovira, A.; Trobajo, R.; Alonso, M.; Duran, C.; Jiménez, P.J.; Munné, A.; Prat, N. Regime shift from phytoplankton to macrophyte dominance in a large river: Top-down versus bottom-up effects. *Sci. Total Environ.* **2012**, *416*, 314–322. [CrossRef] [PubMed]
5. Peñuelas, J.; Poulter, B.; Sardans, J.; Ciais, P.; van der Velde, M.; Bopp, L.; Boucher, O.; Godderis, Y.; Hinsinger, P.; Llusia, J.; et al. Human-induced nitrogen–phosphorus imbalances alter natural and managed ecosystems across the globe. *Nat. Commun.* **2013**, *4*, 2934. [CrossRef] [PubMed]

6. Vollenweider, R.A.; Marchetti, R.; Viviani, R. *Marine Coastal Eutrophication. Proceedings of an International Conference, Bologna, Italy, 21–24 March 1990*; Elsevier: Amsterdam, The Netherlands; London, UK; New York, NY, USA; Tokyo, Japan, 1992; p. 1310. ISBN 978-0-444-89990-3.

7. Howarth, R.; Anderson, D.; Cloern, J.; Elfring, C.; Hopkinson, C.; Lapointe, B.; Malone, T.; Marcus, N.; McGlathery, K.; Sharpley, A.; et al. *Nutrient Pollution of Coastal Rivers, Bays, and Seas*; Issues in Ecology, Ecological Society of America: Washington, DC, USA, 2000; Volume 7, p. 17. ISSN 1092-8987.

8. Yunev, O.A.; Carstensen, J.; Moncheva, S.; Khaliulin, A.; Ærtebjerg, G.; Nixon, S. Nutrient and phytoplankton trends on the western Black Sea shelf in response to cultural eutrophication and climate changes. *Estuar. Coast. Shelf Sci.* **2007**, *74*, 63–76. [CrossRef]

9. Rabouille, C.; Conley, D.J.; Dai, M.H.; Cai, W.-J.; Chen, C.T.A.; Lansard, B.; Green, R.; Yin, K.; Harrison, P.J.; Dagg, M.; et al. Comparison of hypoxia among four river-dominated ocean margin: The Changjiang (Yangtze), Mississippi, Pearl, and Rhône rivers. *Cont. Shelf Res.* **2008**, *28*, 1527–1537. [CrossRef]

10. Oguz, T.; Velikova, V. Abrupt transition of the northwestern Black Sea shelf ecosystem from a eutrophic to an alternative pristine state. *Mar. Ecol. Prog. Ser.* **2010**, *405*, 231–242. [CrossRef]

11. Djakovac, T.; Supić, N.; Bernardi-Aubry, F.; Degobbis, D.; Giani, M. Mechanisms of hypoxia frequency changes in the northern Adriatic Sea during the period 1972–2012. *J. Mar. Syst.* **2014**, *141*, 179–189. [CrossRef]

12. Stachowitsch, M. Coastal hypoxia and anoxia: A multi-tiered holistic approach. *Biogeosciences* **2014**, *11*, 2281–2285. [CrossRef]

13. Bianchi, T.S. The role of terrestrially derived organic carbon in the coastal ocean: A changing paradigm and the priming effect. *Proc. Natl. Acad. Sci. USA* **2011**, *108*, 19473–19481. [CrossRef] [PubMed]

14. Asmala, E.; Autio, R.; Kaartokallio, H.; Stedmon, C.A.; Thomas, D.N. Processing of humic-rich riverine dissolved organic matter by estuarine bacteria: Effects of predegradation and inorganic nutrients. *Aquat. Sci.* **2014**, *76*, 451–463. [CrossRef]

15. Durand, N.; Fiandrino, A.; Fraunié, P.; Ouillon, S.; Forget, P.; Naudin, J.J. Suspended matter dispersion in the Ebro ROFI: An integrated approach. *Cont. Shelf Res.* **2002**, *22*, 267–284. [CrossRef]

16. Berlinsky, N.; Bogatova, Y.; Garkavaya, G. Estuary of the Danube. In *Handbook of Environmental Chemistry*; Water Pollution, 5 (PART H); Wangersky, P.J., Ed.; Springer: Berlin/Heidelberg, Germany, 2006; Volume 5, pp. 233–264. ISBN 978-3-540-00270-3.

17. Syvitski, J.P.M.; Kettner, A.J.; Overeem, I.; Hutton, E.W.H.; Hannon, M.T.; Brakenridge, G.R.; Day, J.; Vörösmarty, C.; Saito, Y.; Giosan, L.; et al. Sinking deltas due to human activities. *Nat. Geosci.* **2009**, *2*, 681–686. [CrossRef]

18. Markaki, Z.; Loÿe-Pilot, M.D.; Violaki, K.; Benyahya, L.; Mihalopoulos, N. Variability of atmospheric deposition of dissolved nitrogen and phosphorus in the Mediterranean and possible link to the anomalous seawater N/P ratio. *Mar. Chem.* **2010**, *120*, 187–194. [CrossRef]

19. The MerMex Group. Marine ecosystems' responses to climatic and anthropogenic forcings in the Mediterranean. *Prog. Oceanogr.* **2011**, *91*, 97–166. [CrossRef]

20. Giani, M.; Djakovac, T.; Degobbis, D.; Cozzi, S.; Solidoro, C.; Fonda Umani, S. Recent changes in the marine ecosystems of the northern Adriatic Sea. *Estuar. Coast. Shelf Sci.* **2012**, *115*, 1–13. [CrossRef]

21. Ludwig, W.; Dumont, E.; Meybeck, M.; Heussner, S. River discharges of water and nutrients to the Mediterranean and Black Sea: Major drivers for ecosystem changes during past and future decades? *Prog. Oceanogr.* **2009**, *80*, 199–217. [CrossRef]

22. Huertas, I.E.; Ríos, A.F.; García-Lafuente, J.; Navarro, G.; Makaoui, A.; Sánchez-Román, A.; Rodriguez-Galvez, S.; Orbi, A.; Ruíz, J.; Pérez, F.F. Atlantic forcing of the Mediterranean oligotrophy. *Glob. Biogeochem. Cycles* **2012**, *26*, GB2022. [CrossRef]

23. Kempe, S.; Pettine, M.; Cauwet, G. Biogeochemistry of European Rivers. In *Biogeochemistry of Major World Rivers*; SCOPE Report 42; Degens, E.T., Kempe, S., Richey, J.E., Eds.; ICSU, UNEP, John Wiley & Sons Ltd.: Chichester, UK, 1991; pp. 169–211. ISSN 1052-7613.

24. Grizzetti, B.; Bouraoui, F.; Aloe, A. Changes of nitrogen and phosphorus loads to European seas. *Glob. Chang. Biol.* **2012**, *18*, 769–782. [CrossRef]

25. Romero, E.; Garnier, J.; Lassaletta, L.; Billen, G.; Le Gendre, R.; Riou, P.; Cugier, P. Large-scale patterns of river inputs in southwestern Europe: Seasonal and interannual variations and potential eutrophication effects at the coastal zone. *Biogeochemistry* **2013**, *113*, 481–505. [CrossRef]

26. DANUBS Final Report. *Nutrient Management in the Danube Basin and Its Impact on the Black Sea.* EVK1-CT-2000-00051. 2005, p. 78. Available online: http://danubs.tuwien.ac.at (accessed on 12 October 2017).

27. Ibáñez, C.; Prat, N.; Duran, C.; Pardos, M.; Munné, A.; Andreu, R.; Caiola, N.; Cid, N.; Hampel, H.; Sánchez, R.; Trobajo, R. Changes in dissolved nutrients in the lower Ebro river: Causes and consequences. *Limnetica* **2008**, *27*, 131–142.

28. Ollivier, P.; Radakovitch, O.; Hamelin, B. Major and trace element partition and fluxes in the Rhône River. *Chem. Geol.* **2011**, *285*, 15–31. [CrossRef]

29. Viaroli, P.; Soana, E.; Pecora, S.; Laini, A.; Naldi, M.; Fano, E.A.; Nizzoli, D. Space and time variations of watershed N and P budgets and their relationships with reactive N and P loadings in a heavily impacted river basin (Po river, Northern Italy). *Sci. Total Environ.* **2018**, *639*, 1574–1587. [CrossRef] [PubMed]

30. Cruzado, A.; Velásquez, Z.; Pérez, M.D.C.; Bahamón, N.; Grimaldo, N.S.; Ridolfi, F. Nutrient fluxes from the Ebro River and subsequent across-shelf dispersion. *Cont. Shelf Res.* **2002**, *22*, 349–360. [CrossRef]

31. Alvisi, F.; Cozzi, S. Seasonal dynamics and long-term trends of hypoxia in the coastal zone of Emilia Romagna (NW Adriatic Sea, Italy). *Sci. Total Environ.* **2016**, *541*, 1448–1462. [CrossRef] [PubMed]

32. Kovats, R.S.; Valentini, R.; Bouwer, L.M.; Georgopoulou, E.; Jacob, D.; Martin, E.; Rounsevell, M.; Soussana, J.-F. Europe. In *Climate Change 2014: Impacts, Adaptation, and Vulnerability. Part B: Regional Aspects. Contribution of Working Group II to the Fifth Assessment Report of IPCC*; Cambridge University Press: Cambridge, UK; New York, NY, USA, 2014; pp. 1267–1326. ISBN 978-1-107-05816-3.

33. Díaz, F.; Naudin, J.J.; Courties, C.; Rimmelin, P.; Oriol, L. Biogeochemical and ecological functioning of the low-salinity water lenses in the region of the Rhone River freshwater influence, NW Mediterranean Sea. *Cont. Shelf Res.* **2008**, *28*, 1511–1526. [CrossRef]

34. Cozzi, S.; Giani, M. River water and nutrient discharges in the Northern Adriatic Sea: Current importance and long-term changes. *Cont. Shelf Res.* **2011**, *31*, 1881–1893. [CrossRef]

35. Tuğrul, S.; Murray, J.W.; Friederich, G.E.; Salihoğlu, T. Spatial and temporal variability in the chemical properties of the oxic and suboxic layers of the Black Sea. *J. Mar. Syst.* **2014**, *135*, 29–43. [CrossRef]

36. Lutz, S.R.; Mallucci, S.; Diamantini, E.; Majone, B.; Bellin, A.; Merz, R. Hydroclimatic and water quality trends across three Mediterranean river basins. *Sci. Total Environ.* **2016**, *571*, 1392–1406. [CrossRef]

37. Zampieri, M.; Scoccimarro, E.; Gualdi, S.; Navarra, A. Observed shift towards earlier spring discharge in the main Alpine rivers. *Sci. Total Environ.* **2015**, *503–504*, 222–232. [CrossRef] [PubMed]

38. Ibáñez, C.; Prat, N. The environmental impact of the Spanish National Hydrological Plan on the lower Ebro River and Delta. *Int. J. Water Resour. Dev.* **2003**, *19*, 485–500. [CrossRef]

39. Falco, S.; Niencheski, L.F.; Rodilla, M.; Romero, I.; González del Río, J.; Sierra, J.P.; Mösso, C. Nutrient flux and budget in the Ebro Estuary. *Estuar. Coast. Shelf Sci.* **2010**, *87*, 92–102. [CrossRef]

40. Torrecilla, N.J.; Galve, J.P.; Zaera, L.G.; Retamar, J.F.; Álvarez, A.N.A. Nutrient sources and dynamics in a Mediterranean fluvial regime (Ebro River, NE Spain) and their implications for water management. *J. Hydrol.* **2005**, *304*, 166–182. [CrossRef]

41. Tornés, E.; Pérez, M.C.; Durán, C.; Sabater, S. Reservoirs override seasonal variability of phytoplankton communities in a regulated Mediterranean river. *Sci. Total Environ.* **2014**, *475*, 225–233. [CrossRef] [PubMed]

42. Higueras, M.; Kerhervé, P.; Sanchez-Vidal, A.; Calafat, A.; Ludwig, W.; Verdoit-Jarraya, M.; Heussner, S.; Canals, M. Biogeochemical characterization of the riverine particulate organic matter transferred to the NW Mediterranean Sea. *Biogeosciences* **2014**, *11*, 157–172. [CrossRef]

43. Denant, V.; Saliot, A. Seasonal variations of nutrients (NO_3^-, NO_2^-, NH_4^+, PO_4^{3-} and $Si(OH)_4$) and suspended matter in the Rhône delta, France. *Oceanol. Acta* **1990**, *13*, 47–52.

44. Moutin, T.; Raimbault, P.; Golterman, H.L.; Coste, B. The input of nutrients by the Rhône River into the Mediterranean Sea: Recent observations and comparison with earlier data. *Hydrobiologia* **1998**, *373/374*, 237–246. [CrossRef]

45. Fruget, J.-F.; Centofanti, M.; Dessaix, J.; Olivier, J.-M.; Druart, J.-C.; Martinez, P.-J. Temporal and spatial dynamics in large rivers: Example of a long-term monitoring of the middle Rhône River. *Ann. Limnol.* **2001**, *37*, 237–251. [CrossRef]

46. Basin Authority of Po River. Characteristics of Po River Basin and Preliminary Assessment of the Environmental Impact of Human Activities on Hydrological Resources. Parma, Italy, 2006; p. 643, [Italian]. Available online: http://www.adbpo.gov.it/it/distretto-del-po/alcuni-dati (accessed on 29 January 2017).

47. Copetti, D.; Carniato, L.; Crise, A.; Guyennon, N.; Palmeri, L.; Pisacane, G.; Struglia, M.V.; Tartari, G. Impacts of Climate Change on Water Quality. In *Regional Assessment of Climate Change in the Mediterranean*; Advances in Global Change Research; Navarra, A., Tubiana, L., Eds.; Springer: Dordrecht, The Netherlands, 2013; Volume 50, pp. 307–332. ISBN 978-94-007-5780-6.

48. Tesi, T.; Miserocchi, S.; Acri, F.; Langone, L.; Boldrin, A.; Hatten, J.A.; Albertazzi, S. Flood-driven transport of sediment, particulate organic matter, and nutrients from the Po River watershed to the Mediterranean Sea. *J. Hydrol.* **2013**, *498*, 144–152. [CrossRef]

49. Humborg, C.; Ittekkot, V.; Cociasu, A.; Bodungen, B.V. Effect of Danube River dam on Black Sea biogeochemistry and ecosystem structure. *Nature* **1997**, *386*, 385–388. [CrossRef]

50. Rice, E.W.; Baird, R.B.; Eaton, A.D.; Clesceri, L.S. *Standard Methods for the Examination of Water and Wastewater*, 22nd ed.; American Public Health Association: Washington, DC, USA, 2012; p. 1360. ISBN 978-087553-013-0.

51. Grasshoff, K.; Kremling, K.; Ehrhardt, M. *Methods of Seawater Analysis. Third Completely Revised and Extended Edition*; Wiley-VCH Verlag GmbH: Weinheim, Germany, 1999; p. 632. ISBN 978-3-527-61399-1.

52. Ott, W.R. *Environmental Statistics and Data Analysis*; Lewis Publishers: Boca Raton, FL, USA; London, UK; New York, NY, USA; Washington, DC, USA, 1994; p. 336. ISBN 9780873718486.

53. Chen, Z.; Grasby, S.E. Impact of decadal and century-scale oscillations on hydroclimate trend analyses. *J. Hydrol.* **2009**, *365*, 122–133. [CrossRef]

54. Rodionov, S.N. A sequential method of detecting abrupt changes in the correlation coefficient and its application to Bering Sea Climate. *Climate* **2015**, *3*, 474–491. [CrossRef]

55. Pujo-Pay, M.; Conan, P.; Oriol, L.; Cornet-Barthaux, V.; Falco, C.; Ghiglione, J.-F.; Goyet, C.; Moutin, T.; Prieur, L. Integrated survey of elemental stoichiometry (C, N, P) from the western to eastern Mediterranean Sea. *Biogeosciences* **2011**, *8*, 883–899. [CrossRef]

56. Dafner, E.V.; Boscolo, R.; Bryden, H.L. The N:Si:P molar ratio in the Strait of Gibraltar. *Geophys. Res. Lett.* **2003**, *30*, 1506. [CrossRef]

57. Ren, L.; Wang, P.; Wang, C.; Chen, J.; Hou, J.; Qian, J. Algal growth and utilization of phosphorus studied by combined mono-culture and co-culture experiments. *Environ. Pollut.* **2017**, *220*, 274–285. [CrossRef]

58. Ivančić, I.; Godrijan, J.; Pfannkuchen, M.; Marić, D.; Gašparović, B.; Djakovac, T.; Najdeka, M. Survival mechanisms of phytoplankton in conditions of stratification-induced deprivation of orthophosphate: Northern Adriatic case study. *Limnol. Oceanogr.* **2012**, *57*, 1721–1731. [CrossRef]

59. Ursella, L.; Poulain, P.M.; Signell, R.P. Surface drifter derived circulation in the northern and middle Adriatic Sea: Response to wind regime and season. *J. Geophys. Res. Oceans* **2007**, *112*, C03S04. [CrossRef]

60. Naudin, J.J.; Cauwet, G.; Fajon, C.; Oriol, L.; Terzic, S.; Devenon, J.-L.; Broche, P. Effect of mixing on microbial communities in the Rhone River plume. *J. Mar. Syst.* **2001**, *28*, 203–227. [CrossRef]

61. Turco, M.; Quintana-Seguí, P.; Llasat, M.C.; Herrera, S.; Gutiérrez, J.M. Testing MOS precipitation downscaling for ENSEMBLES regional climate models over Spain. *J. Geophys. Res.* **2011**, *116*, D18109. [CrossRef]

62. Stagl, J.C.; Hattermann, F.F. Impacts of Climate Change on the Hydrological Regime of the Danube River and Its Tributaries Using an Ensemble of Climate Scenarios. *Water* **2015**, *7*, 6139–6172. [CrossRef]

63. Lionello, P.; Malanotte-Rizzoli, P.; Boscolo, R.; Alpert, P.; Artale, V.; Li, L.; Luterbacher, J.; May, W.; Trigo, R.; Tsimplis, M.; et al. The Mediterranean climate: An overview of the main characteristics and issues. *Dev. Earth Environ. Sci.* **2006**, *4*, 1–26. [CrossRef]

64. Schubert, S.D.; Stewart, R.E.; Wang, H.; Barlow, M.; Berbery, E.H.; Cai, W.; Hoerling, M.P.; Kanikicharla, K.K.; Koster, R.D.; Lyon, B.; et al. Global meteorological drought: A synthesis of current understanding with a focus on SST drivers of precipitation deficits. *J. Clim.* **2016**, *29*, 3989–4019. [CrossRef]

65. Forget, P.; André, G. Can Satellite-derived Chlorophyll Imagery Be Used to Trace Surface Dynamics in Coastal Zone? A Case Study in the Northwestern Mediterranean Sea. *Sensors* **2007**, *7*, 884–904. [CrossRef]

66. D'Ortenzio, F.; Ribera d'Alcalà, M. On the trophic regimes of the Mediterranean Sea. *Biogeosciences* **2009**, *6*, 139–148. [CrossRef]

67. Alekseenko, E.; Raybaud, V.; Espinasse, B.; Carlotti, F.; Queguiner, B.; Thouvenin, B.; Garreau, P.; Baklouti, M. Seasonal dynamics and stoichiometry of the planktonic community in the NW Mediterranean Sea: A 3D modeling approach. *Ocean Dyn.* **2014**. [CrossRef]

68. Harding, L.W.; Degobbis, D.; Precali, R. Production and fate of phytoplankton: Annual cycles and interannual variability. In *Ecosystems at the Land-Sea Margin: Drainage Basin to Coastal Sea*; Coastal and Estuarine Studies; Malone, T.C., Malej, A., Harding, L.W., Smodlaka, N., Turner, R.E., Eds.; American Geophysical Union: Washington, DC, USA, 1999; Volume 55, pp. 131–172. ISBN 0-87590-269-3.

69. Melin, F.; Vantrepotte, V.; Clerici, M.; D'Alimonte, D.; Zibordi, G.; Berthon, J.-F.; Canuti, E. Multi-sensor satellite time series of optical properties and chlorophyll-a concentration in the Adriatic Sea. *Prog. Oceanogr.* **2011**, *91*, 229–244. [CrossRef]

70. Díaz, F.; Raimbault, P.; Boudjellal, B.; Garcia, N.; Moutin, T. Early phosphorus limitation during spring in the Gulf of Lions. *Mar. Ecol. Prog. Ser.* **2001**, *211*, 51–62. [CrossRef]

71. Cristofor, S.; Vadineanu, A.; Ignat, G. Importance of flood zones for nitrogen and phosphorus dynamics in the Danube Delta. *Hydrobiologia* **1993**, *251*, 143–148. [CrossRef]

72. Raimbault, P.; Durrieu de Madron, X. Research activities in the Gulf of Lion (NW Mediterranean) within the 1997–2001 PNEC project. *Oceanol. Acta* **2003**, *26*, 291–298. [CrossRef]

73. Nebra, A.; Alcaraz, C.; Caiola, N.; Muñoz-Camarillo, G.; Ibáñez, C. Benthic macrofaunal dynamics and environmental stress across a salt wedge Mediterranean estuary. *Mar. Environ. Res.* **2016**, *117*, 21–31. [CrossRef] [PubMed]

74. Fonda Umani, S. Pelagic production and biomass in the Adriatic Sea. *Sci. Mar.* **1996**, *60*, 65–77.

75. Giani, M.; Rinaldi, A.; Degobbis, D. Mucilages in the Adriatic and Tyrrhenian Sea: An introduction. *Sci. Total Environ.* **2005**, *353*, 3–9. [CrossRef] [PubMed]

76. Aubrey, D.; Moncheva, S.; Demirov, E.; Diaconu, V.; Dimitrov, A. Environmental changes in the western Black Sea related to anthropogenic and natural conditions. *J. Mar. Syst.* **1996**, *7*, 411–425. [CrossRef]

77. Daskalov, G.M.; Boicenco, L.; Grishin, A.N.; Lazar, L.; Mihneva, V.; Shlyakhov, V.A.; Zengin, M. Architecture of collapse: Regime shift and recovery in an hierarchically structured marine ecosystem. *Glob. Chang. Biol.* **2017**, *23*, 1486–1498. [CrossRef] [PubMed]

78. Bostan, V.; Dominik, J.; Bostina, M.; Pardos, M. Forms of particulate phosphorus in suspension and in bottom sediment in the Danube Delta. *Lakes Reserv. Res. Manag.* **2000**, *5*, 105–110. [CrossRef]

79. Ptacnik, R.; Andersen, T.; Tamminen, T. Performance of the Redfield Ratio and a family of nutrient limitation indicators as thresholds for phytoplankton N vs. P limitation. *Ecosystems* **2010**, *13*, 1201–1214. [CrossRef]

80. Lazăr, L.; Gomoiu, M.-T.; Boicenco, L.; Vasiliu, D. Total Organic Carbon (TOC) of the surface layer sediments covering the seafloor of the Romanian Black Sea coast. *Geoecomarina* **2012**, *18*, 121–132.

Article

The Influence of Flow Regime on Ecological Quality, Bird Diversity, and Shellfish Fisheries in a Lowland Mediterranean River and Its Coastal Area

Oscar Belmar *, Carles Ibáñez, Ana Forner and Nuno Caiola

Marine and Continental Waters Program, IRTA, Carretera Poblenou, km 5.5, Sant Carles de la Ràpita, 43540 Catalonia, Spain; carles.ibanez@irta.cat (C.I.); anaforner13@gmail.com (A.F.); nuno.caiola@irta.cat (N.C.)
* Correspondence: oscar.belmar@irta.cat or oscarbd@um.es; Tel.: +34-977-74-54-27 (ext. 1859)

Received: 12 March 2019; Accepted: 28 April 2019; Published: 1 May 2019

Abstract: Designing environmental flows in lowland river sections and estuaries is a challenge for researchers and managers, given their complexity and their importance, both for nature conservation and economy. The Ebro River and its delta belong to a Mediterranean area with marked anthropogenic pressures. This study presents an assessment of the relationships between mean flows (discharges) computed at different time scales and (i) ecological quality based on fish populations in the lower Ebro, (ii) bird populations, and (iii) two shellfish fishery species of socioeconomic importance (prawn, or *Penaeus kerathurus*, and mantis shrimp, or *Squilla mantis*). Daily discharge data from 2000 to 2015 were used for analyses. Mean annual discharge was able to explain the variation in fish-based ecological quality, and model performance increased when aquatic vegetation was incorporated. Our results indicate that a good ecological status cannot be reached only through changes on discharge, and that habitat characteristics, such as the coverage of macrophytes, must be taken into account. In addition, among the different bird groups identified in our study area, predators were related to river discharge. This was likely due to its influence on available resources. Finally, prawn and mantis shrimp productivity were influenced up to a certain degree by discharge and physicochemical variables, as inputs from rivers constitute major sources of nutrients in oligotrophic environments such as the Mediterranean Sea. Such outcomes allowed revisiting the environmental flow regimes designed for the study area, which provides information for water management in this or in other similar Mediterranean zones.

Keywords: Ebro River; fish community; ecological quality; birds; fisheries; productivity; deltas

---

1. Introduction

Deltas and estuaries are complex ecosystems largely recognized for their productivity and importance, both for the economy and the conservation of nature (e.g., [1–3]). However, the rapid growth of human population has put these areas under increasing pressure that threatens their ecological integrity and economic value [4]. In Mediterranean aquatic ecosystems, the impacts produced by anthropogenic pressures are magnified by their increased, and often extreme, natural hydrological variability [5]. Human response to such hydrological fluctuations includes flow regulation and water extraction that frequently disrupt aquatic ecosystems and produce accentuated environmental stress [6–8].

There is a need to harmonize nature conservation with socioeconomic activities. The provision of environmental flow regimes (hereafter, e-flows) in the context of the Brisbane Declaration [9,10] "to sustain aquatic ecosystems which, in turn, support human cultures, economies, sustainable livelihoods, and well-being" constitutes an essential tool towards this direction. The 2012 European Commission's "Blueprint to Safeguard Europe's Water Resources" proposed the development of a guidance document [11] in the framework of a Common Implementation Strategy (CIS) that would provide a European definition of e-flows and a common understanding of how they should be calculated in general terms. Such a document constituted a complement for the EU Water Framework Directive (WFD) [12], which established the objective of achieving a good ecological status (which categories may be high, good, moderate, poor, or bad) in European water courses.

The ecological status of rivers in the European Union must be assessed through ecological indices (Water Framework Directive; [12]). Those based on fish offer the best sensitiveness to hydromorphological pressures [13,14], such as hydrological variations. The official index in Spain is the Mediterranean Index of Biotic Integrity (IBIMED), based on the Index of Biotic Integrity for Catalan rivers (IBICAT2010) [15] and intercalibrated in Europe [16].

Studies developed in the lower Ebro have linked fish-based ecological quality or alien fish species with flow regimes. Using data from summer 2016, Caiola et al. [17] found that the success (establishment and dispersal) of alien fishes is enhanced by flow reduction, which resulted in decreased flow velocity in the littoral zone. They determined that water velocities lower than 0.4 m/s are associated with an impacted community dominated by alien species (i.e., alien-to-native species ratio greater than 0.5). Belmar et al. [18] used interannual field data sampled between 2006 and 2015 in the lower Ebro and found relationships between hydrological indices associated with the magnitude and variability of flow regimes (using daily and hourly flow records) and ecological quality, assessed using different fish-based indices. They concluded that the index IBICAT2010 [15] is more suitable than IBICAT2b (its variant) [15] and the Improvement and Spatial extension of the European Fish Index (EFI+) [19,20] to detect ecohydrological relationships in the lower Ebro River, depending on the spatial and temporal scales considered. On one side, this further suggested that the time period considered to characterize hydrologic regimes determined the ability to observe relationships between flow indices and ecological quality. On the other side, the ability to detect such ecohydrological relationships depended on the location of the transect, even with those located within the same water unit ("masa de agua"; subdivision of surface waters to implement the WFD in Spain).

The e-flows proposed for the lower Ebro (Table 1) have little in common with the requirements of natural ecosystems or natural flow regimes, as they are not an exception to the common tendency of water management to set minimum flows that are constant for long periods [21]. The Hydrological Plan published in 1996 established such e-flows as the 10% of the natural mean interannual runoff (5% when the mean flow was greater than 80 m^3/s) in the river, maintaining a provision of 100 m^3/s for the mouth. In 2001, the National Hydrological Plan (PHN, in Spanish) developed the Integral Protection Plan of the Ebro Delta (PIPDE, in Spanish) to maintain its "special ecological conditions" but e-flows were defined using the same criteria as in 1996. This simplistic approach was criticized arguing that the classical methods of determining environmental flows in rivers are neither designed nor adequate for the objective of maintaining the deltas and estuaries in a good ecological status [22]. In 2007, the Commission for the Sustainability of the Ebro Land (CSTE, in Catalan), including representatives from the Catalan and Spanish governments, issued a report proposal of monthly e-flows for the river. In addition, the Royal Decree of Hydrological Planning [23] and the Technical Instruction of Hydrological Planning [24] developed normative contents regarding e-flow assessments. The Ebro's Hydrological Plan for 2010–2015, approved in 2014, defined e-flows for the lower part of the river, much lower than those proposed by CSTE in 2007 and 2015 [25]. Finally, the Hydrological Plan for the period 2016–2021 used the same environmental regime (Table 1). Therefore, the e-flows for the lower Ebro are based solely on defining flow magnitudes. Although such magnitudes may change among months and types of year, there are not explicit rules to adjust flow variability (e.g., coefficient of variation).

Table 1. Environmental flow regimes (in m^3/s, except when flows are expressed as percentage of the natural mean interannual runoff) proposed and implemented for the lower Ebro River (and its delta) in the last 20 years (CHE (in Spanish): Ebro's Hydrographic Confederation; PIDPE (in Spanish): Integral Protection Plan of the Ebro Delta; CSTE (in Catalan): Commission for the Sustainability of the Ebro's Land).

Source	Zone	Month											
		October	November	December	January	February	March	April	May	June	July	August	September
CHE, 1996	River	5%–10%	5%–10%	5%–10%	5%–10%	5%–10%	5%–10%	5%–10%	5%–10%	5%–10%	5%–10%	5%–10%	5%–10%
	Mouth	100	100	100	100	100	100	100	100	100	100	100	100
PIDPE, 2001	River	5%–10%	5%–10%	5%–10%	5%–10%	5%–10%	5%–10%	5%–10%	5%–10%	5%–10%	5%–10%	5%–10%	5%–10%
	Mouth	100	100	100	100	100	100	100	100	100	100	100	100
CSTE *, 2007	River	133	214	354	382	428	381	440	505	347	204	151	146
	Mouth	-	-	-	-	-	-	-	-	-	-	-	-
CHE, 2014	River	80	80	91	95	150	150	91	91	81	80	80	80
	Mouth	80	100	100	120	150	155	100	100	100	100	100	80
CSTE *, 2015	River	131	229	275	226	251	297	406	384	314	180	134	150
	Mouth	-	-	-	-	-	-	-	-	-	-	-	-
CHE, 2016	River	80	80	91	95	150	150	91	91	81	80	80	80
	Mouth	80	100	100	120	150	155	100	100	100	100	100	80

* Flows calculated from a weighted mean using the regimes for wet, average, and dry years (these environmental flows have been proposed but not implemented).

Advances in measuring flow-ecology relationships allow inferring conclusions on the ecological impacts of specific flow regimes on communities, including e-flows, but further research is still necessary in order to determine the patterns of habitat complexity that may explain differences in ecohydrological relationships (*sensu* [26]) among river transects (and water units). Exceptional rises in magnitude, together with a general reduction in flow magnitude and variability, are common in Mediterranean main stems (such as the lower Ebro) because of dam management (with consequences on river habitats) [27]. Given the relevance of flow magnitude in e-flow regimes (Table 1), basing such flow-ecology relationships on mean discharges results relevant from a water management perspective (see examples in [28,29]). In addition, the sustainability of deltas cannot be guaranteed only with the allocation of e-flows for the fish inhabiting the low section of the river, which is the current practice in Spain and most countries. E-flows must be determined not only for the river ecosystem but also for the associated coastal and marine systems, which represents a challenge for scientists and water managers [22]. The possible effects of water quality (e.g., nutrient content) have also to be considered. In this context, a relationship between freshwater inputs from the Ebro River and coastal fishery species such as anchovy (*Engraulis encransicolus*) has been highlighted [30]. Major river outflows are one of the nutrient enrichment processes that maintain fishery production worldwide [31], and similar results have been obtained in other areas in Mexico [32] and Australia [33]. In this context, the littoral of the Ebro delta is suitable to study the relationship between river outflows and two shellfish fishery species: prawn (*Penaeus kerathurus*) and mantis shrimp (*Squilla mantis*). Due to their commercial value and their dependence on deltaic habitats, the delta is the only fishing ground of these species exploited by the corresponding fishermen's society (the most important in Catalonia in terms of income). Therefore, significant relationships between river outflows and these species would have socioeconomic implications relevant for water management at catchment scale (particularly, for e-flow assessment). In addition, the delta is characterized as one of the main feeding and breeding areas of several endangered bird species, amongst which Audouin's gull (*Larus audouinii*) stands out. Determining relationships between flow regimes and bird communities in the delta and in similar Mediterranean areas, as it has been done in (scarce) studies developed in other areas [34], may be essential for nature conservation.

The present study, developed in the lower Ebro and its delta, aimed to (i) deepen into the ecohydrological relationships found in Belmar et al. [18] in order to obtain additional conclusions for water managers; and (ii) complete such conclusions using other groups of organisms relevant for ecological conservation and socioeconomic activities at the delta and its littoral zone [22]. This will also provide conclusions useful for water management. Specific tasks were planned to:

i. Determine the relationship between mean discharge (instead of the original set of hydrological indices in Belmar et al. [18]) averaged along the same time periods and ecological status in order to use the obtained models to calculate e-flows to preserve the good ecological status in the lower Ebro (assessing the suitability of the proposals presented up-to-date). Fish communities were characterized using not only ecological quality indices (IBICAT2010, IBICAT2b, and EFI+) but also the ratio of alien species (both in terms of richness and abundance). This allowed determination of the relationship between ecological quality and alien species, in order to validate the 0.5 threshold used by Caiola et al. [17] to consider a community as impacted.

ii. Introduce habitat and riparian characteristics in the models to determine the influence of habitat complexity.

iii. Identify potential relationships among discharge, water quality, and two (shellfish) fishery species with socioeconomic relevance (prawn, or *Penaeus kerathurus*, and mantis shrimp, or *Squilla mantis*), as well as between flows and bird populations (ecological relevance) at the delta.

2. Materials and Methods

2.1. Study Area

The Ebro Delta is one of the largest wetlands in the western Mediterranean and one of the most important estuarine zones in Europe [35,36]. Declared a Natural Park in 1983, a Special Protection Area (SPA) under the Birds Directive [37] in 2006 and a World Biosphere Reserve in 2013, more than 8000 ha are protected by the Spanish legislation. The delta has a great diversity of habitats, with endemic faunal (ornithological and ichthyologic) and halophilic floral composition [38], together with important human activities such as rice agriculture, fisheries, aquaculture, and tourism [22].

The lower Ebro River is located upstream of the delta, between the Flix Reservoir and the Tortosa Estuary (Figure 1). The lower Ebro's hydrology, geomorphology, and ecology are strongly impacted by the operation of three dams and two weirs. This river section is divided in four water units (Figure 1) according to the current Hydrological Plan: ES091463 (from the Xerta weir to Tortosa), ES091461 (from Ascó Weir to Xerta Weir), ES091460 (from Flix dam to Ascó Weir), and ES091459 (Flix Meander). The water unit ES091461 is by far longer than the others (Figure 1). The rest of the basin is strongly regulated by nearly 200 dams, most of them built from 1940 to 1970 [39,40].

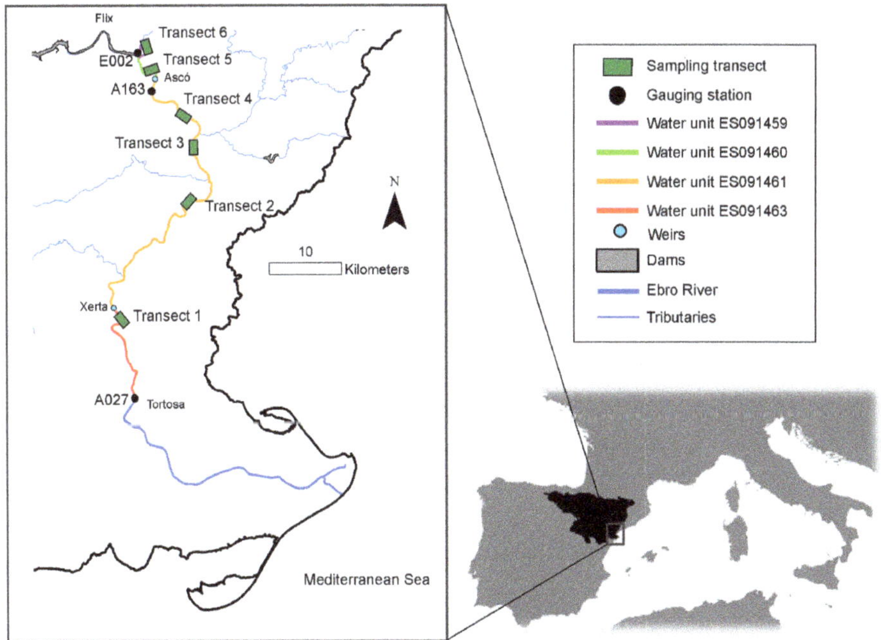

Figure 1. Study area showing the sampling transects and gauging stations located on the water units (masses) of the low Ebro River (dams and weirs are also showed).

2.2. Data

2.2.1. Fish Data at the Lower Ebro River

Fishes were sampled from six sampling transects in the lower Ebro River (Figure 1) from 2006 to 2015, during summer or early autumn, using electrofishing. The dataset belongs to a long-term sampling developed with generalists' objectives, where transects were selected to represent different hydromorphological typologies in the river section (for more details, see [17]) and to capture as much variability as possible. They also provided a weighed representation of the water units in the study area (as the greatest number of transects was located in the longest water unit): ES091463 (transect 1; from Xerta weir to Tortosa), ES091461 (transects 2, 3, and 4; from Ascó to Xerta), ES091460 (transect 5; from Flix to Ascó), and ES091459 (transect 6; Flix meander), respectively. One survey per year was carried out between summer and early autumn. At each transect, fish were caught with boat-based electrofishing gear that generated up to 400 V and 10 A pulsed DC, working from the downstream to the upstream direction. Following the CEN's ("Comité Européen de Normalisation") standards for fishing with electricity [41], each transect was 2 km long and was sampled in ten equidistant points in the littoral zone, either on the left or the right bank. At each sampling point, habitat variables (presence of aquatic vegetation and riparian land cover types; Table 2) were additionally recorded. All fish catches from the sampling points of each transect were aggregated to ensure an adequate sampling effort [17]. The habitat descriptors were expressed at the transect level calculating the proportion of the corresponding categories (aquatic vegetation and riparian land cover types) using the ten points included in each transect.

Table 2. Habitat variables recorded at each sampling point. All point measurements were converted to percentages to describe their relative distribution at each transect.

Factor	Variable
Aquatic vegetation	Coverage
	Potamogeton pectinatus
	Potamogeton natans
	Potamogeton crispus
	Ceratophyllum demersum
	Myriophyllum spicatum
	Cladophora spp.
	Lemna minor
	Others
Riparian land cover types	Riparian forest
	Reed
	Riprap
	Rock
	Grass
	Wall
	Without vegetation

All diadromous species were removed, as the Xerta weir prevents their movement upstream and these species can only be found in the lowermost sampling transect (Figure 1). By doing so, we ensured that this transect was comparable with the rest. Then, fish-based ecological quality was characterized through the indices IBICAT2010, IBICAT2b, and EFI+ in each sampling transect. Captures in transects 2, 3, and 4 were combined to account for the increased length of water unit ES091461, thus obtaining results comparable to those in Belmar et al. [18]. Finally, the ratio between alien and native species was computed, using richness and abundance (CPUEs; Catches Per Unit Effort).

2.2.2. Bird and Fishery Data at the Delta

The census of birds and the records of fish landings were obtained from the Ebro's Delta Natural Park and the Verge del Carme fishermen's society (Tarragona, Spain), respectively. The bird census consisted of species abundances in winter each year in the delta (including nesting species), which were grouped by families. Fisheries data consisted of the number of boats and the biomass fished, by means of trawling and artisanal techniques. Prawn (*Penaeus kerathurus*) and mantis shrimp (*Squilla mantis*) catches were expressed as kilograms per number of boats. They were complemented with physicochemical data (Table 3) obtained from different regional and national organizations: Ebro's Hydrographic Confederation (CHE, in Spanish), Tarragona's Water Consortium (CAT, in Catalan), and Catalan Water Agency (ACA, in Catalan).

The abundances of bird populations (around 140 species) showed a growing tendency since the 70's that did not cease until the 00's, which was probably influenced by protection measures after the creation of the Natural Park. We selected, for analyses, a posterior time series (2005–2015) to ensure that the variability in bird populations was mainly caused by environmental change. Such period also allowed using coetaneous data of fish (river) and birds. Fishery data (delta) covered the same period (2005–2015).

Table 3. Physicochemical variables obtained from Ebro's Hydrographic Confederation (CHE, in Spanish), Tarragona's Water Consortium (CAT, in Catalan), and Catalan Water Agency (ACA, in Catalan).

Variable	Units
Chlorophyll	µg/L
Phytoplankton	Relative frequency
Zooplankton	Relative frequency
Dissolved Inorganic Nitrogen	mg/L
Dissolved Inorganic Phosphorous	mg/L
Total Organic Carbon	mg/L
Turbidity	Nephelometric units
Suspended Matter	mg/L

2.2.3. Hydrological Records

Daily discharge data for the Tortosa (A027), Ascó (A163), and Flix (E002) stations (Figure 1) were obtained from the automatic network of gauging stations (SAIH, Automatic System of Hydrologic Information, in Spanish) in the Ebro Basin. Tortosa (A027) was assigned to the water unit ES091463 (transect 1), Ascó (A163) was assigned to the water units ES091461 (transects 2, 3, and 4), and ES091460 (transect 5), and Flix (E002) was assigned to the water unit ES091459 (transect 6), considering the location of the weirs and transects in the study area. Finally, Tortosa was also assigned to the delta and its littoral zone. The period 2000–2015 was selected in order to encompass the period with biological data (2005–2015; Figure 2) and at least the four previous years, to average flows along the same periods as in Belmar et al. [18].

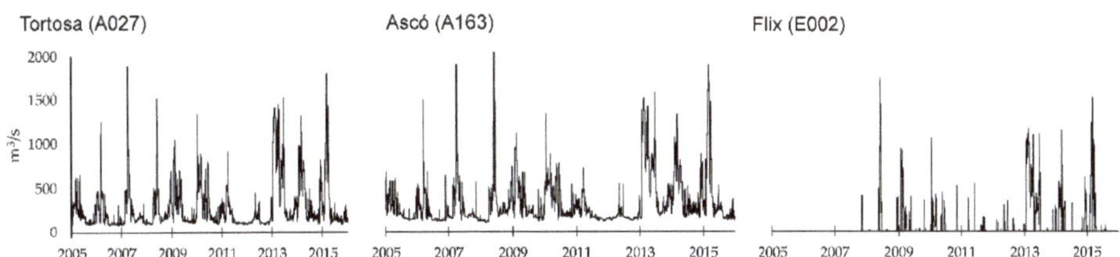

Figure 2. Flow series recorded in Tortosa (A027), Ascó (A163), and Flix (E002) from 2005 to 2015.

2.3. Analyses

For river fish, daily records from the three gauging stations (Tortosa, Ascó and Flix) were used to compute mean discharge values in the lower Ebro for periods of 1, 3, 6, 9, 12, 24, 36, and 48 months prior to sampling, similarly to the process applied in Belmar et al. [18]. In addition, the Pearson correlation coefficient between total annual runoff and the runoff generated only by sudden flow rises each year was computed to guide the interpretation of the effects of mean discharge on river communities. For birds, mean annual flows on a hydrological-year basis (October–September) were computed from the data of the Tortosa gauge (A027). This temporal scale has been used previously to assess relationships between flow regimes and birds [34]. For fisheries, annual flows using the hydrological year were also employed, as the hydrologic year practically matches the starting of the trawl-fishing season. Relationships among flows, fish-based ecological quality, birds, and fisheries in the low Ebro River and its delta were assessed as described below.

First, a set of general linear models (GLMs) was employed to investigate relationships between the mean discharge values, computed for the different time periods, and ecological quality (fish indices) at the lower Ebro. Given the different hydromorphological typologies represented by the transects, these models were developed to understand the ecohydrological relationships of each specific transect using temporal series, which may allow to provide water managers with useful information. Aquatic vegetation and riparian land cover were considered as independent habitat variables representing the patterns of complexity that may explain differences in ecohydrological relationships among transects. GLMs were also used to investigate the relationships between ecological quality and the ratio of alien species. Then, using the equations of the models, the values to obtain moderate or good ecological qualities (*sensu* the WFD) were computed.

A second set of general linear models (GLMs) was employed to test the significance of the relationships between mean annual discharge values, on one side, and birds and fisheries, on the other.

For all bird taxa, a two-step procedure was followed. (i) A between-years dissimilarity matrix was constructed using Bray–Curtis distances. This matrix was used to develop a permutational multivariate analysis of variance (PERMANOVA; [42]) for fitting linear models to the distance matrix using a permutation test with pseudo-F ratios. (ii) The Pearson correlation was used to search for relationships between mean annual discharge values and bird families and species counts, developing GLMs between discharge values and those taxa with a correlation (discharge–taxa) greater than 70% [18]. In addition, using sequentially longer periods, an iterative process was carried out to determine the possible effect of the temporal period selected on the results of two endangered species: *Larus audouinii* and *Larus genei* (the two species of *Larus* with enough data to use this approach).

For fisheries, the second set of GLMs was developed with the two selected species (and fishing methods) as dependent variables. Mean annual discharge values (on a hydrological-year basis) and physicochemical variables (to represent nutrient availability) were used as independent variables. Mean discharges computed using the natural year and the months of autumn, spring, and March (the month with maximum flow values) for each year were also used as independent variables. This last set of mean discharges was computed using also the flow records of the one and two years before each year. Stepwise procedures were employed to discard variables in the GLMs [43]. The assumptions of Gaussian models were verified (data series were also tested to discard the presence of autocorrelation). In addition, monthly data were used to run an integrated, autoregressive, moving average (ARIMA; [44]) time series model using flows and physicochemical variables as covariables. ARIMAs are fitted models to time series data either to better understand the data or to predict future points in these series (forecasting). The use of covariables allows determining if they improve the model, which is indicated by a reduction in the Akaike Information Criterion (AIC). All analyses were developed in R [45].

3. Results

3.1. Relationships between Flows and Fish-Based Ecological Quality in the Lower Ebro River

Total annual runoff and the runoff generated by sudden flow rises each year were highly correlated (Pearson correlation coefficient: 0.98; $p < 0.001$; Figure S1, Supplementary Material). Statistically significant correlations were also observed between the 9-, 12-, 24-, 36-, and 48-month average discharge values and the IBICAT2010 index, with the strongest correlation for the 12-month average (Table S1a, Supplementary Material). Transect 4 had the greatest number of significant cases and the highest R^2 values observed. Transects 5 and 6 also showed statistically significant relationships between mean flows and IBICAT2010. On the contrary, transects 1, 2, and 3 did not show statistically significant results. Finally, the results obtained when transects within the water unit ES091461 (transects 2, 3, and 4) were combined were similar to those obtained in transect 4 (i.e., the means computed using periods between nine and 36 months produced statistically significant results).

In all transects, which showed an ecological status between poor and bad, the minimum flow needed to guarantee a moderate or good ecological status decreased as the period used to compute the mean flow increased (Figure 3; Table S1b, Supplementary Material). Interestingly, the greatest mean discharges were those provided by the combination of the transects 2, 3 and 4. Such flows were by far above the current mean discharge (around 300 m^3/s), which is why their use was discarded from subsequent analyses (i.e., addition of habitat variables, which may in fact result unrealistic at combined scale). Transects that showed statistically significant relationships between fish indices and discharge also showed them between fish indices and the ratio of alien species (with the exception of the combination of transects 2, 3, and 4). The indices IBICAT2b and EFI+ showed less statistically significant relationships with flows, but a greater number of relationships with alien fishes was detected using IBICAT2b. All transects showed statistically significant relationships with the richness of alien species (Table S2a, Supplementary Material). A proportion of 0.5 would imply a status below moderate in almost all of them and below good in all of them (Figure 3; Table S2b, Supplementary Material). In addition, whereas transect 6 showed a relative tolerance to alien individuals to achieve a moderate ecological status, transect 5 showed a very low tolerance (Figure 3).

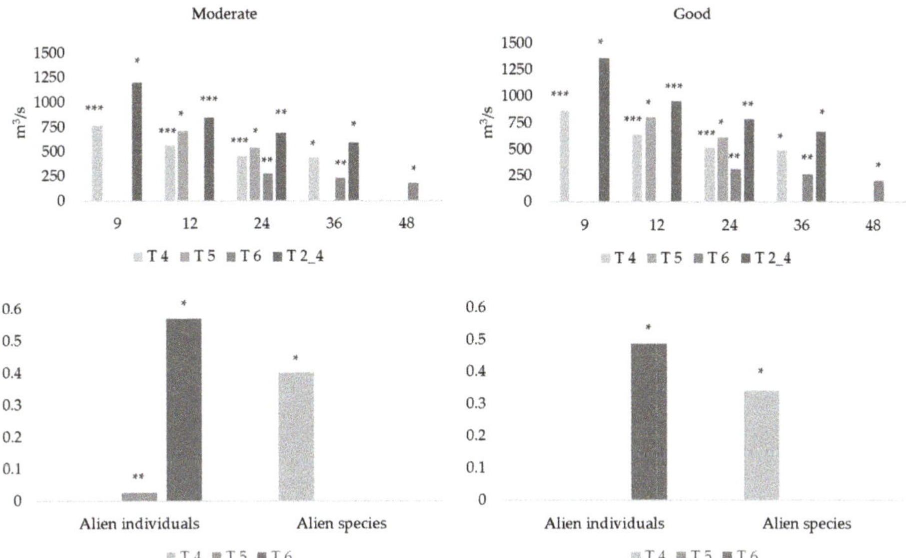

Figure 3. Mean discharges (m^3/s) computed with different periods (9, 12, 24, 36, and 48 months) and maximum proportion of alien individuals and species acceptable to ensure moderate (left) and good (right) ecological status in the transects with statistically significant results (significance levels: * p < 0.05; ** p < 0.01; *** p < 0.001).

When habitat variables (aquatic vegetation and riparian land cover) were added to mean flow (12-month period, as it provided the greatest coefficients of determination) to explain variations in ecological quality (IBICAT2010, given its greater sensitiveness to flows), results improved for all transects, with coefficients of determination that reached even almost 90% (Table 4). Whereas the use of mean flows did not provide statistically significant results in transects 2 and 3, the overall coverage of aquatic vegetation and the coverage of specific aquatic macrophytes (*Ceratophyllum*) produced significant p-values. In addition, by considering a coverage of 0% (to simulate the suppression of macrophytes), the flows necessary to achieve a moderate or good ecological status in transect 5 diminished (Figure 4; Table S3, Supplementary Material).

Table 4. Coefficients of determination (R^2) of the general linear models (GLMs) for the different river transects (T) using ecological quality (IBICAT2010) as dependent variable and mean discharge (12-month period; "qmean_12") and habitat variables as independent variables, through a stepwise selection procedure.

Transect	R^2	Independent Variable	Sign of Coefficient
T 1	0.23	qmean_12	+
		Coverage	−
		Potamogeton pectinatus	−
		Ceratophyllum demersum	−
		Riparian forest	+
T 2	*** 0.75	qmean_12 *	+
		Coverage **	−
T 3	*** 0.70	Coverage *	−
		Ceratophyllum demersum *	−
T 4	*** 0.79	qmean_12 ***	+
T 5	*** 0.88	qmean_12 **	+
		Coverage *	−
		Potamogeton pectinatus	+
		Myriophyllum spicatum *	−
T 6	0.85	qmean_12	+
		Coverage	−
		Potamogeton natans	+
		Potamogeton crispus	−

* $p < 0.05$; ** $p < 0.01$; *** $p < 0.001$.

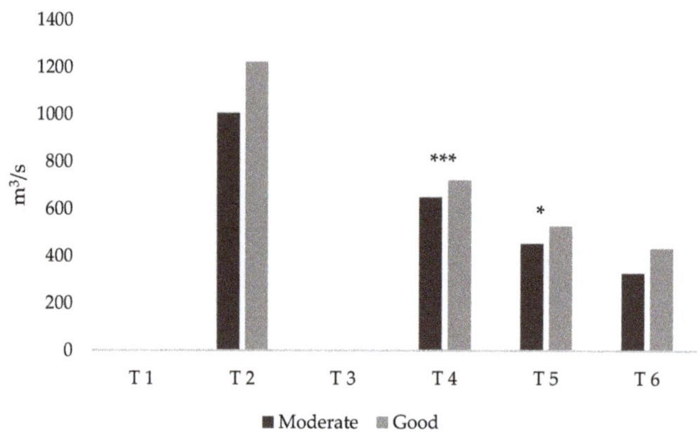

Figure 4. Mean discharges computed with a 12-month period necessary to achieve a moderate and good ecological status in the transects (T) in the absence of aquatic vegetation (significance levels: * $p < 0.05$; *** $p < 0.001$).

3.2. Relationships between Flows, Birds, and Fisheries at the Ebro Delta

Although the PERMANOVA analysis revealed that the bird community was not related to annual flows ($p > 0.05$), some species did show correlations. Such was the case of *Circus aeruginosus*, *Accipiter nisus*, *Grus grus*, and *Tringa stagnatilis*. After inspecting the GLMs, the latter was discarded because it did not fit the assumptions of Gaussian models. Regarding taxonomic groups, only predators were significantly correlated with mean annual flows (Table 5). No statistically significant relationship was found between nesting species and mean flows ($p > 0.05$). Finally, the iteration process carried out to determine the effect of the temporal framework on *Larus* species revealed that the results were not significant in practically any case. The coefficients of determination obtained were never greater than 0.25, independently of the time series used (Table 6).

Table 5. Coefficients of determination (R^2) of the general linear models (GLMs) using bird species and groups as dependent variable and mean annual discharge as independent variable. Taxa and groups with non-significant values are not shown.

Species or Group	R^2	Sign of Coefficient
Circus aeruginosus	*** 0.68	+
Accipiter nisus	* 0.42	+
Grus grus	*** 0.81	+
Predators	*** 0.60	+

* $p < 0.05$; *** $p < 0.001$.

Table 6. Coefficients of determination (R^2) of the general linear models (GLMs), using two endangered bird species (*Larus audouinii* and *Larus genei*) as the dependent variable and mean annual discharge as the independent variable, within an iteration process with different years.

Period	*Larus audouinii*	*Larus genei*
1995–2014	0.01	0.01
1995–2013	0.01	0.00
1995–2012	0.12	0.00
1995–2011	0.19	0.00
1995–2010	* 0.25	0.00
1995–2009	0.25	0.00
1995–2008	0.24	0.00
1995–2007	0.22	0.00
1995–2006	0.21	0.00
1995–2005	0.20	0.02

* $p < 0.05$.

The studied fisheries in the delta seemed to respond to annual flows in the case of prawns (*Penaeus kerathurus*; Figure 5), but they were significantly related only for those fished through traditional techniques (Table 7a). The models did not improve using physicochemical variables. The only exception was the case of mantis shrimps fished through trawling, but the model discarded annual flows (Table 7b). On the contrary, at a monthly scale, the ARIMA analysis showed that only the use of physicochemical covariables improved the models: dissolved inorganic phosphorous, phytoplankton, and zooplankton mainly (Table 8). Monthly flows did not have the ability to reduce the AIC. However, the flows of the previous two years, spring, March, and the previous two autumns did provide some additional statistically significant results according to the corresponding GLMs (Table 9).

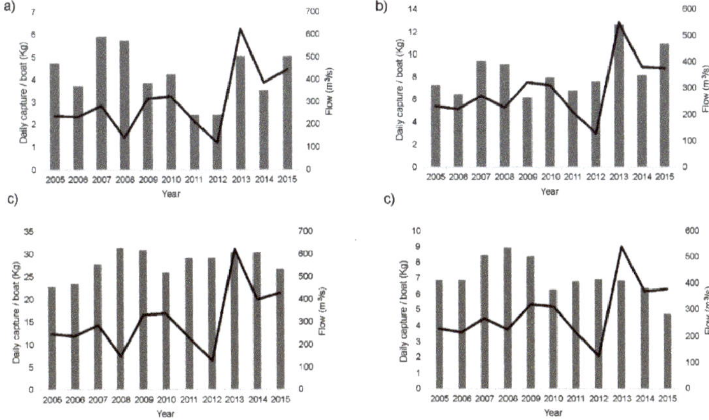

Figure 5. Annual average captures by number of boats (columns, first axis) and flows (line, secondary axis) and for the fishery species selected and the two techniques used: (**a**) Prawn and trawling, (**b**) prawn and other techniques; (**c**) mantis shrimp and trawling; and (**d**) mantis shrimp and other techniques.

Table 7. (a) Coefficients of determination (R^2) of general linear models (GLMs) using the selected fishery taxa and techniques (a: trawling; b: other techniques) as the dependent variable and annual flows based on hydrological year (Flow) as the independent variable. (b) Results using annual flows and physicochemical variables (see Table 3 for details) as independent variables, through a stepwise selection procedure.

(a)			
Dependent Variable	R^2	**Independent Variable**	**Sign of Coefficient**
Pennaeus_a	0.07	Flow	+
Pennaeus_b	* 0.40	Flow	+
Squilla_a	0.04	Flow	+
Squilla_b	0.08	Flow	−
(b)			
Dependent Variable	R^2	**Independent Variable**	**Sign of Coefficient**
Pennaeus_a	0.22	Flow	+
		Phytoplankton	+
		Dissolved Inorganic Phosphorous	−
Pennaeus_b	0.46	Flow	+
		Phytoplankton	−
		Zooplankton	+
		Dissolved Inorganic Phosphorous	−
		Turbidity	+
		Suspended Matter	−
Squilla_a	* 0.50	Phytoplankton	−
		Suspended Matter	−
Squilla_b	n/a	n/a	n/a

* $p < 0.05$; n/a: not applicable.

Table 8. Akaike Information Criterion (AIC) for different covariables (see details in Table 3), obtained employing integrated, autoregressive, moving average (ARIMA) time series models on the selected fishery taxa and technique (a: trawling; b: other techniques). The co-variables that reduced the AIC are in bold.

Co-Variable	Penaeus_a	Penaeus_b	Squilla_a	Squilla_b
None	510	739	630	514
Flow	512	742	631	515
Chlorophyll	491	728	611	507
Phytoplankton	452	727	611	502
Zooplankton	493	723	605	500
Dissolved Inorganic Nitrogen *	512	740	625	516
Dissolved Inorganic Phosphorous	333	419	371	389
Total Organic Carbon	497	727	614	505
Turbidity	512	743	631	515
Suspended Matter	413	570	478	406

* only for Squilla_a.

Table 9. Coefficients of determination (R^2) and signs of coefficients of general linear models (GLMs) developed using the selected fishery taxa and technique (a: trawling; b: other techniques) as the dependent variable and flows averaged along different periods as the independent variables ("Indep. var."; "-2" indicates that it is the value from two years ago), through a stepwise selection procedure. Only variables with relevant results are shown.

Indep. Var.	Pennaeus_a	Sign	Pennaeus_b	Sign	Squilla_a	Sign	Squilla_b	Sign
year	0.16	+	** 0.52	+	0.11	+	0.09	−
year-2	0.01	−	0.01	−	* 0.47	−	0.23	−
March	0.05	+	0.30	+	0.01	−	** 0.57	−
Spring	* 0.41	+	* 0.37	+	0.33	+	0.13	+
Autumn-2	0.00	−	0.10	−	*** 0.84	−	0.13	−

* $p < 0.05$; ** $p < 0.01$; *** $p < 0.001$.

4. Discussion

This study shows relationships between water discharge, riverine fishes, bird diversity, and coastal (shellfish) fisheries. Based on a dataset with more than ten years of records (which is scarce in bibliography), it constitutes a contribution to validating sustainable management strategies in the lower Ebro River and its delta (and in similar areas), and; therefore, to water management. The environmental flow regimes proposed up-to-date would result as insufficient in the river, according to the models developed. However, controlling the development of macrophytes may allow for the achievement of the same objectives of quality with lower volumes. This is coherent with previous research that highlighted that some consequences of flow alterations can be remedied by physical habitat alterations, such as encroaching aquatic vegetation removed by mechanical clearing in absence of scouring flows, but pointed out that the long-term effectiveness and sustainability of such actions are in many cases likely to be low [15]. This study contributes to the understanding of the connection among flows, macrophytes, and fish, as well as ecological (i.e., birds) and socioeconomical (i.e., fisheries) components of the delta, in order to provide water managers with valuable information to guide their decisions.

4.1. Fish-Based Ecological Quality and Flows in the Lower Ebro

Fish-based ecological quality may respond to mean discharge, depending on the spatial and temporal scale considered, but water management and e-flow assessment must pay attention to the specific mechanisms that explain such variability. Like in the study developed by Belmar et al. [18], ecological quality (IBICAT2010) showed positive relationships with flow magnitude in the lower Ebro River and a 12-months span provided the best results, which is likely related to the relationship between total annual runoff and the runoff caused by sudden flow rises (given their ecological relevance). Such results were conditioned by spatial heterogeneity (*sensu* [46]), as local conditions (e.g., aquatic vegetation) influenced fish communities. Fish-based indices are not focused on one "objective" species, contrarily to habitat suitability models, which means that ecohydrological relationships are more difficult to establish using only hydrological information. Nevertheless, approaches at community level like this one are necessary because the use of "objective" species ignore vital ecological linkages [47]. Our results indicate that combining transects to obtain results representative at a greater spatial scale for a long water unit [18] may condition the applicability of the results, given that this may provide too high discharge thresholds to achieve a predefined ecological status or not be realistic (as aquatic vegetation must also be considered). Additional research will allow better understanding this output, which seems to evidence the importance of considering the influence of multiple environmental factors (i.e., not just flow) and spatial scale on the ability of species to complete their life cycles and thus persist as members of a community, as Poff [47] highlighted. This helps to anticipate under what circumstances flow interventions will be successful in sustaining socially valued ecological characteristics. In this sense, the choice of a suitable quality index to establish meaningful ecohydrological relationships may depend on the specific objective. IBICAT2b showed greater sensitiveness to the ratio of the richness of alien species (as the relationships were statistically significant in all transects). The fact that a proportion above 0.5 would imply an ecological status below good in all of them seems to confirm the suitability of such criterion to define a community as impacted within the approach to biologically validate environmental flows proposed by Caiola et al. [17].

Ecological quality may be improved both with relatively high mean flows during a short period and with lower flows during longer periods, given that the relationship between time span and flow magnitude is inverse. Such inverse relationship must be interpreted with caution, as it has been observed within the temporal framework established by Belmar et al. [18], who did not find relationships between daily flow records and ecological quality out of the temporal range observed here (between 9 and 48 months). Additional factors, such as aquatic habitats, may help to better understand this inverse relationship (see text about macrophytes in the paragraph below). Within the stated temporal range, longer periods (more months) involve greater uncertainty, as they are associated with lower coefficients of determination. In general, both a good and a moderate ecological status are unreachable only through increases in mean flows, as the required flow values are way greater than the current mean values. The annual mean for the series considered in this study is around 300 m^3/s (period 2005–2015), whereas the historical annual mean discharge (423 m^3/s for the period 1912–2014) is greater, similar to the flow necessary to achieve a moderate ecological status when the span considered increases up to 36 months (435 m^3/s) and also similar to the value necessary to achieve a good status (492 m^3/s). This evidences that the current ecological status of the river is the result of a process of degradation caused by a gradual increase in hydrologic stress (mostly caused by water demands) and also that flow regimes closer to the natural regime would be capable to provide a better ecological status. Nevertheless, we must emphasize the necessity to interpret our results with caution. Given that all samples showed an ecological status between poor and bad, the discharge needed to achieve a moderate and good status had to be extrapolated, which involves incertitude. Despite this limitation, the dataset employed in this study results very valuable because biological series with more than 10 years of data are scarce in bibliography.

Measures additional to water provisioning are needed to improve the ecological quality in the lower Ebro River, such as controlling aquatic vegetation (macrophyte development). Such vegetation showed to represent spatial heterogeneity effectively, as it improved the models in those transects where flows provided poor results on their own (practically all of them except transect 4). The coverage of aquatic vegetation seems to be in general inversely related to ecological quality (IBICAT2010), which was also the case of *Ceratophilum* and *Myriophyllum* (associated with stagnant waters). Therefore, in those transects where both discharge and macrophytes influence ecological quality, reducing their coverage would imply lower discharges to achieve the same ecological status (e.g., transect 5). This could be related to the inverse relationship between water discharge and macrophytes observed in the study area by Ibáñez et al. [48]. They showed that flows with enough magnitude and duration reduced macrophyte abundance, given that lower discharges can allow vegetation to encroach into river channels. Our results also suggest that increased macrophyte coverage is accompanied by an increase in alien species. During the fieldwork conducted, alien species were more abundant than native species in habitats rich in aquatic macrophytes, which have spread in the study area as the degree of regulation has increased during the last years together with the corresponding reductions in mean discharge [48] (given that both alien plants and animals are more vulnerable to flood scour [49]). The effects of flows on ecological quality (fish communities) seem thus to be direct and indirect. Direct effects would be related to the suitability of the conditions generated by flows (e.g., depth, sediment transport, etc.) for fish, whereas indirect effects would be associated with other factors that may also influence fish communities (i.e., macrophytes). The ecological index employed (which uses metrics based on alien species) and the fact that flow discharge is currently below the natural regime may help to understand the inverse relationship between time span and flow magnitude observed to achieve a certain ecological status. Greater flows (more close to the natural regime) may produce flood scour in a shorter period, with a positive effect on ecological quality (given its negative effect on alien species). Further research is necessary to fully understand the relationships among these factors, as well as to understand the ecological responses to individual hydrologic events or sequences of events at other ecological resolution such as process rates and species traits [47].

4.2. Effects of River Flows on Birds and Fisheries in the Delta

The relationship between flows and bird communities showed as being restricted to a few taxa. Only some species typical of lacustrine or riverine environments (*Accipiter nisus* is associated with forests, but can also inhabit riparian understory; [50]), and the group of predators, were related to mean annual discharge. Lowland species that feed on submerged prey or macrophytes from the water surface are vulnerable to changes in flow regimes, because such species may display negative relationships with high flow frequency and duration [34], as foraging efficiency is likely to be severely compromised under conditions of elevated water velocity, depth and turbidity [51]. However, we found a positive relationship between flows and predators. This might be due to an indirect relationship between river flow and abundance of prey birds in the delta caused by an increase in their availability in rainy years, but this issue requires further research. Marine birds such as seagulls did not provide correlations greater than 70%. The iterative process developed on two *Larus* species revealed a small effect of the temporal framework on the results, as the coefficients of determination were always low (and usually non-significant). However, the last iterations (from 1995–2012 to 1995–2014) showed a reduction in the values of the coefficients that also deserves further investigation. Such reduction may indicate a weakening in the dependence of these species on flows, probably caused by other factors associated with the decreasing trends in populations detected in the last years such as changes in fishing practices by boats (e.g., prohibition of returning fishing discards to sea).

The significant relationships between discharge (together with the additional indices calculated for fisheries) and some fishery groups detected evidence that flow magnitude may have implications for socioeconomic activities, although other factors must be considered. The influence of water physicochemical characteristics associated with river nutrients on populations at a monthly scale is relevant for seasonal water management, as nutrients determine the availability of nourishment in an oligotrophic environment such as the Mediterranean Sea [52]. For example, monitoring variations in the parameters highlighted in this study (chlorophyll, phytoplankton, zooplankton, dissolved inorganic phosphorous, total organic carbon, and suspended matter) may allow anticipating monthly variations in the fishery groups selected. Fishery management must also take into consideration that the effect of discharge on populations depends on the temporal scale, species, and fishing technique considered. For example, only prawns caught with traditional techniques showed a statistically significant annual response. In addition, the flows experienced in March, spring, and during the two previous years are good candidates to be taken into consideration by water managers in order to anticipate changes in fish productivity. Such relationships are also related to nutrient enrichment processes caused by discharge [31–33], but further research is necessary to explain the specific mechanisms that link river flows and fish productivity in coastal areas like this one. In this context, the relationships observed in this study involve a reduced number of factors (i.e., hydrologic regime and nutrient content), which results as relevant for water management at the catchment scale to assess environmental flow regimes. More accurate models could be developed to account for more variables associated with other factors, such as water pollution (which was out of the scope and would require a greater number of cases). In addition, analyses based on other species may allow a broader knowledge in terms of how fisheries respond to river flows. Distinguishing among reproductive strategies and species' location in the water column will result as essential.

Supplementary Materials: The following are available online at http://www.mdpi.com/2073-4441/11/5/918/s1, Figure S1: Plot showing the relationship between annual runoff and the runoff caused by sudden flow rises (hm^3) each year, Table S1: (a) Coefficients of determination (R^2) of the General Linear Models (GLMs) for the different river transects (T) developed using IBICAT2010 as the dependent variable and (i) the mean discharge (qmean) computed with different months, (ii) the proportion of alien individuals (CPUEI_I_T) and (iii) the proportion of alien species (R_I_T), as independent variables (Indep. var.) (b) Modelled mean discharges (m^3/s) to achieve the moderate and good ecological status for each transect at each period, Table S2: (a) Coefficients of determination (R^2) of the General Linear Models (GLMs) for the different river transects (T) developed using IBICAT2b and EFI+ as dependent variables (Dep. var.) and (i) the mean discharge (qmean) computed with different months, (ii) the proportion of alien individuals (CPUEI_I_T) and (iii) the proportion of alien species (R_I_T) as independent variables (Indep. var.). (b) Modelled mean discharges (m^3/s) to achieve moderate and good ecological status for each transect at each period, Table S3: Coefficients of determination (R^2) using only mean annual flow (12-month period) as the independent variable to explain ecological quality using the IBICAT2010 index.

Author Contributions: Conceptualization, C.I. and N.C.; data curation, O.B. and A.F.; formal analysis, O.B.; methodology, O.B.; supervision, C.I. and N.C.; writing—original draft, O.B.; writing—review and editing, C.I. and N.C.

Funding: This study was developed through an agreement on the ecological status of the low Ebro River and its estuary between IRTA ("Institut de Recerca i Tecnologia Agroalimentàries") and ACA ("Agència Catalana de l'Aigua"), which acted as financial supporter.

Acknowledgments: We wish to thank "Confederación Hidrográfica del Ebro" for providing their flow records, Antoni Curcó for its guidance on bird data and Carles Alcaraz, Núria Vila-Martínez, David Mateu and Lluis Jornet for the fieldwork conducted. The authors also acknowledge support from the CERCA Programme/Generalitat de Catalunya.

Conflicts of Interest: The authors declare no conflicts of interest. The funders had no role in the design of the study; in the collection, analyses, or interpretation of data; in the writing of the manuscript, or in the decision to publish the results.

References

1. McLusky, D.S. Estuarine benthic ecology: A European perspective. *Aust. Ecol.* **1999**, *24*, 302–311. [CrossRef]
2. Peirson, W.; Bishop, K.; Van Senden, D.; Horton, P.; Adamantidis, C. *Environmental Water Requirements to Maintain Estuarine Processes*; Environment Australia: Canberra, Australia, 2002.
3. Russell, M.J.; Montagna, P.A.; Kalke, R.D. The effect of freshwater inflow on net ecosystem metabolism in Lavaca Bay, Texas. *Estuar. Coast. Shelf Sci.* **2006**, *68*, 231–244. [CrossRef]
4. Dauer, D.M.; Ranasinghe, J.A.; Weisberg, S.B. Relationships between benthic community condition, water quality, sediment quality, nutrient loads, and land use patterns in Chesapeake Bay. *Estuaries* **2000**, *23*, 80–96. [CrossRef]
5. Gasith, A.; Resh, V.H. Streams in Mediterranean climate regions: Abiotic influences and biotic responses to predictable seasonal events. *Annu. Rev. Ecol. Syst.* **1999**, *30*, 51–81. [CrossRef]
6. Caiola, N.; Vargas, M.J.; de Sostoa, A. Feeding ecology of the endangered Valencia toothcarp, Valencia hispanica (Actinopterygii: Valenciidae). *Hydrobiologia* **2001**, *448*, 97–105. [CrossRef]
7. Caiola, N.A.; Vargas, M.J.; de Sostoa, A. Life history pattern of the endangered Valencia toothcarp, Valencia hispanica (Actinopterygii: Valenciidae) and its implications for conservation. *Arch. Hydrobiol.* **2001**, *150*, 473–489. [CrossRef]
8. Ferreira, T.; Oliveira, J.; Caiola, N.; De Sostoa, A.; Casals, F.; Cortes, R.; Economou, A.; Zogaris, S.; Garcia-Jalon, D.; Ilhéu, M. Ecological traits of fish assemblages from Mediterranean Europe and their responses to human disturbance. *Fish. Manag. Ecol.* **2007**, *14*, 473–481. [CrossRef]
9. Arthington, A.H.; Bhaduri, A.; Bunn, S.E.; Jackson, S.E.; Tharme, R.E.; Tickner, D.; Young, B.; Acreman, M.; Baker, N.; Capon, S.; et al. The Brisbane Declaration and Global Action Agenda on Environmental Flows (2018). *Front. Environ. Sci.* **2018**, *6*, 45. [CrossRef]
10. Brisbane Declaration. The Brisbane Declaration: Environmental flows are essential for freshwater ecosystem health and human well-being. In Proceedings of the 10th International River Symposium, Brisbane, Australia, 3–6 September 2007.
11. European Commission. *A Blueprint to Safeguard Europe's Water Resources*; European Commission: Brussels, Belgium, 2012.
12. European Commission. *Ecological Flows in the Implementation of the Water Framework Directive. Guidance Document No. 31*; European Commission: Brussels, Belgium, 2015.
13. Ormerod, S.J. Current issues with fish and fisheries: editor's overview and introduction. *J. Appl. Ecol.* **2003**, *40*, 204–213. [CrossRef]
14. Schinegger, R.; Pucher, M.; Aschauer, C.; Schmutz, S. Configuration of multiple human stressors and their impacts on fish assemblages in Alpine river basins of Austria. *Sci. Total Environ.* **2018**, *616*, 17–28. [CrossRef] [PubMed]
15. Sostoa, A.D.; Caiola, N.; Casals, F.; García-Berthou, E.; Alcaraz Cazorla, C.; Benejam Vidal, L.; Maceda-Veiga, A.; Solà, C.; Munné, A. *Ajust de L'índex d'Integritat Biòtica (IBICAT) Basat en L'ús Dels Peixos Com a Indicadors de La Qualitat Ambiental Als Rius de Catalunya*; Generalitat de Catalunya: Barcelona, Spain, 2010.
16. Segurado, P.; Caiola, N.; Pont, D.; Oliveira, J.M.; Delaigue, O.; Ferreira, M.T. Comparability of fish-based ecological quality assessments for geographically distinct Iberian regions. *Sci. Total Environ.* **2014**, *476*, 785–794. [CrossRef] [PubMed]
17. Caiola, N.; Ibáñez, C.; Verdú, J.; Munné, A. Effects of flow regulation on the establishment of alien fish species: A community structure approach to biological validation of environmental flows. *Ecol. Indic.* **2014**, *45*, 598–604. [CrossRef]

18. Belmar, O.; Vila-Martínez, N.; Ibáñez, C.; Caiola, N. Linking fish-based biological indicators with hydrological dynamics in a Mediterranean river: Relevance for environmental flow regimes. *Ecol. Indic.* **2018**, *95*, 492–501. [CrossRef]

19. Pont, D.; Hugueny, B.; Beier, U.; Goffaux, D.; Melcher, A.; Noble, R.; Rogers, C.; Roset, N.; Schmutz, S. Assessing river biotic condition at a continental scale: A European approach using functional metrics and fish assemblages. *J. Appl. Ecol.* **2006**, *43*, 70–80. [CrossRef]

20. Pont, D.; Hugueny, B.; Rogers, C. Development of a fish-based index for the assessment of river health in Europe: The European Fish Index. *Fish. Manag. Ecol.* **2007**, *14*, 427–439. [CrossRef]

21. Renöfält, B.M.; Jansson, R.; Nilsson, C. Effects of hydropower generation and opportunities for environmental flow management in Swedish riverine ecosystems. *Freshw. Biol.* **2010**, *55*, 49–67. [CrossRef]

22. Ibáñez, C.; Prat, N. The environmental impact of the Spanish national hydrological plan on the lower Ebro river and delta. *Int. J. Water Resour. Dev.* **2003**, *19*, 485–500. [CrossRef]

23. MAM. Real Decreto 907/2007, de 6 de julio, por el que se aprueba el Reglamento de la Planificación Hidrológica. In *Boletín Oficial del Estado*; Ministerio de Medio Ambiente: Madrid, Spain, 2007; Volume 162, pp. 1–53.

24. MARM. Orden ARM 2656/2008, de 10 de septiembre, por la que se aprueba la Instrucción de Planificación Hidrológica. In *Boletín Oficial del Estado*; Ministerio de Medio Ambiente, y Medio Rural y Marino: Madrid, Spain, 2008; Volume 229, pp. 38472–38582.

25. CSTE. *Revisió i Actualització de la Proposta de Règim de Cabals Ecològics al Tram Final del Riu Ebre, Delta i Estuari*; CSTE, Generalitat de Catalunya: Barcelona, Spain, 2015.

26. Livingston, R.J.; Diaz, R.J.; White, D.C. *Field Validation of Laboratory-Derived Multispecies Aquatic Test Systems*; United States Environmental Protection Agency, Office of Research and Development, Environmental Research Laboratory: Gulf Breeze, FL, USA, 1985.

27. Belmar, O.; Bruno, D.; Martínez-Capel, F.; Barquín, J.; Velasco, J. Effects of flow regime alteration on fluvial habitats and riparian quality in a semiarid Mediterranean basin. *Ecol. Indic.* **2013**, *30*, 52–64. [CrossRef]

28. Feyrer, F.; Healey, M.P. Fish community structure and environmental correlates in the highly altered southern Sacramento-San Joaquin Delta. *Environ. Biol. Fish.* **2003**, *66*, 123–132. [CrossRef]

29. Costa, M.J.; Vasconcelos, R.; Costa, J.L.; Cabral, H.N. River flow influence on the fish community of the Tagus estuary (Portugal). *Hydrobiologia* **2007**, *587*, 113–123. [CrossRef]

30. Lloret, J.; Palomera, I.; Salat, J.; Sole, I. Impact of freshwater input and wind on landings of anchovy (*Engraulis encrasicolus*) and sardine (*Sardina pilchardus*) in shelf waters surrounding the Ebre (Ebro) River delta (north-western Mediterranean). *Fish. Oceanogr.* **2004**, *13*, 102–110. [CrossRef]

31. Caddy, J.; Bakun, A. A tentative classification of coastal marine ecosystems based on dominant processes of nutrient supply. *Ocean. Coast. Manag.* **1994**, *23*, 201–211. [CrossRef]

32. Deegan, L.A.; Day, J.W.; Gosselink, J.G.; Yáñez-Arancibia, A.; Chávez, G.S.; Sánchez-Gil, P. Relationships among physical characteristics, vegetation distribution and fisheries yield in Gulf Mexico estuaries. In *Estuarine Variability*; Wolfe, D.A., Ed.; Academic Press: Cambridge, MA, USA, 1986; pp. 83–100.

33. Loneragan, N.R. River flows and estuarine ecosystems: Implications for coastal fisheries from a review and a case study of the Logan River, southeast Queensland. *Aust. J. Ecol.* **1999**, *24*, 431–440. [CrossRef]

34. Royan, A.; Hannah, D.M.; Reynolds, S.J.; Noble, D.G.; Sadler, J.P. River birds' response to hydrological extremes: New vulnerability index and conservation implications. *Biol. Conserv.* **2014**, *177*, 64–73. [CrossRef]

35. Colomé, J.V.F.; Sancho, J.C.; Comín, F.A. Los medios acuáticos del Delta del Ebro y su capacidad de producción. *Rev. Obras Publ.* **1997**, *3368*, 67–71.

36. Day, J.W., Jr.; Maltby, E.; Ibáñez, C. River basin management and delta sustainability: A commentary on the Ebro Delta and the Spanish National Hydrological Plan. *Ecol. Eng.* **2006**, *26*, 85–99. [CrossRef]

37. European Parlament. Directive 2009/147/EC of the European Parliament and of the Council of 30 November 2009 on the conservation of wild birds. *Off. J. Eur. Union* **2010**, *20*, 7–25.

38. Ibáñez, C.; Prat, N.; Canicio, A.; Curcó, A. *El delta del Ebro: Un Sistema Amenazado*; Bakeaz: Bilbao, Spain, 1999.

39. Ibáñez, C.; Alcaraz, C.; Caiola, N.; Rovira, A.; Trobajo, R.; Alonso, M.; Duran, C.; Jiménez, P.J.; Munné, A.; Prat, N. Regime shift from phytoplankton to macrophyte dominance in a large river: Top-down versus bottom-up effects. *Sci. Total Environ.* **2012**, *416*, 314–322. [CrossRef]

40. Nebra, A.; Caiola, N.; Ibáñez, C. Community structure of benthic macroinvertebrates inhabiting a highly stratified Mediterranean estuary. *Sci. Mar.* **2011**, *75*, 577–584.

41. CEN. *Water Quality—Sampling of Fish with Electricity. European Standard EN*; CEN: Brussels, Belgium, 2003.

42. McArdle, B.H.; Anderson, M.J. Fitting multivariate models to community data: A comment on distance-based redundancy analysis. *Ecology* **2001**, *82*, 290–297. [CrossRef]

43. Venables, W.; Ripley, B. Random and mixed effects. In *Modern Applied Statistics with S*; Springer: Berlin, Germany, 2002; pp. 271–300.

44. Box, G.E.; Jenkins, G.M. *Time Series Analysis: Forecasting and Control*; Holden-Day: San Francisco, CA, USA, 1976.

45. R Core Team. *R: A Language and Environment for Statistical Computing*; R Core Team: Vienna, Austria, 2017.

46. Livingston, R.J. Field sampling in estuaries: The relationship of scale to variability. *Estuaries* **1987**, *10*, 194–207. [CrossRef]

47. Poff, N.L. Beyond the natural flow regime? Broadening the hydro-ecological foundation to meet environmental flows challenges in a non-stationary world. *Freshw. Biol.* **2018**, *63*, 1011–1021. [CrossRef]

48. Ibáñez, C.; Caiola, N.; Rovira, A.; Real, M. Monitoring the effects of floods on submerged macrophytes in a large river. *Sci. Total Environ.* **2012**, *440*, 132–139. [CrossRef]

49. Power, M.E.; Dietrich, W.E.; Finlay, J.C. Dams and downstream aquatic biodiversity: Potential food web consequences of hydrologic and geomorphic change. *Environ. Manag.* **1996**, *20*, 887–895. [CrossRef]

50. SEO/BirdLife. Guía de Aves. Available online: https://www.seo.org/listado-aves/ (accessed on 29 April 2019).

51. Vilches, A.; Arizaga, J.; Salvo, I.; Miranda, R. An experimental evaluation of the influence of water depth and bottom color on the common kingfisher's foraging performance. *Behav. Process.* **2013**, *98*, 25–30. [CrossRef]

52. Margalef, R. Introduction to the Mediterranean. In *Key Environments: Western Mediterranean*; Pergamon Press: Oxford, UK, 1985.

Article

A Comment on Chinese Policies to Avoid Negative Impacts on River Ecosystems by Hydropower Projects

Miao Wu [1], Ang Chen [2], Xingnan Zhang [1,3,4,*] and Michael E. McClain [5,6]

[1] College of Hydrology and Water Resources, Hohai University, Nanjing 210098, China; 3water@hhu.edu.cn
[2] Yangtze Ecology and Environment Co., Ltd., Wuhan 430062, China; angteenchen@gmail.com
[3] National Cooperative Innovation Center for Water Safety & Hydro-Science, Hohai University, Nanjing 210098, China
[4] National Engineering Research Center of Water Resources Efficient Utilization and Engineering Safety, Hohai University, Nanjing 210098, China
[5] Department of Water Resources and Ecosystems, IHE Delft Institute for Water Education, 2611AX Delft, The Netherlands; m.mcclain@un-ihe.org
[6] Faculty of Civil Engineering and Geosciences, Delft University of Technology, 2628CN Delft, The Netherlands
* Correspondence: xingnan.zhang@outlook.com or zxn@hhu.edu.cn

Received: 24 February 2020; Accepted: 18 March 2020; Published: 20 March 2020

Abstract: The rapid economic development of river basins depends on the excessive use of water resources. China experienced a rapid development of hydropower projects in the last two decades and thus faces many ecological and environmental issues, especially in ecologically sensitive areas. Environmental flow is an important management tool that requires attention in the environmental impact assessment of hydropower projects. Environmental flows are of great significance for maintaining river structures and protecting the health of both aquatic ecosystems and human sustainable livelihoods. Although the government authorities have done much work in this area and attempted to consider technical requirements to address the negative externalities of hydropower projects, there are still defects in the basic procedures, calculation methods, and ultimately implementation process from policy to operationalization in terms of environmental flows. The official standards for environmental flows assessment mainly appear in two documents: 1. specification for calculation of environmental flow in rivers and lakes; and 2. code for calculation ecological flow of hydropower projects. This paper reviewed the overarching framework of the two documents and then summarized their fitness in terms of environmental flows implementation in hydropower projects. The research status of environmental flows and future directions for China were also proposed in this paper.

Keywords: environmental flows; ecological flow; China; hydropower; river ecosystem

---

1. Introduction

Hydropower in China has been developing rapidly for the past 20 years and the government encourages the development and utilization of renewable energy. Nearly one hundred thousand of multi-objective dams have been constructed, hydropower stations account for a half [1,2]. Hydropower stations are supposed to protect the ecosystems by clean electricity production and take into account the requirements of flood control, water supply, irrigation, shipping and fisheries [3]. The unguaranteed inflow is the direct cause of the insufficient ecological flow of some rivers and lakes [4]. Over-exploitation of water resources has brought serious threats to the ecosystem, such as the hydropower project has greatly changed river flow regimes [5–7]. To prevent more human disturbance from harming the ecosystem, the government of China (GoC) has made great efforts to promote ecological sustainability [8].

Unnatural flow alternations negatively influence ecosystems [9–11]. Since 2006, GoC has issued many relevant guidelines or standards to promote the sustainable development of water resources (Table 1). In these guidelines, the purpose of river environmental base flow is to prevent the river flow from being cut off and to avoid minimum flows that might cause irreversible damage of river aquatic communities. Water conservancy projects and hydropower projects are required to take measures to protect the environmental base flow both in the construction and operation period [12,13]. It is clear that most of the key disturbances in the flow state come from in-stream water works, which are designed to store water during the rainy season and transfer it downstream as needed. Maintaining the environmental flows as designed is an effective way to protect river ecosystems [14]. The environmental flows management of dammed rivers in China can be implemented in accordance with relevant official guidelines to ensure the health of the ecosystem.

Table 1. Typical guidelines for environmental flows issued by Chinese authorities.

Issuing Authority	Guideline	Related Content
State Environmental Protection Administration	Technical guide for environmental impact assessment of river ecological flow, cold water, and fish passage facilities for water conservation construction projects (trial) EIA Letter (2006) No.4	The lower limit of the ecological base flow should not be less than 10% of the average annual natural runoff.
Ministry of Water Resources	Guidelines for assessment of rivers and lakes eco-water demands (SL/Z 479-2010)	Ecological base flow generally takes the minimum monthly average flow of 90% of the control nodes; the minimum ecosystems water demand in the river, and the average annual runoff of the control nodes in the north is generally 10–20%, 20–30% in the south.
Ministry of Water Resources	Technical specification for the analysis of supply and demand balance of water resources (SL 429-2008)	The eco-hydrological elements of water project planning and design should consider the ecological base flow and sensitive ecological water demand at the river basin scale, river corridor scale and river scale, and further standardize.

Due to the significant regional differences in climate and geography in China, how to determine the environmental flows in a logical process is challenging for water managers. It is imperative that environmental flows be quantified and implemented. Considerable research has been done on the definition of environmental flows in China, the development of calculation methods, and implementation of environmental flows, including the selection of indicators for monitoring [15–18]. Given this level of research, no specific provisions on environmental flows have been formulated [3]. Where they do appear, environmental flows are grouped with other demands and given unclear priority. For example, the new "Water Law of the People's Republic of China" promulgated in 2016 states that "the development and utilization of water resources should first meet the requirements of urban and rural residents, and take into account the needs of agriculture, industry, ecosystems and shipping". In the development and utilization of water resources in semi-arid areas, the ecosystems needs be fully considered [6]. In the related norms, there are some recommend methods to calculate environmental flows. Previous studies [19,20] have conducted research into the advantages and disadvantages of environmental flows calculation methods and applicable conditions for a specific area or region. In order to maximize the use of environmental flows calculation specifications, it is time to comprehensively sort out GoC's calculation of environmental flows, analyzing the relevant normative standards, the environmental flows calculation principles and recommended methods proposed in the different normative guidelines. In the rest of this comment, we review the related policies on environmental flows release from hydropower projects in China and summarize the defects in the current policy. This provides not only effective suggestions for environmental flows research, but also some empirical reference for environmental flows practices.

2. Description of the Two Standards for Development of Water Use Projects

There are two main standards for development of water use projects for the environmental flows, the 2015 "Specification for calculation of environmental flow in rivers and lakes" (SERL) [21] which applies in all water conservancy projects and the 2017 "Code for calculation of ecological flow of hydropower projects" (CEHP) [22] which applies specifically to hydropower projects. They are the two latest and important specifications of GoC on the principles and calculation methods of environmental flows. The standards are state-of-the art in requiring that flows be set to protect a broad range of ecological processes, including migrations, spawning, and other habitat requirements. In general, SERL is the reference basis for CEHP, and CEHP is a sector-specific (hydropower) analysis of environmental flow calculations for rivers. Therefore, in the actual operation process, the manager must make the discharge flow meet the requirements of SERL, and also meet the specific regulations in CEHP.

About the compilation process of preparing the standards, on the basis of in-depth research, the compiler summarized the practical experience of relevant environmental flows calculations, incorporated the scientific and technological achievements of relevant research, and solicited opinions from relevant design units and scientific research experts. Although they are both standards for development of water use projects of the People's Republic of China for the calculation of environmental flows, they also have many differences (Table 2). The two standards come from different government authorities and the objects are different. Additionally, there are differences in the objectives and purposes of their implementation, but they do not exist independently. From the point of view of the title, SERL covers a wider range and content than CEHP.

Table 2. Comparison of the two standards for development of water use projects for calculating environmental flows in China.

Guideline	SERL	CEHP
Issued authority	Ministry of Water Resources of the People's Republic of China	National Energy Administration of China
Purpose	Technical requirements, basic procedures and calculation methods for regulating the ecosystems of rivers and lakes in order to protect and restore the ecosystems of rivers and lakes	In order to standardize the conditions, contents and methods for calculating the ecological flow of hydropower projects [1], and to unify the technical requirements, this specification is formulated.
Object	It is applicable to the ecosystems water demand calculation and water project [2] planning, design and management of water environment integration and professional planning for watershed and regional ecosystems water demand calculation	It is suitable for the analysis and calculation of ecological flow of hydropower projects [1].

[1] Hydropower projects: the first main use is hydropower generation. [2] Water conservancy projects: the first main use is not hydropower generation.

3. Comparison Analysis of the Two Standards for Development of Water Use Projects

3.1. Basic Principles

The SERL stipulates that environmental flows should be considered at a river basin scale and not only for individual river reaches. On the basis of water resources development, environmental protection and social and economic development, the environmental flows can be determined scientifically and reasonably [23–25]. The environmental flow method selected by the manager depends on the ecological conditions and data availability in the downstream, such as the rare species, spawning area, wetland or historic landscape existed in the downstream. These requirements also apply to CEHP. It is worth noting that CEHP emphasizes the consideration of river fish resources, water quality deterioration, saltwater intrusion, algal blooms and other ecological and environmental issues, but also needs to consider the downstream of hydropower projects construction and operation. Eco-environmental problems are also to be considered, such as changes in hydrological conditions in the river section, insufficient water supply to connected wetlands, and weakened aquatic habitats. From the basic principle, the two standards are similar, but the impact of the operation of water development projects is emphasized in CEHP.

3.2. Components and Definition of Terms

Environmental flows in rivers and lakes, in-stream environmental flow, off-stream environmental flow, in-stream fundamental environmental flow, in-stream targeted environmental flow are defined in the SERL. In CEHP, ecological flow, aquatic ecological flow needs, aquatic ecological base flow, riparian wetland flow needs, environment flow needs, scenery flow needs, estuary ecological flow needs and in-stream groundwater recharge needs are defined (Table 3). In terms of components, the names of the definitions are inconsistent (Table 4), but the purpose is to protect the river ecosystem in a healthy condition. Additionally, in order to facilitate management, both guidelines divide the environmental flows into two parts, which are basic (fundamental/base) flow and target flow. Considering the scope of application, the environmental flows are divided into two categories that are in-stream environmental flows and off-stream environmental flows. Through the analysis of the definitions of each term in the two standards for development of water use projects, we refer to the subject discussed in this paper as environmental flows that is the in-stream environmental flows impacted by the hydropower projects.

Table 3. The comparison of application of the two standards for development of water use projects.

Category	SERL	CEHP
In-stream environmental flow	River, Lake, Swamp	River, Riparian wetland, Scenery, Estuary, In-stream groundwater,
Off-stream environmental flow	Urban green space Urban sanitation Ecological grassland Rivers, lakes and marshes replenish water	Riparian wetland, Scenery, Estuary

Table 4. Comparison of key terms definitions of the two standards for development of water use projects.

Guideline	Main Terms	Definition
SERL	In-stream environmental flow	In order to maintain the ecological and environmental protection goals of rivers, lakes and marshes.
	In-stream fundamental environmental flow	The minimum amount of water in the river.
	In-stream targeted environmental flow	The amount of water retained in the river to maintain the ecological and environmental functions.
	Off-stream environmental flow	Artificial water supply in order to achieve certain ecological goals.
CEHP	Ecological flow	In order to ensure the flow for ecosystem in the downstream reach of hydropower project.
	Aquatic ecological flow needs	Suitable flow to guarantee the basic stability of the aquatic ecosystem in the downstream sections of hydropower projects.
	Aquatic ecological base flow	Minimum flow to guarantee the basic quality of aquatic habitat in the downstream sections of hydropower projects.

3.3. Analysis of the Calculation Process

The two documents both summarize the assessment of environmental flows into several similar and effective steps. The first step is to collect the basic data of the river. The second step is to determine the ecological objectives (sensitive ecosystems protection objectives: nature reserves, important wetlands, natural forests, rare and endangered wildlife nature reserves, natural spawning grounds, etc.). The third step is to analyze and calculate the environmental flows by different methods. The fourth step is the comprehensive analysis of environmental flows in the specific river.

In the CEHP, there is a clear formulation to calculate the environmental flows, as follows:

$$Q_{st}(t) = Max(Q_{ss}(t), Q_{sh}(t), Q_{jg}(t), Q_{hk}(t)) + Q_{hl}(t) + Q_{dx}(t) \tag{1}$$

where

$Q_{st}(t)$ is the environmental flows (m³/s);

$Q_{ss}(t)$ is the aquatic ecological flow needs (m³/s);

$Q_{sh}(t)$ is the environmental flow (m³/s) which is considered by the water quality and water environmental functions (shipping);

$Q_{jg}(t)$ is the water required for landscape and ecology (m³/s);

$Q_{hk}(t)$ is the estuary environmental flow needs (m³/s); *dx*

$Q_{hl}(t)$ is the river riparian wetland flow needs (m³/s);

$Q_{dx}(t)$ is the in-stream groundwater recharge needs (m³/s).

In the SERL, there is no deterministic equation to calculate the environmental flows. A framework is explicitly presented to guide managers on how to determine the environmental flows (Figure 1). Because the calculation of environmental flows is a complicated process, various factors need be considered at the same time. Thus, this frame diagram has a certain reference, but its operability is not strong.

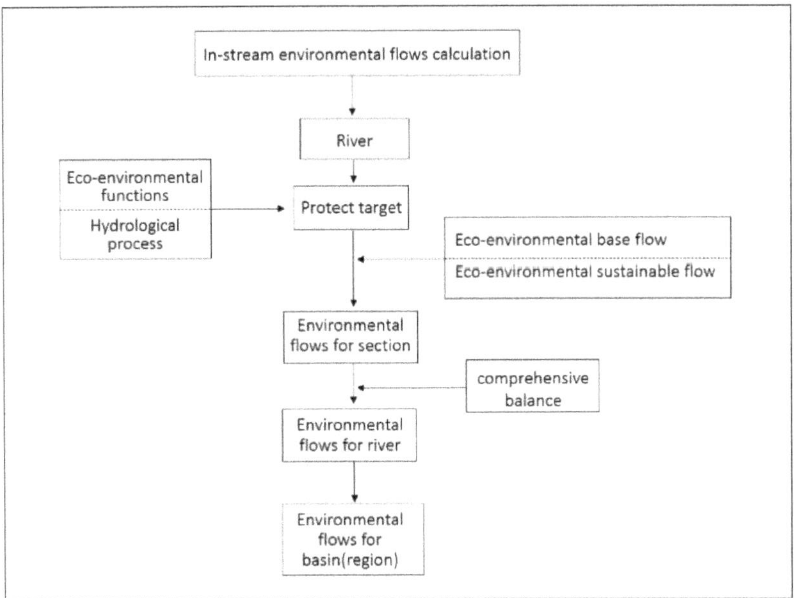

Figure 1. Framework for determining the in-stream environmental flows in SERL.

3.4. Recommended Methods

The environmental flows calculations we discuss here, according to the analysis, correspond to the in-stream environmental flows and aquatic ecological flow needs in the two standards (may vary due to different definitions of terms). Both standards mention that in the stage of method selection, it is necessary to choose the appropriate method according to different objects. Although many types of methods are recommended in the standards, there is no specific explanation on the methods selection principle (Table 5). In the process of how to determine the eco-environmental protection objectives, SERL mentions that the protection objectives should be reasonably determined according to the development and utilization of water resources in the basin. In CEHP, it is mentioned that the eco-environmental protection objectives should be reasonably determined by the downstream hydrological conditions and the aquatic ecology. Obviously, there are significant differences between the two standards in determining the protection objectives. CEHP is more specific on hydropower protection objectives and the SERL is broader to cover all rivers and lakes.

Table 5. Recommended methods of the two standards for development of water use projects.

Guideline	Objects	Methods
SERL	In-stream fundamental environmental flow	Qp [1], 7Q10, The most dry month method (MDM) [2], Flow duration curve method, Tennant, Frequency curve method, Wetted perimeter method, Habitat analysis method, River bed morphology analysis method.
	In-stream targeted environmental flow	Tennant, Frequency curve method, Habitat analysis method.
CEHP	Aquatic ecological flow needs	Wetted perimeter method, R2-CROSS method, Eco-hydraulic method, Habitat analysis method, Tennant, 7Q10.
	Aquatic ecological base flow	Wetted perimeter method, R2-CROSS method, Eco-hydraulic method, Tennant, 7Q10.

[1] Qp: This method uses different flow percentile (It depends on the water manager) as the result. [2] MDM: This method uses the driest month flow as the corresponding month result.

Additionally, during the calculating process, there are many differences in the two documents. In the SERL, the in-stream environmental flows include the in-stream fundamental environmental flow and the in-stream targeted environmental flow. The in-stream fundamental environmental flow needs to be fully assessed by the minimum value, the values during different time periods of the year and the annual value. Among them, the minimum value and the annual value are calculated by the hydrological methods, and the hydraulic method and the wet perimeter method can be selected under the conditions of sufficient data during different periods of the year. The in-stream targeted environmental flow is to maintain the water demand requirement at the normal level according to the ecosystems function corresponding to the protection target. Then, the in-stream targeted environmental flow will be calculated in different periods and as an annual value. For more accurate and reasonable results, the SERL recommends that managers could use a variety of methods to compare the results and analysis to determine in-stream environmental flow.

It is mentioned in CEHP that the calculation of aquatic ecological flow needs to consider the hydrological characteristics and aquatic ecological protection objectives. This requirement is basically the same as in SERL. It is clearly stated in CEHP that when there is no important fish spawning area in the downstream, the aquatic ecological base flow should be analyzed and calculated. When there are important fish spawning fields, it is necessary to analyze and calculate the hydrological processes required for the aquatic ecological base flow and fish breeding period. The hydraulics, ecological hydraulics, and hydrology methods are recommended for the calculation of aquatic ecological base flow. The aquatic ecological flow needs for the fish breeding period can be calculated by the habitat analysis method. Additionally, the standard requires that at least two methods be used to calculate the aquatic ecological base flow and then based on the experts to choose a better one. During the fish breeding period, the requirements of different fish should be all considered, the specific hydrological process should be calculated separately, and the maximum should be taken as the result. Finally, the water manager should consider the recommended environmental flows threshold (Table 6), according to the size of the river area, the climatic conditions of the geographical location and the current water resources of the river to determine the final environmental flows.

Table 6. Recommended environmental flows thresholds [1] for different rivers in China.

River Type		High		Medium		Low	
		Base Flow	Suitable Flow	Base Flow	Suitable Flow	Base Flow	Suitable Flow
L [2]	Nor [5]	10–20	40–50	15–25	45–55	≥25	≥60
	Sou [6]	20–30	65–80	25–35	70–80	≥35	≥80
M [3]	Nor [5]	10–15	40–50	10–25	40–55	≥25	≥55
	Sou [6]	15–30	60–70	20–35	65–75	≥35	≥75
S [4]	Nor [5]	5–10	40–45	10–20	40–50	≥20	≥50
	Sou [6]	15–25	50–60	20–30	55–65	≥30	≥65

[1] Environmental flow threshold = environmental flows/Surface available water resources × 100%. [2] L means large river (area > 100,000 km^2). [3] M means medium river (10,000 km^2 < area < 100,000 km^2). [4] S means small river (3000 km^2 < area < 10,000 km^2). [5] Nor means northern of China. [6] Sou means southern of China.

4. Weaknesses and Recommendations

Our examination of the two overarching standards governing environmental flows in China included comparisons of term definitions, calculation processes and calculation methods. Our findings of deficiencies in the two standards summarized below and can provide suggestions for water resources managers.

1. Because the standards come from two different authorities, the definition of related terms is not clear. This may lead to misunderstandings during implementation. Water managers may have errors in textual understanding when using the standards. More seriously, due to differences in understanding, there may be increased misunderstandings between stakeholders.

2. In the process of data collection, both standards mention the collection of watershed hydrology, river topography, aquatic ecology, relevant planning and research results. Although the process of evaluating environmental flows in the standards is scientific, it is difficult to evaluate and implement environmental flows according to the requirements of the standards considering the actual situation.

3. Both standards classify environmental flows into base environmental flows (minimum flows) and targeted environmental flows (variable environmental flows). The purpose of this consideration is to rationally use water resources and then create economic value under the premise of protecting ecosystems. How to trade-off the relationship between environment value and the economic value? There is no operational suggestion for water manager.

4. In the specific calculation process, both standards recommend a variety of calculation methods. The standard also clarifies how to choose the method for base environmental flow and targeted environmental flow. However, most of the methods are hydrological methods, and the results of hydrological calculations often lack consideration of ecological processes. Additionally, most of the rivers lack sufficient data, which can impede the implementation of environmental flows.

5. In the ecological process, both standards recommend habitat analysis methods and consider the relationship between hydrological-ecological responses. However, in the actual process, only the large watersheds in China currently have relevant ecological data, which brings difficulties to river managers. In CEHP, it is recommended that water managers consider the hydrological process of different fish breeding seasons. Due to the difficulty of data collection and long cycle monitoring, this is not feasible in practical work.

6. During the process of verifying the rationality of the results, both standards emphasize the need to use multiple methods for calculations and comparison of results to determine a reasonable hydrological process. This is positive but undoubtedly increases the cost of river management.

7. A reasonable range of parameters is given in the two standards, except for habitat analysis methods. Because habitat analysis methods are always based on a rigorous aquatic ecology survey and expert advice, a reasonable range of habitat parameters can be determined. Due to the many remaining problems in the history of watershed management, the habitat analysis method

is not available in most rivers in China. Moreover, much upfront capital investment is needed which cannot be used in medium or small river.

8. The unknown long-term change of climate and hydrological conditions will bring unpredictable impacts on the ecosystem, such as the invasion of non-native species caused by the change of environmental conditions (rainfall and temperature under changing conditions) that lead to the failure of traditional calculation methods (provided in the two standards) in practical water resources management.

9. Most of the environmental flow calculation methods recommended in the two standards are from foreign countries, and the portability and preconditons for some of them needs further research.

5. Conclusions

Hydropower development has brought huge economic benefits, but the complex water resources and intensive human activities have led to serious ecological and environmental problems. [2]. This also urges GoC to resolve to do some necessary remedial work, such as issuing relevant policies and encouraging relevant research. In China, some rivers have a high degree of development and utilization, and a large amount of economic and social water use. Environmental flows is one of the important management tools to maintain the ecological function of rivers and control the intensity of water resources development [26]. Strengthening the management of river and lake environmental flows is related to the overall situation of water security and realizing China's goal of an ecological civilization. Based on rich experience on environmental flows around the world [27–29], how to minimize the bad influence on the ecosystem becomes more important in the sustainable world. In this study, based on the GoC issued for the environmental flows policy to do a systematic analysis.

The two major standards for development of water use projects (SERL and CEHP) were developed in the context of the national strategy for protecting the ecosystems (ecological civilization) in order to alleviate the pressure of human activities on river ecosystems in China. They all have strict scientific logic and consider all aspects of the factors. However, as the units set up are different administrative authorities, there are certain deviations in the interpretation and implementation of different managers. Our assessment identified defects in the current guidelines and standards. At present, there are no unified provisions on the concept, connotation and evaluation index of environmental flows in the aspects of laws, regulations, policy systems and planning standards. It leads to confusion in the formulation and management of relevant policies and also has a great impact on the public's understanding and implementation of environmental flows. Therefore, unreasonable environmental flow calculations may seriously affect the sustainable development of the ecosystem [30].

Weakness with SERL and CEHP need to be addressed in several ways. Aim at the defects of the basic concepts and recommend calculation methods in standards. First of all, government needs to encourage research institutions to conduct systematic research on the theoretical problems of environmental flows and promote the development of China's environmental flow research technology to the direction of methodology. Let each industry standardize the basic concept of environmental flows. As for the calculation method, the original method introduced from abroad should be changed. Researchers are encouraged to discriminate between methods, assess their applicability and choose the most appropriate method. Most of the previous [31–33] studies have carried out environmental flow calculations for specific projects. How to reduce the uncertainty and improve the rationality and accuracy of the evaluation results is worth further discussion.

The evaluation and implementation of environmental flows is a complex problem that needs to consider not only the rationality of scientific research, but also the feasibility of engineering [9,34]. As it involves economic benefits, it also needs to consider social and economic benefits. In general, the GoC needs to promote the establishment of a complete set of standards for development of water use projects system, unified connotation and understanding. The relevant government authorities need to establish sound policies to ensure the environmental flow of rivers and lakes in accordance with the standards. Promoting the hierarchical management of river and lake environmental flows guarantees and clarifies the responsibilities of authorities at all levels for river and lake environmental flow implementation. We should give priority to ensuring the environmental flow of rivers and lakes and strengthen the unified allocation of water resources in river basins. For different rivers, the requirements of environmental flows need to be considered in a targeted way and water resources need to be allocated in a unified way according to local conditions. The research and implementation of environmental flows in rivers and lakes is a complex and comprehensive work involving multiple authorities, fields and links. Processes for stakeholder engagement should be established. The involvement of society at large in decisions on water resources management is an inevitable consequence of social development, and it is important that the general public be encouraged to participate in such decision-making processes. Eco-environmental impact assessments of hydropower projects are significant and should be comprehensive and consider all of the short- and long-term benefits and drawbacks. Therefore, it is urgent to study river environmental flows assessment frameworks and establish a set of general assessment processes in order to achieve differentiated management for different rivers.

Author Contributions: Conceptualization, M.W. and M.E.M.; Funding acquisition, X.Z.; Methodology, M.W.; Resources, A.C.; Supervision, X.Z. and M.E.M.; Writing—original draft, M.W.; Writing—review and editing, M.W., A.C., X.Z. and M.E.M. All authors have read and agreed to the published version of the manuscript.

Funding: This research was funded by the National Key Research and Development Program (Grant No. 2019YFC0409000) approved by the Ministry of Science and Technology of China and National natural science foundation of China (51420105014). The first author was supported by fellowship from the China Scholarship Council for her visit to IHE Delft Institute for Water Education, Delft, The Netherlands.

Conflicts of Interest: The authors declare no conflict of interest.

References

1. Li, X.; Chen, Z.; Fan, X.; Cheng, Z. Hydropower development situation and prospects in China. *Renew. Sustain. Energy Rev.* **2018**, *82*, 232–239. [CrossRef]

2. Xingang, Z.; Lu, L.; Xiaomeng, L.; Jieyu, W.; Pingkuo, L. A critical-analysis on the development of China hydropower. *Renew. Energy* **2012**, *44*, 1–6. [CrossRef]

3. Chen, A.; Wu, M. Managing for Sustainability: The Development of Environmental Flows Implementation in China. *Water* **2019**, *11*, 433. [CrossRef]

4. Benjankar, R.; Jorde, K.; Yager, E.M.; Egger, G.; Goodwin, P.; Glenn, N.F. The impact of river modification and dam operation on floodplain vegetation succession trends in the Kootenai River, USA. *Ecol. Eng.* **2012**, *46*, 88–97. [CrossRef]

5. Wang, Y.; Zhang, N.; Wang, D.; Wu, J.; Zhang, X. Investigating the impacts of cascade hydropower development on the natural flow regime in the Yangtze River, China. *Sci. Total Environ.* **2018**, *624*, 1187–1194. [CrossRef]

6. Chen, A.; Wu, M.; Chen, K.; Sun, Z.Y.; Shen, C.; Wang, P.Y. Main issues in research and practice of environmental protection for water conservancy and hydropower projects in China. *Water Sci. Eng.* **2016**, *9*, 312–323. [CrossRef]

7. Yu, B. The ecological damage compensation for hydropower development based on trade-offs in river ecosystem services. In Proceedings of the IOP Conference Series: Earth and Environmental Science, Ordos, China, 14–16 April 2017; Volume 64, p. 12047.

8. Hansen, M.H.; Li, H.; Svarverud, R. Ecological civilization: Interpreting the Chinese past, projecting the global future. *Glob. Environ. Chang.* **2018**, *53*, 195–203. [CrossRef]

9. Bejarano, M.D.; Sordo-Ward, A.; Gabriel-Martin, I.; Garrote, L. Tradeoff between economic and environmental costs and benefits of hydropower production at run-of-river-diversion schemes under different environmental flows scenarios. *J. Hydrol.* **2019**, *572*, 790–804. [CrossRef]

10. Tranmer, A.W.; Marti, C.L.; Tonina, D.; Benjankar, R.; Weigel, D.; Vilhena, L.; McGrath, C.; Goodwin, P.; Tiedemann, M.; Mckean, J.; et al. A hierarchical modelling framework for assessing physical and biochemical characteristics of a regulated river. *Ecol. Model.* **2018**, *368*, 78–93. [CrossRef]

11. Zhang, Y.; Arthington, A.H.; Bunn, S.E.; Mackay, S.; Xia, J.; Kennard, M. Classification of Flow regimes for environmental flow assessment in regulated rivers: The huai river basin, china. *River Res. Appl.* **2012**, *28*, 989–1005. [CrossRef]

12. Popa, F.; Dumitran, G.E.; Vuta, L.I.; Tica, E.I.; Popa, B.; Neagoe, A. Impact of the ecological flow of some small hydropower plants on their energy production in Romania. *J. Phys. Conf. Ser.* **2020**, *1426*, 12043. [CrossRef]

13. Lu, S.; Dai, W.; Tang, Y.; Guo, M. A review of the impact of hydropower reservoirs on global climate change. *Sci. Total Environ.* **2020**, *711*, 134996. [CrossRef] [PubMed]

14. Webb, J.A.; Watts, R.J.; Allan, C.; Conallin, J.C. Adaptive Management of Environmental Flows. *Environ. Manag.* **2018**, *61*, 339–346. [CrossRef] [PubMed]

15. Ban, X.; Diplas, P.; Shih, W.; Pan, B.; Xiao, F.; Yun, D. Impact of Three Gorges Dam operation on the spawning success of four major Chinese carps. *Ecol. Eng.* **2019**, *127*, 268–275. [CrossRef]

16. Xue, J.; Gui, D.; Zhao, Y.; Lei, J.; Zeng, F.; Feng, X.; Mao, D.; Shareef, M. A decision-making framework to model environmental flow requirements in oasis areas using Bayesian networks. *J. Hydrol.* **2016**, *540*, 1209–1222. [CrossRef]

17. Baoligao, B.; Xu, F.; Chen, X.; Wang, X.; Chen, W. Acute impacts of reservoir flushing on fishes in the Yellow River. *J. Hydro-Environ. Res.* **2016**, *13*, 26–35. [CrossRef]

18. Yan, Y.; Yang, Z.; Liu, Q.; Sun, T. Assessing effects of dam operation on flow regimes in the lower Yellow River. *Procedia Environ. Sci.* **2010**, *2*, 507–516. [CrossRef]

19. Tan, G.; Yi, R.; Chang, J.; Shu, C.; Yin, Z.; Han, S.; Feng, Z.; Lyu, Y. A new method for calculating ecological flow: Distribution flow method. *AIP Adv.* **2018**, *8*, 45118. [CrossRef]

20. Kumar, A.U. Assessment of environmental flows using hydrological methods for Krishna River, India. *Adv. Environ. Res.* **2018**, *7*, 161–175.

21. China MOWR. *Specification for Calculation of Environmental Flow in Rivers and Lakes*; China MOWR: Beijing, China, 2015.

22. China NEA. *Code for Calculation of Ecological Flow of Hydropower Projects*; China NEA: Beijing, China, 2016.

23. Zhang, W.; Peng, H.; Jia, Y.; Ni, G.; Yang, Z.; Zeng, Q. Investigating the Simultaneous Ecological Operation of Dam Gates to Meet the Water Flow Requirements of Fish Spawning Migration. *Pol. J. Environ. Stud.* **2019**, *28*, 1967–1980. [CrossRef]

24. Li, W.; Chen, Q.; Cai, D.; Li, R. Determination of an appropriate ecological hydrograph for a rare fish species using an improved fish habitat suitability model introducing landscape ecology index. *Ecol. Model.* **2015**, *311*, 31–38. [CrossRef]

25. Yi, Y.; Tang, C.; Yang, Z.; Chen, X. Influence of Manwan Reservoir on fish habitat in the middle reach of the Lancang River. *Ecol. Eng.* **2014**, *69*, 106–117. [CrossRef]

26. Arthington, A.H.; Bunn, S.E.; Poff, N.L.; Naiman, R.J. The challenge of providing environmental flow rules to sustain river ecosystems. *Ecol. Appl.* **2006**, *16*, 1311–1318. [CrossRef]

27. Acreman, M.C.; Ferguson, A.J.D. Environmental flows and the European Water Framework Directive. *Freshwater Biol.* **2010**, *55*, 32–48. [CrossRef]

28. Hecht, J.S.; Lacombe, G.; Arias, M.E.; Dang, T.D.; Piman, T. Hydropower dams of the Mekong River basin: A review of their hydrological impacts. *J. Hydrol.* **2019**, *568*, 285–300. [CrossRef]

29. Yi, Y.; Cheng, X.; Yang, Z.; Wieprecht, S.; Zhang, S.; Wu, Y. Evaluating the ecological influence of hydraulic projects: A review of aquatic habitat suitability models. *Renew. Sustain. Energy Rev.* **2017**, *68*, 748–762. [CrossRef]

30. Pang, A.; Li, C.; Sun, T.; Yang, W.; Yang, Z. Trade-Off Analysis to Determine Environmental Flows in a Highly Regulated Watershed. *Sci. Rep.* **2018**, *8*, 1–11. [CrossRef] [PubMed]

31. Książek, L.; Woś, A.; Florek, J.; Wyrębek, M.; Młyński, D.; Wałęga, A. Combined use of the hydraulic and hydrological methods to calculate the environmental flow: Wisloka river, Poland: Case study. *Environ. Monit. Assess.* **2019**, *191*, 254. [CrossRef] [PubMed]

32. Beaussier, T.; Caurla, S.; Bellon-Maurel, V.; Loiseau, E. Coupling economic models and environmental assessment methods to support regional policies: A critical review. *J. Clean. Prod.* **2019**, *216*, 408–421. [CrossRef]

33. Kuriqi, A.; Pinheiro, A.N.; Sordo-Ward, A.; Garrote, L. Influence of hydrologically based environmental flow methods on flow alteration and energy production in a run-of-river hydropower plant. *J. Clean. Prod.* **2019**, *232*, 1028–1042. [CrossRef]

34. Kuriqi, A.; Pinheiro, A.N.; Sordo-Ward, A.; Garrote, L. Flow regime aspects in determining environmental flows and maximising energy production at run-of-river hydropower plants. *Appl. Energy* **2019**, *256*, 113980. [CrossRef]

Article

Analysis of an Ecological Flow Regime during the Ctenopharyngodon Idella Spawning Period Based on Reservoir Operations

Jie Li [1,2], Hui Qin [1,2,*], Shaoqian Pei [1,2], Liqiang Yao [3], Wei Wen [4], Liang Yi [4], Jianzhong Zhou [1,2] and Lingyun Tang [1,2]

[1] School of Hydropower and Information Engineering, Huazhong University of Science and Technology, Wuhan 430074, China; jieli94@hust.edu.cn (J.L.); a825299702@163.com (S.P.); jz.zhou@hust.edu.cn (J.Z.); M201873751@hust.edu.cn (L.T.)
[2] Hubei Key Laboratory of Digital Valley Science and Technology, Wuhan 430074, China
[3] Changjiang River Scientific Research Institute of Changjiang Water Resources Commission, Wuhan 430015, China; liqiang85327@gmail.com
[4] Central-southern Safety and Environment Technology Institute Co LTD, Wuhan 430051, China; lss864@foxmail.com (W.W.); stbcyl@163.com (L.Y.)
* Correspondence: hqin@hust.edu.cn

Received: 15 June 2019; Accepted: 24 September 2019; Published: 29 September 2019

Abstract: The study of fish habitats is important for us to better understand the impact of reservoir construction on river ecosystems. Many habitat models have been developed in the past few decades. In this study, a fuzzy logic-based habitat model, which couples fuzzy inference system, two-dimensional laterally averaged hydrodynamic model, and two-dimensional shallow water hydrodynamic model, is proposed to identify the baseline condition of suitable habitat for fish spawning activities. The proposed model considers the reservoir and the downstream river channel, and explores the comprehensive effects of water temperature, velocity, and water depth on habitat suitability. A real-world case that considers the *Ctenopharyngodon idella* in the Xuanwei Reservoir of Qingshui River is studied to investigate the effect of in- and outflow of reservoir on fish habitat and the best integrative management measure of the model. There were 64 simulations with different reservoir in- and outflows employed to calculate the weighted usable area and hydraulic habitat suitability. The experimental results show that the ecological flow for *Ctenopharyngodon idella* spawning can satisfy the basic demand when the reservoir inflow is greater than 60 m^3/s and the reservoir outflow is greater than 100 m^3/s. The habitat ecological suitability is the best when the reservoir outflow is 120 m^3/s. A more reasonable and reliable ecological flow range can be obtained based on the habitat model in this paper, which provides the best scenario for water resources planning and management in the Qingshui River Basin.

Keywords: water temperature; fuzzy logic; habitat model; spawning period; ecological flow

---

1. Introduction

Reservoirs play key roles in the social and economic development as a water conservancy project that prevents or controls flood and regulates water flow [1], however, the construction of a reservoir changes natural flow regimes of a river and has an impact on the river. These effects are mainly manifested as follows: (1) blocking the migration path of fish to make it impossible to complete its life history process, resulting in a decline in genetic diversity of fish populations [2]; (2) affecting sediment transport, resulting in sedimentation at the bottom of the reservoir and river blockage [3]; (3) affecting nitrogen, phosphorus, and other substances migration, thereby inducing eutrophication [4]; (4) leading to a decrease in the temperature of the outflow during the spawning period of the fish which delays the breeding season, and thus affects the growth of the individual born in the current year. Furthermore, the resulting discharge process changes the river characteristics and related fish habitats quantity and quality [5]. As a result, reservoir construction may cause a decrease in the biological abundance of fish species and even leads to the disappearance of some spawning grounds [6]. With the rapid development of the economy, the problem of energy shortage is becoming more and more serious. Hydropower is attracting more and more attention as a clean and sustainable energy source. In recent years, more and more reservoirs have been built. According to statistics, more than 100,000 reservoirs have been built in China with more than 100,000 m^3 of storage capacity [7,8]. Watershed water resources management and planning and the development of rational flow plans also require in-depth research and exploration.

Fish are a common indicator of aquatic habitat richness [9,10]. Their habitats include the waters necessary to complete the entire life history process, such as spawning grounds, feeding grounds, wintering grounds, and migratory passages connecting waters of different life stages [11,12]. Fish habitats not only provide living space for fish, but also provide all environmental factors such as water temperature, substrate, flow rate, pH value, and dissolved oxygen that satisfy the survival, growth, and reproduction of fish. Fish habitats are mainly affected by the hydrological conditions of the river, where flow and water temperature are one of the two main important factors affecting fish spawning [13]. Thermal stratification is common in narrow channel reservoirs and greatly influences the aquatic animal habitat environment [14]. It causes a significant difference between the reservoir outflow temperature and the inflow temperature [15]. The hysteresis of the water temperature in the downstream of the thermal stratification reservoir is affected to some extent by the discharge. Meanwhile, the discharge of the reservoir also affects the flow velocity and water depth distribution of the downstream river channel [16]. Therefore, it is necessary to predict the quality and quantity of fish habitats by simulating the spatial and temporal distribution of various hydraulic factors such as water temperature, velocity, and water depth in the reservoir and the downstream river. The development of a rational watershed planning and management plan (develop a suitable ecological flow scenarios) is significant for the survival, reproduction, and development of fish [17].

Recently, several classical approaches were used to estimate ecological flow, including hydrologic [18], hydraulic [19], and a habitat suitability modeling approach [20]. The hydrological approach determines ecological flow based on historical hydrological data, of which one representative method is the Tennant method [21]. The hydraulic approach determines the ecological flow according to the wetted perimeter of the cross section of the river [22]. Hydrological and hydraulic approaches are favored because of their simplicity and ease of calculation, but both of them lack the biological mechanisms and biological requirements. Fortunately, the habitat suitability modeling approach combines the knowledge of hydraulics and biology to establish the relationship between habitat and hydraulic factors [23]. It has certain advantages in the evaluation of ecological flow and has attracted more and more researchers' attention. The most classical habitat suitability modeling approach is the instream flow incremental methodology (IFIM) and its physical habitat simulation component (PHABSIM) [24], which includes the suitability of habitat target species to a series of hydraulic factors such as flow velocity and water depth to build a habitat suitability index model. The habitat suitability modeling approach provides the best range of hydraulic factors such as flow for habitat target species, which has certain reference significance for guiding reservoir operation [25]. In past decades, habitat suitability modeling approaches have been applied to rainbow trout *Oncorhynchus mykiss* at the Colorado River [26], brown trout *Salmo trutta* at the Calore Irpino River [27], and Indian Carp fish [28], which have successively assessed the relationship between habitat weighted usable area and outflow. The computer aided simulation model for instream flow requirements (CASiMiR) is a habitat simulation tool for aquatic organisms with a focus on fish and macroinvertebrates [29], which uses a multivariate fuzzy logic approach to link abiotic attributes with habitat requirements of aquatic species, resulting in a habitat suitability index (HSI) [29]. The fuzzy model considers the uncertainties of habitat hydraulics variables and enables expressing nonlinear relationships between habitat hydraulic variables in a transparent manner [30]. It uses the three physical characteristics of rivers that allow for the determination of the quality of habitats for fish species, i.e., velocity, water depth, and water temperature [31]. Each parameter is classified by an overlapping membership function described by a fuzzy language. Since the boundaries of two consecutive fuzzy sets are overlapping, an object may partially belong to two consecutive fuzzy sets [32]. The expert knowledge is embedded in fuzzy if–then rules to determine the relationship between these physical parameters and biological responses, and to define fuzzy rules in combination. It uses language descriptions such as "low", "moderate", and "high" for the quantification of hydraulics variables and combines the knowledge of aquatic ecology experts to translate these descriptions into a data processing framework [33]. In recent years, fuzzy models such as a fuzzy logic-based *Spinibarbus hollandi* habitat suitability model [34], data-driven fuzzy habitat suitability models for brown trout [32], and a fuzzy habitat model based on Chinese sturgeon (*Acipenser sinensis*) [35] have been used to calculate habitat ecological flow. In the same way, this study uses fuzzy models to explore the effects of water temperature, velocity, water depth, and other hydraulic factors on *Ctenopharyngodon idella* habitat. Through the simulation results of habitat suitable ecological flow, the best flow management scenario suitable for the study area is given. The main contributions of this study are as follows:

1. We consider the thermal stratification of reservoirs and analyze the important influence of the hysteresis of the water temperature on the spawning of *Ctenopharyngodon idella* in habitats.
2. A fuzzy logic-based habitat suitability model is established to evaluate the influence of hydraulic factors such as water temperature, velocity, and water depth on the suitability of a *Ctenopharyngodon idella* habitat.
3. By simulating a variety of different reservoir inflow and outflow conditions, the functional relationship between flow and habitat suitability is established, which provides the best management plan for water resources planning and construction.

The remaining parts of this paper are arranged as follows: In Section 2, we introduce the study area and target ecological species. In Section 3, we describe the construction of hydraulic habitat suitability (HHS) for target ecological species, a method for evaluating the weighted usable area (WUA) based on hydraulic simulation and the fuzzy logic-based habitat model. The results analysis and discussion are presented in Section 4. Finally, in Section 5, we state our conclusion for this study.

2. Study Area and Species

2.1. Study Area

The Qingshui River, located in the upper reaches of the Yuanjiang River, belongs to the Yangtze River system in China. It is located in the southeast of Guizhou Province and southwest of Hunan Province, with a drainage area of 15,000 km^2. Hills and mountainous regions compose most of the Qingshui River Basin. The basin has abundant rainfall, warm climate, rich oil, and diverse types of aquatic habitats. The Xuanwei reservoir is located in the upper reaches of the mainstream of the Qingshui River, with a catchment area of 1337 km^2. The design parameters of the reservoir include a normal water level of 681 m, a dead water level of 657 m, a total storage capacity of 139.80 million m^3, and an average annual storage flow of 903.39 million m^3. Among the fish in the study area, the largest number of cypriniformes (34 species) accounted for 65.4% of the total number of fish species by the scene investigation and sampling in the field, followed by 17.3% perciformes (9 species), 15.4% siluriformes (8 species), and 1.9% synbranchiformes (1 species). There is a spawning ground from the reservoir to the downstream river channel with a length of 3 km. The main fish species in the spawning ground is *Ctenopharyngodon idella* which produces drifting eggs. The bottom of the riverbed is an open rocky beach formed by pebbles and river sand. By considering the 3 km river channel downstream of the reservoir as a study case, the influence of hydraulic factors on fish spawning is simulated and analyzed. The location of the study area is shown in Figure 1.

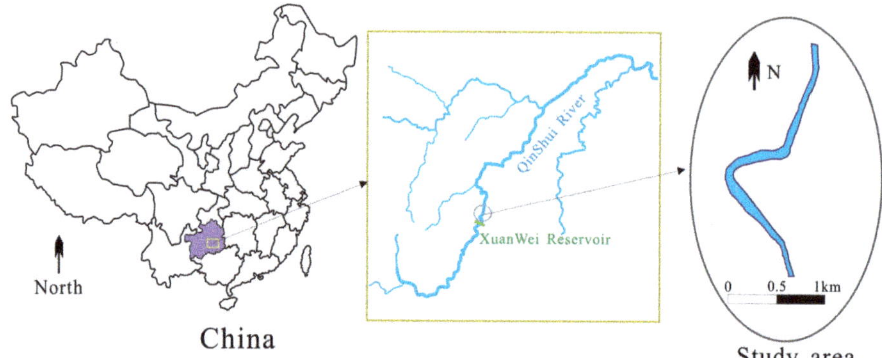

Figure 1. Study area.

2.2. Ctenopharyngodon idella (C. idella)

Ctenopharyngodon idella (*C. idella*) is a freshwater species native mainly to China. It plays an important role in aquaculture, being the second highest in world fish production. *C. idella*, as a food fish or as a biological controller of aquatic vegetation, has been introduced into more than 100 countries in various continental regions, including East Asia, North America, and Europe [36]. The natural reproduction of *C. idella* has certain demands for water temperature [37], landform, hydrology, and hydrodynamic conditions, especially related to flow and water temperature. *C. idella* migrates from the middle and lower reaches of the Qingshui River to the upstream for spawning and breeding during the summer flood season. In the autumn, it swims to the deep waters of the lower reaches of the Qingshui River for overwintering. Its spawning occurs during the flood period from mid-May to mid-July each year when the water temperature reaches 18 °C. According to the survey data of *C. idella* in the Qingshui River, it is most suitable for adult *C. idella* spawning when the water temperature is between 21 °C and 24.5 °C, the velocity is between 0.3 m/s and 0.9 m/s, and the water depth is between 3 m and 7.5 m. This migratory fish is the target species of reservoir ecological scheduling because the construction of the dam greatly threatens its survival. The ecological scheduling of such reservoirs generally involves releasing a suitable flow for fish migration during the migratory period of the fish and combining certain fish facilities to help the fish reach the spawning ground [38]. Therefore, the natural reproduction of *C. idella* is required to incorporate the water environment requirements of the spawning habitat into the reservoir scheduling objective. The construction of multi-objective ecological scheduling is one of the ways to effectively protect the river's important economic fish species resources [39]. However, a key step in the implemention of a reservoir ecological scheduling scenario that takes into account the conservation of rare aquatic organisms is to determine the ecological water demand and to clarify the relationship between reservoir scheduling and the suitability of water environment for protecting aquatic habitats.

Practical topographic data of the Xuanwei reservoir and the downstream river channel was acquired from the Hydraulic Exploration Design Institute of the Guizhou Province. Meteorological data such as temperature, relative humidity, wind speed, and cloudage come from the Duyun meteorological station, since 1970. Hydraulic data such as reservoir inflow and water level are provided by the Xiasi hydrological station. The *C. idella* information of the Qingshui River comes from the sampling data of the past five years in the field. According to the data of the Duyun hydrological station in the study area, the data of reservoir inflow and water temperature of typical years are obtained as follows: wet year (75% guarantee rate of runoff), normal year (50% guarantee rate of runoff), and dry year (25% guarantee rate of runoff). According to the inflow data of the reservoir, the annual large-scale flow is mainly concentrated from mid-May to early-July. The highest water temperature in the year occurred in August, and the lowest water temperature occurred in February. According to the survey, the spawning period of *C. idella* in the study area is mostly concentrated from mid-May to mid-July. The suitable flow range and suitable water temperature range determine the spawning period of the fish. The annual reservoir inflow rate and water temperature of the study area is shown in Figure 2.

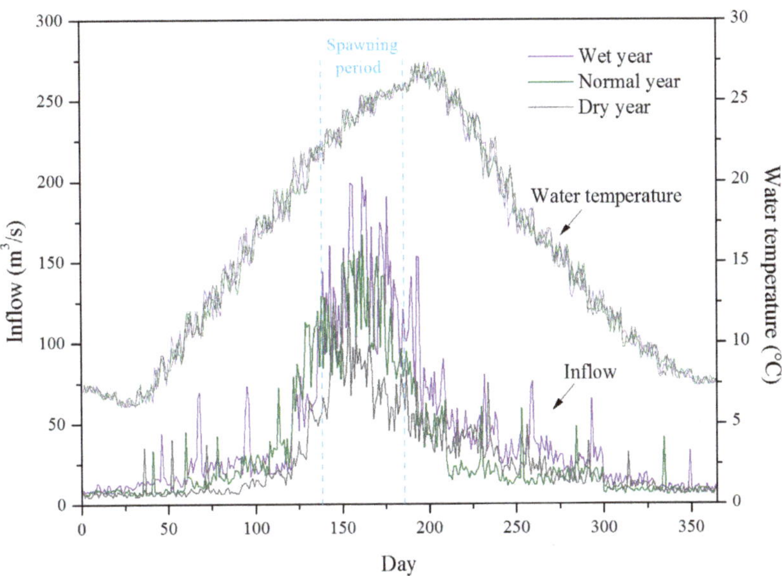

Figure 2. A graph of annual reservoir inflow and water temperature in the study area.

C. idella has an important economic value and is the target species in the study area for this study. On the basis of constructing the hydrodynamic mathematical model and habitat suitability evaluation model of the spawning ground in the downstream of the Xuanwei Reservoir, the correlation between the key indicators of reservoir scheduling and the suitability of *C. idella* spawning habitat have been studied. The research results provide the basis for constructing reservoir ecological scheduling of *C. idella* protection

3. Methods

3.1. Model Framework

In this study, the *C. idella* in the study area was used as the target species to study the flow and water temperature effects of *C. idella* during spawning. This study specifically analyzed the following three main factors affecting the spawning of *C. idella*: velocity, water depth, and water temperature. Two hydrodynamic models were used to simulate the hydraulic conditions of reservoir and the downstream habitats of reservoir. MIKE21 simulated the spatial distribution of velocity and water depth under different flows in the habitat. CE-QUAL-W2 simulated the water temperature stratification of reservoirs under different inflows and outflows. The simulation results of velocity, water depth, and water temperature were used to construct a fuzzy model [31,40] to comprehensively analyze the effects of the three factors on *C. idella* spawning. The simulation results of the fuzzy logic-based habitat suitability model were analyzed and evaluated by WUA and HHS, and the best flow management scenario in the study area was finally obtained. The overall framework of the habitat model is shown in Figure 3.

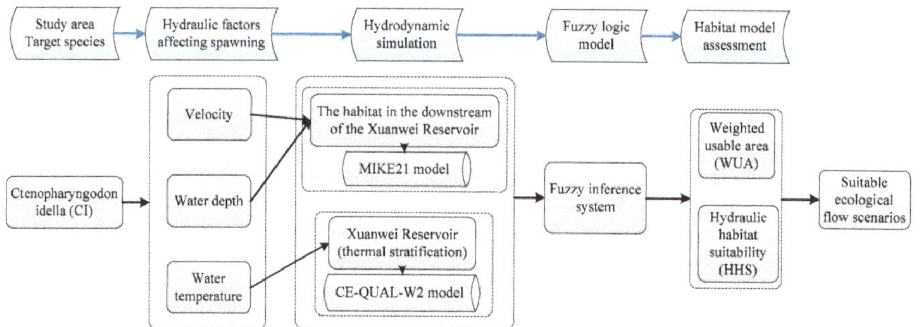

Figure 3. A graph of habitat model framework.

3.2. Hydrodynamic Simulation

The hydrodynamic modeling was divided into two parts. The first part of the reservoir was simulated by CE-QUAL-W2 model, which mainly explored the influence of water temperature change caused by reservoir thermal stratification on downstream habitat. The second part of the downstream channel was simulated by MIKE21 model, which mainly explored the velocity and water depth distribution of the downstream habitat.

3.2.1. CE-QUAL-W2 Model

The CE-QUAL-W2 model is a two-dimensional (longitudinal/vertical) laterally averaged hydrodynamic model for surface water systems [41]. The model is based on the assumption that lateral homogeneity and the flow in a reservoir with a distinct flow direction can be computed from the laterally integrated Navier–Stokes equation [42]. This is particularly suited for relatively long and narrow water bodies exhibiting longitudinal and vertical water temperature gradients. The W2 model was successfully applied to stratified water systems, including reservoirs, lakes, and estuaries.

On the basis of the Boussinesq assumption, the model integrates the homogeneous continuity equation, the momentum equation, and the caloric equation of the three-dimensional turbulent buoyant flow to obtain the equations of the two-dimensional laterally averaged temperature model as follows:

$$\frac{\partial}{\partial x}(Bu) + \frac{\partial}{\partial z}(Bw) = 0 \tag{1}$$

$$\frac{\partial}{\partial t}(Bu) + u\frac{\partial}{\partial x}(Bu) + w\frac{\partial}{\partial z}(Bw) = \\ 2\frac{\partial}{\partial x}\left(Bv_e\frac{\partial u}{\partial x}\right) + \frac{\partial}{\partial z}\left(Bv_e\frac{\partial u}{\partial z}\right) + \frac{\partial}{\partial z}\left(Bv_e\frac{\partial w}{\partial x}\right) - \frac{B}{\rho_a}\frac{\partial p}{\partial x} \tag{2}$$

$$\frac{\partial}{\partial t}(Bw) + u\frac{\partial}{\partial x}(Bw) + w\frac{\partial}{\partial z}(Bw) = \\ 2\frac{\partial}{\partial z}\left(Bv_e\frac{\partial w}{\partial z}\right) + \frac{\partial}{\partial x}\left(Bv_e\frac{\partial w}{\partial x}\right) + \frac{\partial}{\partial x}\left(Bv_e\frac{\partial u}{\partial z}\right) - \frac{B}{\rho_a}\frac{\partial p}{\partial z} + \beta\Delta T g B \tag{3}$$

$$\frac{\partial}{\partial t}(BT) + u\frac{\partial}{\partial x}(BT) + w\frac{\partial}{\partial z}(BT) = \frac{\partial}{\partial x}\left(\frac{Bv_e}{\sigma_T}\frac{\partial T}{\partial x}\right) + \frac{\partial}{\partial z}\left(\frac{Bv_e}{\sigma_T}\frac{\partial T}{\partial z}\right) \tag{4}$$

where B is the river width, u is the longitudinal velocity, w is the vertical velocity, v is the coefficient of kinematic viscosity, v_t is the turbulent viscosity coefficient, $v_e = v + v_t$ is the effective viscosity coefficient, T is the water temperature, T_a is the reference temperature of water, $\Delta T = T - T_a$, ρ is the water density, ρ_a is the reference density of water, β is the coefficient of thermal expansion, and σ_T is the Prandtl number of temperature.

The Xuanwei Reservoir was represented as 63 longitudinal segments with lengths of 250 m and 48 vertical layers with thicknesses of 1 m according to the topographic data recorded in 2018. The CE-QUAL-W2 turbulence closure algorithm was used. There is only one inflow entrance upstream of the mainstream and there are no other inflow entrances from the tributaries. The boundary data was input with a daily resolution, including reservoir inflow and temperature, outflow, air temperature, wind speed and direction, solar radiation, cloud cover, precipitation, and evaporation. Reservoir grid model is shown in Figure 4.

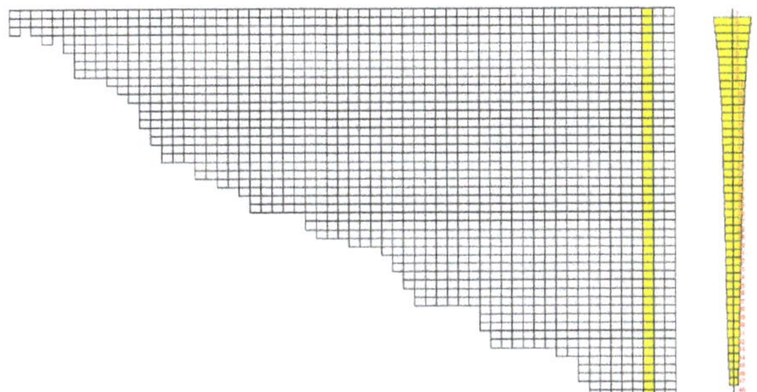

Figure 4. Reservoir grid model.

3.2.2. MIKE21 Model

The fish habitat in the downstream of the Xuanwei Reservoir was simulated by MIKE21, and the spatial distribution of velocity and water depth under different outflows was calculated. The hydrodynamic module is based on the solution of two-dimensional shallow water equations obtained from the Navier–Stokes equations, which are integrated over the water depth including the depth and horizontal momentum equation of the flow just as follows [43]:

$$\frac{\partial T}{\partial t} + \frac{\partial U}{\partial x} + \frac{\partial V}{\partial y} = S \tag{5}$$

where t indicates the time; x and y are Cartesian coordinates; T, U, V, and S indicate vectors of the conserved variables, fluxes in the x- and y-direction, and source terms, respectively.

$$T = \begin{bmatrix} h \\ hu \\ hv \end{bmatrix}, \; U = \begin{bmatrix} hu \\ hu^2 + gh^2/2 \\ huv \end{bmatrix}, \; V = \begin{bmatrix} hv \\ huv \\ hv^2 + gh^2/2 \end{bmatrix}, \; S = \begin{bmatrix} 0 \\ gh\left(S_{0x} - S_{fx}\right) \\ gh\left(S_{0y} - S_{fy}\right) \end{bmatrix} \tag{6}$$

where h is the water depth; $g = 9.81$ m/s^2 is the gravity acceleration; u and v is the depth-averaged velocity components in the x-direction and y-direction, respectively; S_{0x} and S_{0y} are the bed slopes in the x-direction and y-direction, respectively; S_{fx} and S_{fy} are the friction slopes in the x-direction and y-direction, respectively.

Tests of numerical experiments and results of the practical applications showed that this model can simulate the water flow movement over complex terrain accurately [44]. The MIKE21 model was used to divide the downstream habitat of the Xuanwei Reservoir into 796 triangular grids, with a maximum area of 523.49 m^2 and a minimum area of 130.33 m^2. The measured outflow time curve of the reservoir was taken as the upper boundary of the model. The measured water level time curve at 3 km from the reservoir was taken as the lower boundary of the model. The model simulated the velocity and water depth distribution in the habitat under different outflows. Habitat grid model is shown in Figure 5.

3.3. Fuzzy Logic-Based Habitat Modeling

Fuzzy logic can copy the judgment of the uncertainty concept and the way of reasoning thinking of the human brain by using fuzzy sets and fuzzy rules [45,46]. Meanwhile, the model considers the inherent uncertainty of ecological variables and enables expressing nonlinear relations between ecological variables in a transparent way [47,48]. Fuzzy species distribution models transform fuzzy modeling into a technique suitable for species distribution modeling, which can reflect the specific characteristics of ecological problems. This study was based on fuzzy inference for habitat suitability (HS) simulation. The inference process consisted of three steps as follows: (1) fuzzification, in which fuzzifier is used to convert crisp input to fuzzy set using membership functions; (2) fuzzy inference, in which a fuzzy inference system is applied to convert fuzzy input sets to fuzzy output sets; and (3) defuzzification, in which fuzzy output sets are converted to crisp output by a defuzzifier. The modeling for the fuzzy inference process is shown in Figure 6.

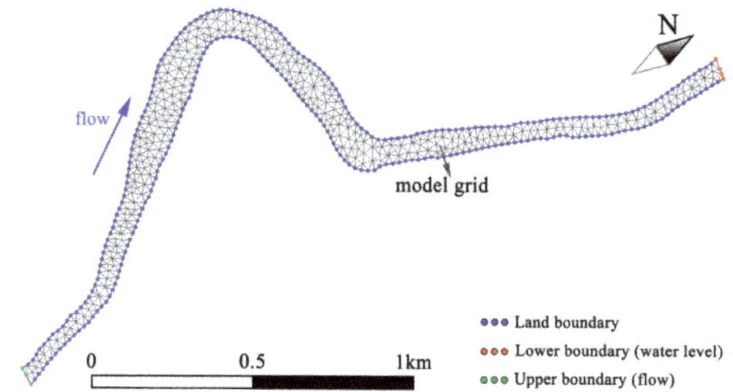

Figure 5. Habitat grid model.

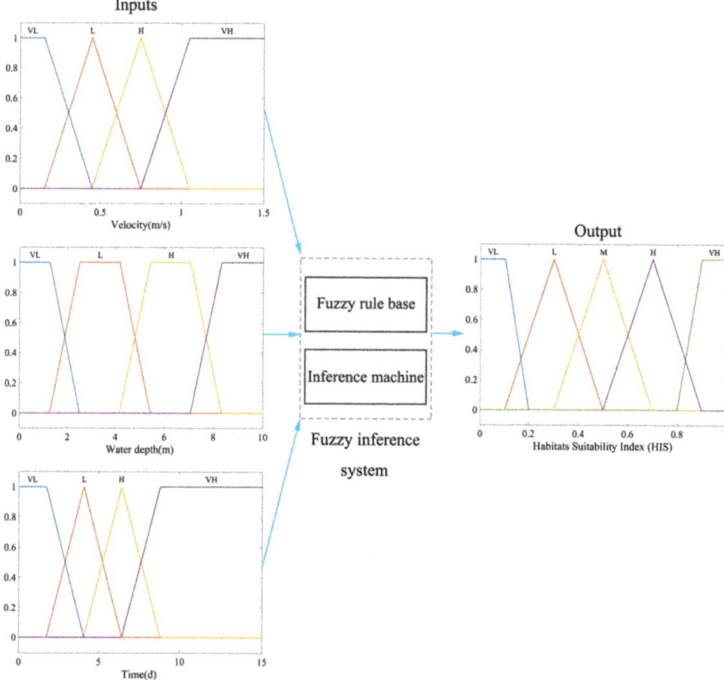

Figure 6. The modeling for the fuzzy inference process.

3.3.1. Fuzzification and Defuzzification

(1)   Fuzzification

In this study, for the habitat modeling, the membership functions of trapezoidal and triangular shapes are used to convert crisp input values of velocity, water depth, and water temperature lag time to fuzzy sets expressed by one or more linguistic values with membership degrees. Fuzzy sets are usually expressed by linguistic description such as "very low", "low", "high", and "very high". The membership functions of the fuzzy sets have overlapping boundaries and the linguistic statement "the water temperature lag time is quite short but tending to be long" can be translated into a time which has a membership degree of 0.8 to the fuzzy set "high" and of 1 to the fuzzy set "very high".

(2)   Fuzzy Inference

A fuzzy inference system is composed of a fuzzy rule base and an inference machine. The fuzzy rule consists of if–then fuzzy rules, such as "if x IS A AND y IS B, THEN z IS C", by links the input variables to output. In the same way, the fuzzy rules as a total of 64 (velocity with four linguistic values, water depth with four linguistic values, and water temperature lag time with four linguistic values) describing HS for the spawning period of *C. idella* are created based on literature reviews and expert experience [49–51]. The fuzzy rules for the spawning of *C. idella* is shown in Table 1. The water temperature lag time is used as an evaluation index of the fuzzy model. The water temperature lag time indicates the time required for the outflow water temperature to reach the same water temperature when the reservoir inflow water temperature (21 °C) is suitable for *C. Idella* to spawn. By using this fuzzy rule base, the Mamdani–Assilian inference was used in the current application. An example of a fuzzy inference process is described in Figure 7. The first line and second line show three inputs and a single output. The third line depicts how three input values are partial members of three linguistic variables for "velocity", "water depth" and "water temperature lag time". This leads three fuzzy rules to be activated. The implication operator determines the partial membership to the output. These are aggregated into an output surface.

Table 1. Fuzzy rule base representing the habitat suitability index (HSI) for the spawning of *Ctenopharyngodon idella* (*C. idella*).

Velocity	Water Depth	Time	HSI	Velocity	Water Depth	Time	HSI
VL	VL	VL	M	VL	VL	L	L
L	VL	VL	H	L	VL	L	M
H	VL	VL	H	H	VL	L	M
VH	VL	VL	M	VH	VL	L	L
VL	L	VL	H	VL	L	L	M
L	L	VL	VH	L	L	L	VH
H	L	VL	VH	H	L	L	VH
VH	L	VL	H	VH	L	L	M
VL	H	VL	H	VL	H	L	M
L	H	VL	VH	L	H	L	VH
H	H	VL	VH	H	H	L	VH
VH	H	VL	H	VH	H	L	M
VL	VH	VL	M	VL	VH	L	L
L	VH	VL	H	L	VH	L	M
H	VH	VL	H	H	VH	L	M
VH	VH	VL	M	VH	VH	L	L
VL	VL	H	L	VL	VL	VH	VL
L	VL	H	L	L	VL	VH	VL
H	VL	H	L	H	VL	VH	VL
VH	VL	H	L	VH	VL	VH	VL

Table 1. *Cont.*

Velocity	Water Depth	Time	HSI	Velocity	Water Depth	Time	HSI
VL	L	H	L	VL	L	VH	VL
L	L	H	H	L	L	VH	L
H	L	H	H	H	L	VH	L
VH	L	H	L	VH	L	VH	VL
VL	H	H	L	VL	H	VH	VL
L	H	H	H	L	H	VH	L
H	H	H	H	H	H	VH	L
VH	H	H	L	VH	H	VH	VL
VL	VH	H	L	VL	VH	VH	VL
L	VH	H	L	L	VH	VH	VL
H	VH	H	L	H	VH	VH	VL
VH	VH	H	L	VH	VH	VH	VL

Note: Fuzzy sets are usually expressed by linguistic description such as "very low", "low", "high" and "very high". VL, L, H, VH indicate "very low", "low", "high" and "very high", respectively.

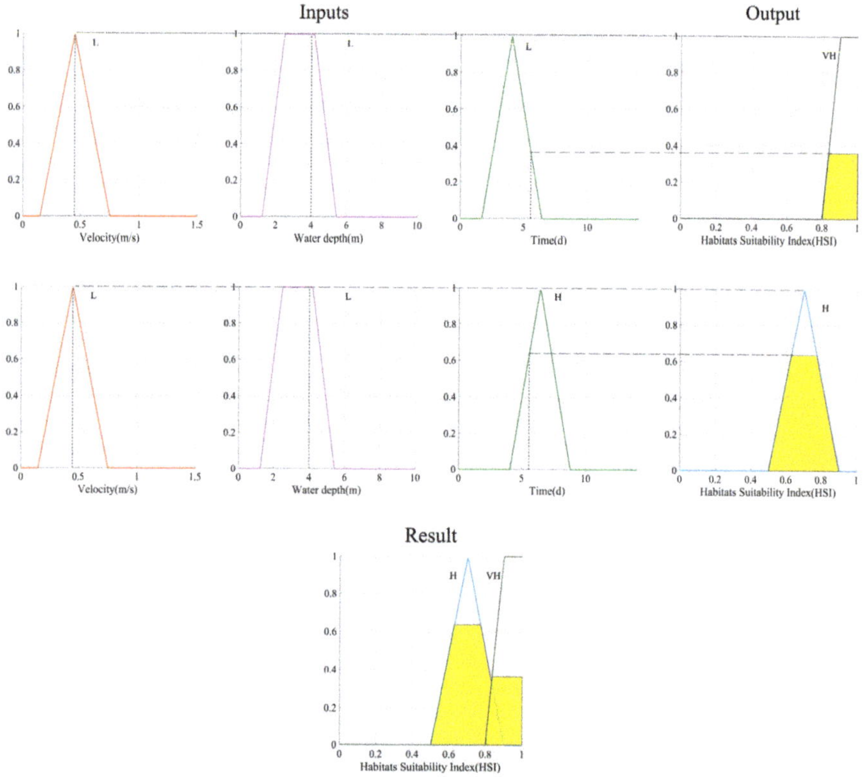

Figure 7. Illustration of the fuzzy inference process.

(3) Defuzzification

The fuzzy inference system uses implication and aggregation operators to scale the membership functions of output linguistic variables before performing defuzzification. Afterwards, a numerical output value of the fuzzy inference process is obtained with defuzzification methods, center of gravity. In this study, the center of gravity defuzzification method is used as follows:

$$CG = \frac{\int_{x_{min}}^{x_{max}} f(x) \times x\,dx}{\int_{x_{min}}^{x_{max}} f(x)\,dx} \tag{7}$$

where CG is the center gravity of the area, x is the value of the linguistic variable, and x_{max} and x_{min} indicate the range of the linguistic variable. The CG defuzzification method effectively calculates the best compromise between multiple output linguistic terms. This crisp output value calculated by this method denotes HS for *C. idella*, described as the habitat suitability index (HSI).

3.3.2. Criteria for Assessing Habitat Quality

In order to make a better and accurate evaluation for availability and suitability of habitat, some related evaluation methodology such as weighted usable area (WUA), hydraulic habitat suitability (HHS), and habitat suitability maps were calculated in the study. Habitat suitability maps were generated to visualize the habitat suitability within the river under a steady flow condition. The WUA was calculated by integrating habitat quality over the cells of the studied area, which can be widely used to show the habitats available for species. It was assumed that WUA is positively associated with fish biomass in IFIM. The HHS was obtained by dividing the WUA by the total area, which assigns relative values between 0 and 1. It describes the suitability of hydrodynamic variables for the fish species considered. The calculation methods of WUA and HHS are as follows:

$$WUA = \sum_{i=1}^{n} HSI_i \times A_i \tag{8}$$

$$HHS = \frac{1}{\sum_{i=1}^{n} A_i} \sum_{i=1}^{n} HSI_i \times A_i \tag{9}$$

where A_i indicates the i-th grid area, n indicates the total number of grids, and HSI_i indicates the i-th habitat suitability index.

The best flow management scenario is determined by comparing and analyzing the simulation results under WUA and HHS for each scenario. A larger the WUA value, the larger total habitat weighted usable area for fish spawning. A larger WUA value is more favorable for fish spawning. HHS is a normalized value between 0 and 1. The larger the HHS value, the larger the proportion of habitats area for fish spawning. A larger HHS value is more favorable for fish spawning.

4. Results and Discussion

4.1. Result of CE-QUAL-W2 Model Simulation

According to the research, the Xuanwei Reservoir has thermal stratification, and therefore it is necessary to analyze the negative effect on the *C. idella* spawning in downstream of the reservoir caused by the releasing of water with low temperature. In this study, the CE-QUAL-W2 model is used to simulate the time taken to reach an appropriate ecological water temperature under different inflow and outflow conditions. First of all, the initial conditions and boundary conditions of the model are set. The historical runoff data (water level and flow) of the hydrological station and the historical meteorological data (wind speed, wind direction, relative humidity, temperature, and cloudage) of the meteorological station are collected to simulate the daily change process of the water temperature of the Xuanwei Reservoir. The vertical distribution of the water temperature of the reservoir in mid-May is obtained by calculating the simulation results, and the results are shown in Figure 8a. The vertical distribution of water temperature in mid-May was taken as the initial conditions of the model, when the inflow water temperature is 21 °C, and the outflow water temperature is 15.8 °C. According to the sampling survey in the study area, the spawning period of *C. idella* occurs mostly from mid-May to mid-July of the year, and the optimal water temperature range for spawning is between 21 °C to 24.5 °C.

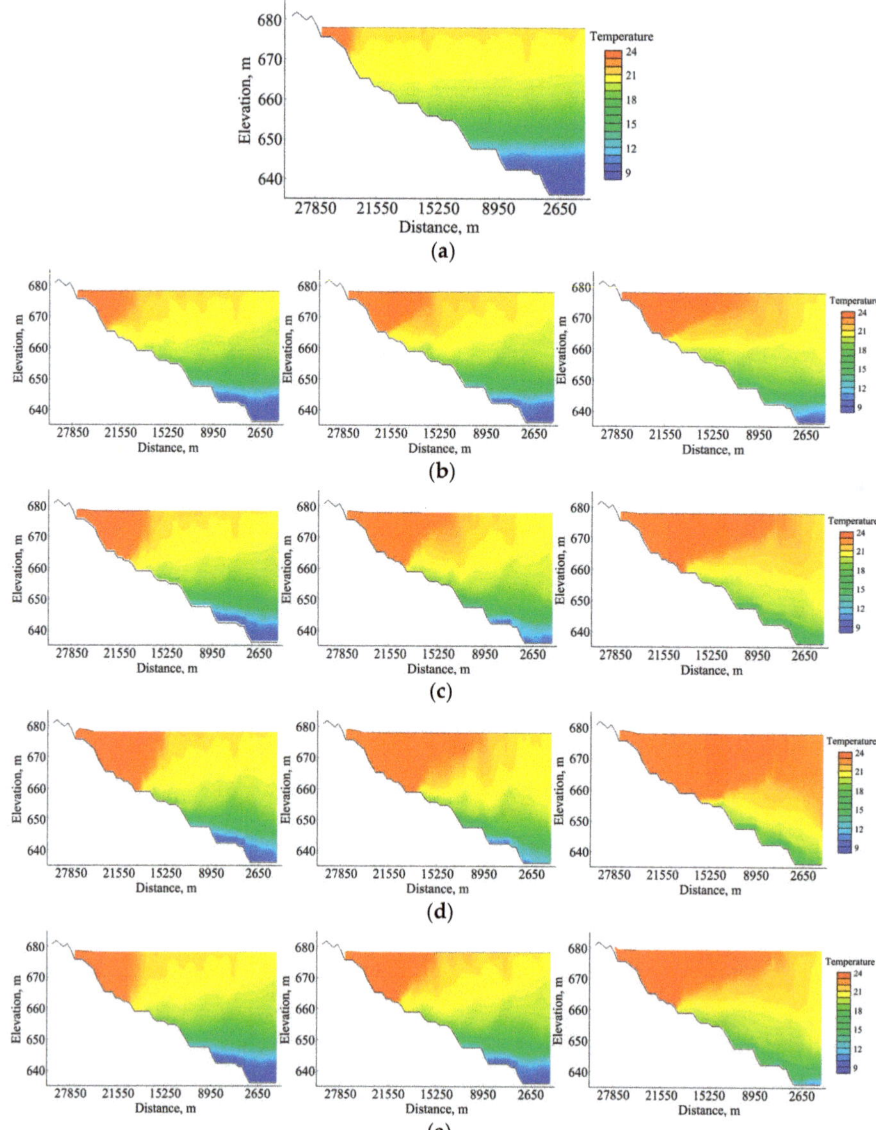

Figure 8. *Cont.*

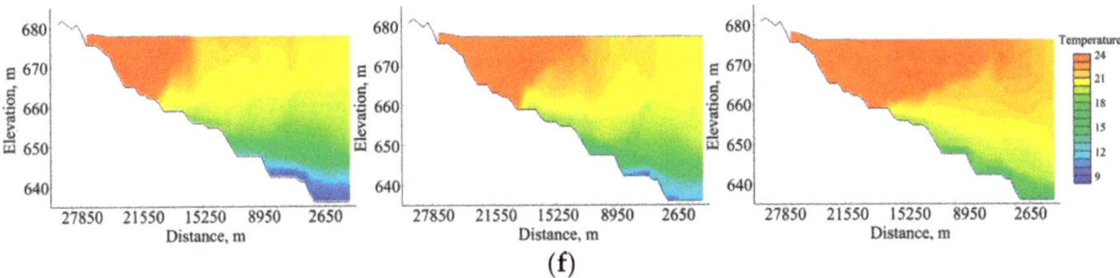

Figure 8. Results of thermal stratification simulation: (**a**) Initial thermal stratification of the reservoir; (**b**) inflow, 40 m^3/s and outflow, 40 m^3/s (from left to right: 1 day, 3 days, 5 days); (**c**) inflow, 100 m^3/s and outflow, 100 m^3/s (from left to right: 1 day, 3 days, 5 days); (**d**) inflow, 140 m^3/s and outflow, 140 m^3/s (from left to right: 1 day, 3 days, 5 days); (**e**) inflow, 100 m^3/s and outflow, 80 m^3/s (from left to right: 1 day, 3 days, 5 days); (**f**) inflow, 100 m^3/s and outflow, 120 m^3/s (from left to right: 1 day, 3 days, 5 days).

The inflow range is between 21 m^3/s to 163 m^3/s according to the historical runoff data from the hydrological station in May, since 1970. The inflow and outflow are each divided into eight different scenarios by using 20 m^3/s as the flow interval. Different inflows and different outflows are combined into 64 sets of scheduling scenarios. The inflow water temperature and meteorological data are based on the multi-year average data in mid-May. By simulating 64 different scenarios, the results obtained are shown in Figure 8 and Table 2.

Table 2. The time taken to reach suitable ecological water temperature under different inflow and outflow scenarios.

Time (day)		Outflow (m^3/s)							
		20	40	60	80	100	120	140	160
	20	13	10.6	8	6.4	6.2	/	/	/
	40	10.8	9.8	7.2	6	5.7	5.3	/	/
	60	/	7.7	6.5	5.6	5.2	4.8	4.5	/
	80	/	6.3	6	5.4	4.8	4.5	4.3	4.2
Inflow (m^3/s)	100	/	/	5.6	5	4.6	4.3	4.1	4
	120	/	/	4.8	4.6	4.3	4	3.9	3.8
	140	/	/	/	4.5	4.1	3.9	3.7	3.6
	160	/	/	/	/	4	3.8	3.6	3.5

Note: / indicates that this scheduling scenario does not meet the requirements. According to this scenario, the water level is higher or lower than the water level limit.

In Figure 8, the vertical distribution of water temperature in the reservoir for one day, three days, and five days is shown by five different scheduling scenarios. The vertical distribution of the reservoir water temperature, respectively, is show in Figure 8b–d, when the reservoir inflow is equal to reservoir outflow. The comparison results show that the large inflow of the reservoir makes the water body mix well, which leads to the decrease of the thermal stratification stability. Therefore, the temperature difference between the inflow and the outflow is relatively small. Similarly, the small inflow of the reservoir makes the water body mix slowly, resulting in enhanced thermal stratification stability. Therefore, the temperature difference between the inflow and the outflow is relatively high. The vertical distributions of the reservoir water temperature are shown in Figure 8c, e, and f, where the reservoir inflow is 100 m^3/s and reservoir outflows are 100 m^3/s, 80 m^3/s, and 120 m^3/s, respectively. The comparison results show that the larger outflow of the reservoir makes the water body mix more fully for the same reservoir inflow, which leads to the decrease of the thermal stratification stability.

When the inflow water temperature is 21 °C, the time taken for the outflow water temperature to reach 21 °C is shown in Table 2. It can be seen from Table 2 that the larger outflow of the reservoir makes the time to reach the appropriate water temperature shorter for the same reservoir inflow.

Due to the thermal stratification of the reservoir, there appears to exist a time lag phenomenon between inflow water temperature and outflow water temperature. The results of the model simulation show that the larger inflow and outflow of the reservoir will make the water body mix more intensely, resulting in shorter time delay of water temperature. At the same time, the fish reach the appropriate water temperature for spawning in a short time, which is more beneficial to *C. idella*. Conversely, a long lag time of water temperature delays the spawning of *C. idella*, which is less favorable for spawning.

4.2. Result of MIKE21 Model Simulation

The MIKE21 model is used to simulate the velocity and water depth distribution of fish habitats under different outflows of the reservoir. The outflow range uses the value simulated by the CE-QUAL-W2 model, which is divided into eight different scenarios. The minimum outflow is 20 m³/s and the maximum outflow is 160 m³/s with an outflow interval of 20 m³/s. The upper boundary of the habitat model uses a flow boundary, and the lower boundary uses a water level boundary. The topographic data is provided by the Hydraulic Exploration Design Institute of the Guizhou Province in China. On the basis of the hydraulic simulation, the spatial distribution of water depth and velocity under eight different modeling scenarios is shown in Figure 9. As can be seen from the figure, the velocity is relatively large near the reservoir location, and the water depth is deeper at the river bend.

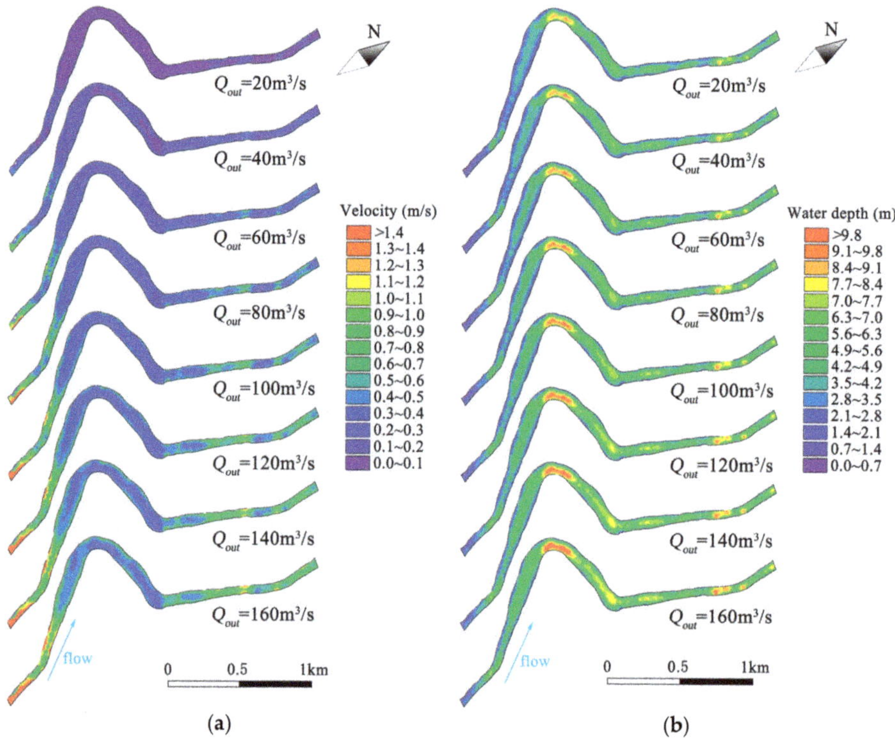

Figure 9. The spatial distribution of velocity and water depth under different reservoir outflows: (a) velocity and (b) water depth.

4.3. Suitable Ecological Flow Calculation

The simulation results of the CE-QUAL-W2 model and the MIKE21 model are applied to the suitability evaluation of representative species in the study area. The effects of flow, water depth, and water temperature hysteresis on habitats are reflected in the suitability evaluation. The implementation of the fuzzy inference system represented in the previous section gives the outputs for all possible inputs, as shown in Figure 10. The combined effects of each of the two indicators on habitat suitability are shown in Figure 10, with a total of three sets of analysis results. The results show that when velocity and water depth are excessively large or small, habitat suitability is so poor that the hydraulic condition cannot meet the spawning requirements of *C. idella*. According to the survey data of *C. idella* in the Qingshui River (Section 2, *2.2. Ctenopharyngodon idella*) it is most suitable for adult *C. idella* spawning when the water temperature is between 21 °C and 24.5 °C, velocity is between 0.3 m/s and 0.9 m/s, and water depth is between 3 m and 7.5 m. The results derived from the 3D fuzzy surface graph are basically the same as the field measured data, so the fuzzy logic-based habitat model should be reliable and adoptable.

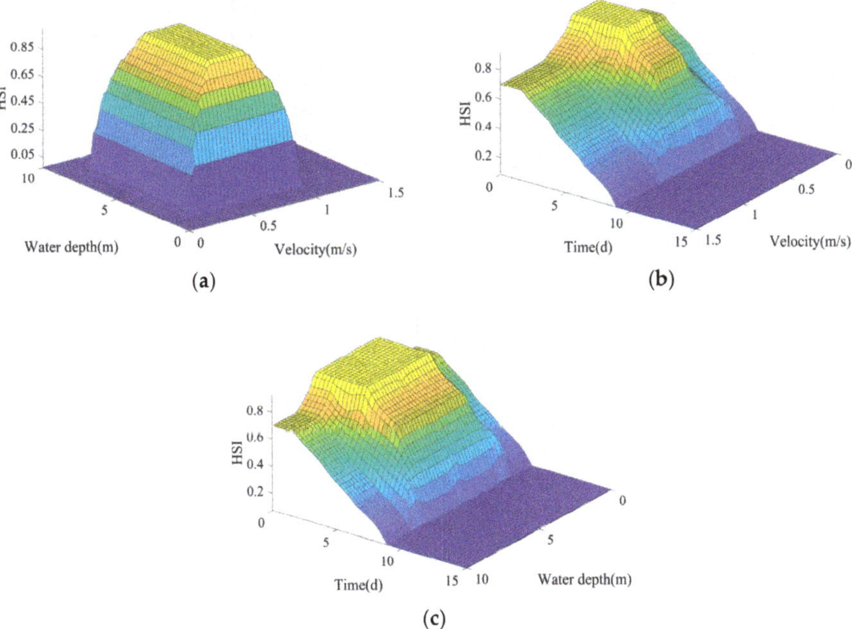

Figure 10. Output for habitat suitability index (HSI) after defuzzification. (**a**) HSI of velocity and water depth, (**b**) HSI of velocity and water temperature lag time, and (**c**) HSI of water depth and water temperature lag time.

On the basis of the hydraulic simulation and the fuzzy logic-based habitat model, the weighted usable area (WUA) and hydraulic habitat suitability (HHS) are evaluated. The specific operational steps are to simulate the distribution of velocity, water depth, and water temperature lag time under different inflows and outflows through two hydraulic models. Then, the fuzzy-based habitat model is used to construct the relationship between hydraulic factors and habitat suitability index. Finally, the WUA and the HHS under different reservoir inflows and reservoir outflows are obtained by fuzzy calculation. This study simulates a combination of eight different reservoir inflows and eight different reservoir outflows, including a total of 64 scenarios. The results of 64 scenarios are analyzed, but 19 of them do not meet the scheduling requirements. According to these scenarios, the water level is higher or lower than the water level limit. The simulation results under 45 scenarios are shown in Figure 11 and Table 3.

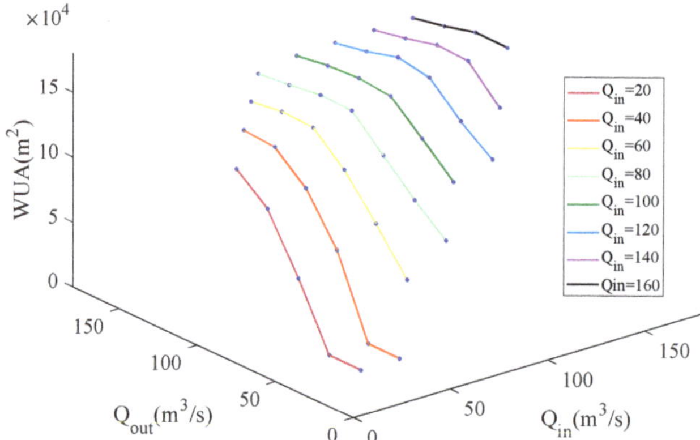

Figure 11. Weighted usable area (WUA) curves for *C. idella* in the study area.

Table 3. Hydraulic habitat suitability (HHS) under different inflow and outflow scenarios.

HHS		Outflow (m³/s)							
		20	40	60	80	100	120	140	160
	20	0.077	0.079	0.300	0.497	0.586	/	/	/
	40	0.077	0.079	0.361	0.529	0.623	0.631	/	/
	60	/	0.264	0.414	0.553	0.651	0.655	0.640	/
Inflow (m³/s)	80	/	0.364	0.455	0.563	0.669	0.675	0.659	0.647
	100	/	/	0.478	0.581	0.681	0.693	0.686	0.669
	120	/	/	0.517	0.601	0.705	0.726	0.696	0.674
	140	/	/	/	0.608	0.723	0.729	0.700	0.678
	160	/	/	/	/	0.728	0.731	0.702	0.680

Note: / indicates that this scheduling scenario does not meet the requirements. According to this scenario, the water level is higher or lower than the water level limit.

The WUA and HHS are calculated from the habitat quality maps by using Equations (8) and (9). As can be seen from Figure 11 and Table 3, the change tendencies of the WUA and HHS are almost consistent. The suitability of hydraulic habitats under different reservoir inflows and outflows are shown in Table 3. The best flow management scenario is determined by comparing and analyzing the simulation results under the WUA and HHS for each scenario as follows:

1. The effects of different outflow on the WUA and HHS for the same reservoir inflow. It can be seen from the table that an increase of the reservoir outflow will make the HHS value gradually increase for the same reservoir inflow. When the reservoir outflow reaches 120 m³/s, the HHS value is the largest. However, the HHS value gradually decreases when the reservoir outflow is greater than 120 m³/s. At the same time, it can be seen from the table that when the reservoir outflow reaches 100 m³/s or more, the comprehensive index of HHS can generally reach 0.6 or more.

2. The effect of the WUA and HHS when the reservoir inflow is equal to outflow. A total of eight scenarios (20, 40, 60, 80, 100, 120, 140, and 160 m³/s) were simulated for the same reservoir inflow and outflow. The habitat ecological suitability is the best when the reservoir outflow is 120 m³/s. When the flow is greater than 100 m³/s, the HHS index can reach 0.6 or more, which is beneficial to *C. idella* spawning.

3. Determination of the suitable ecological flow range based on 64 scenarios. According to the results of 64 scenarios, it shows that the ecological flow for *C. idella* spawning can satisfy the basic

demand when the reservoir inflow is greater than 60 m^3/s and the reservoir outflow is greater than 100 m^3/s. The habitat ecological suitability is the best when the reservoir outflow is 120 m^3/s. Therefore, compared with historical data, the suitable ecological flow from 100 m^3/s to 160 m^3/s for *C. idella* propagation is reliable, and the ecological flow can meet the spawning requirements of the fish species.

4.4. Discussion

4.4.1. Suitable Ecological Flow Discussion

For further analysis of the simulation results, several scenarios are selected for analysis. The spatial distribution of habitat suitability for several scenarios is shown in Figures 12 and 13. According to the simulation results of velocity and water depth in Figures 12 and 13, the velocity is larger at the dam site, but the excessive velocity leads to poor habitat suitability. The water depth at the bend of the river is deep, and the deepwater depth leads to poor habitat suitability.

The spatial distribution of habitat suitability indicators is shown in Figure 12, where the reservoir inflow is equal to outflow. It can be seen from the results that the large inflow of the reservoir makes the water body mix well, which leads to a decrease of the thermal stratification stability. Therefore, the temperature difference between the inflow and the outflow is relatively small, which is beneficial to *C. idella* spawning. Similarly, the small inflow of the reservoir makes the water body mix slowly, resulting in an enhanced thermal stratification stability. Therefore, the temperature difference between the inflow and the outflow is relatively high, which is unfavorable for *C. idella* spawning. Comparing the spatial distribution characteristics of the four different scenarios in Figure 12, the overall suitability is optimal at 120 m^3/s. Habitat suitability is extremely poor when the reservoir outflow is below 40 m^3/s. This shows that too much or too little flow will be detrimental to *C. idella* spawning. The results show that when reservoir inflow and reservoir outflow are excessively large or small, habitat suitability is so poor that the hydraulic condition cannot meet the spawning requirements of *C. idella*.

Figure 13 shows the spatial distribution of habitat suitability index when the reservoir inflow is 120 m^3/s and reservoir outflows are 80 m^3/s, 100 m^3/s, 120 m^3/s, 140 m^3/s, and 160 m^3/s, respectively. It can be seen from the results that the larger outflow of the reservoir makes the water body mix well for the same reservoir inflow, which leads to the decrease of the thermal stratification stability. Therefore, the temperature difference between the reservoir inflow and the outflow is relatively small, which is beneficial to *C. idella* spawning, however, increasing the outflow of the reservoir leads to excessive velocity in the local area of the habitat, and also leads to deeper water depth. A too high velocity and too deepwater depth can negatively affect *C. idella* spawning. On the basis of the spatial distribution characteristics of habitats under five different reservoir outflow scenarios, the overall suitability is optimal at 120 m^3/s. This further validates the results in the previous section. It is most beneficial to the reproduction of *C. idella* when the reservoir outflow reaches 120 m^3/s.

Figure 14 shows the spatial distribution of habitat suitability index when the reservoir outflow is 120 m^3/s and the reservoir outflows are 40 m^3/s, 60 m^3/s, 80 m^3/s, 100 m^3/s, 120 m^3/s, 140 m^3/s, and 160 m^3/s, respectively. From the analysis in Figure 13, 120 m^3/s is a suitable reservoir outflow. From the results of Figure 14, it can be seen that under this suitable outflow, the larger reservoir inflow will make the overall habitat suitability index better. The larger reservoir inflow makes the water body mix well for the same reservoir outflow, which leads to a decrease of the thermal stratification stability. Therefore, the temperature difference between the reservoir inflow and the outflow is relatively small, which is beneficial to *C. idella* spawning. On the basis of the spatial distribution characteristics of habitats under seven different reservoir inflow scenarios, the overall suitability is optimal at 160 m^3/s. This shows that reservoir inflow has a certain impact on *C. idella* spawning. The reservoir inflow provides a space for rational allocation of water resources.

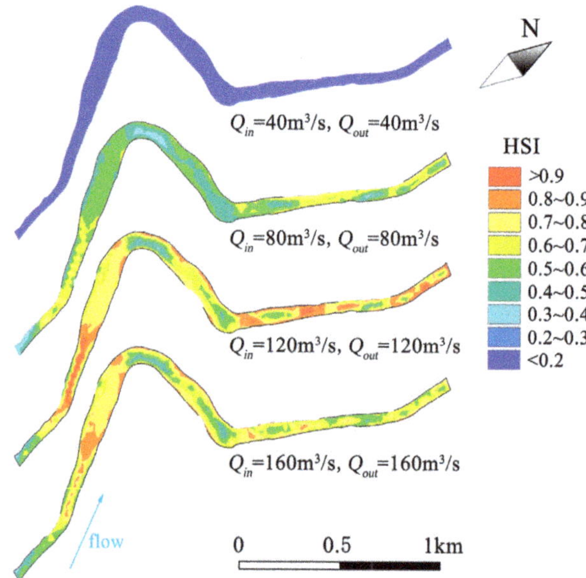

Figure 12. The spatial distribution of HSI under different inflow and outflow scenarios (inflow is equal to outflow).

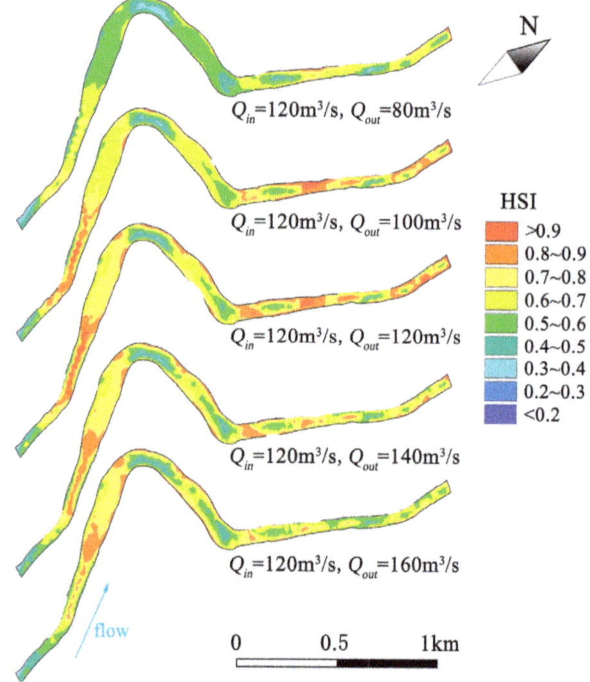

Figure 13. The spatial distribution of HSI under different outflow scenarios.

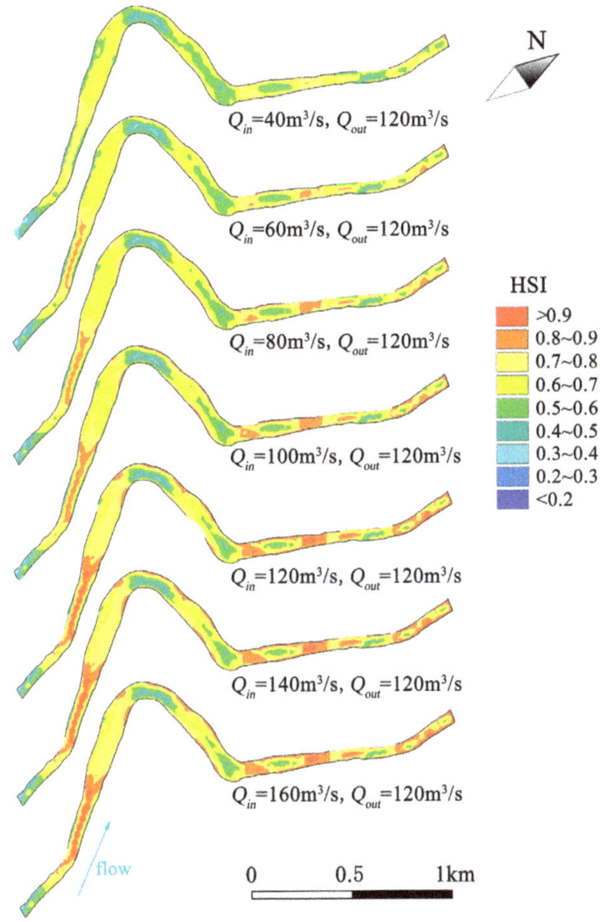

Figure 14. The spatial distribution of HSI under different inflow scenarios.

4.4.2. Suitable Spawning Areas Discussion

In order to discuss suitable spawning activity areas for *C. Idella*, this study calculates and analyzes the weighted usable area with a suitability of 0.6 or more. As shown in Table 4, the following results can be seen from the table: (1) under the seven scenarios with low flow, the weighted usable area with a suitability of 0.6 or more is 0; (2) when the reservoir inflow is 100 m³/s and outflow is 160 m³/s, the weighted usable area with a suitability of 0.6 or more is the largest. The maximum value is 135,396 m³, accounting for 62.6% of the total habitat area of the study area; and (3) when the reservoir inflow is greater than 60 m³/s and outflow is greater than 100 m³/s, the *C. idella* has more than 100,000 m³ area suitable for spawning activity.

The main spawning area of *C. idella* spawning is shown in Figure 15 (select a scenario where the reservoir inflow is 100 m³/s and the outflow is 160 m³/s). According to the survey data, the *C. idella* historical sampling points are mainly concentrated in three areas, which is close to the best suitability area of the simulation results. The model has a certain reliability under less data samples and is a better choice.

Table 4. WUA (HSI > 0.6) under different inflow and outflow scenarios.

WUA (HSI>0.6) (m^3)		Outflow (m^3/s)							
		20	40	60	80	100	120	140	160
Inflow (m^3/s)	20	0	0	0	27764	61698	/	/	/
	40	0	0	0	35069	86714	99344	/	/
	60	/	0	9308	42825	108387	111137	103931	/
	80	/	2979	14038	47581	119950	119168	109860	101351
	100	/	/	18432	55961	124693	124188	117030	106967
	120	/	/	27236	69335	130415	131471	118182	107612
	140	/	/	/	74387	134015	132086	119230	108050
	160	/	/	/	/	135396	132667	120162	108712

Note: / indicates that this scheduling scenario does not meet the requirements. According to this scenario, the water level is higher or lower than the water level limit.

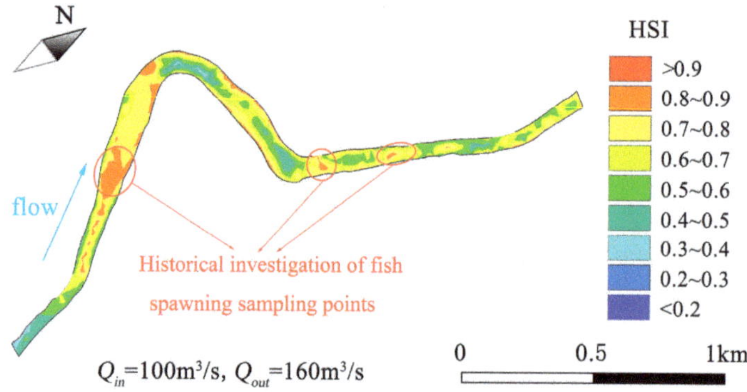

Figure 15. The main distribution area of *C. idella* spawning.

4.4.3. Future Work

There are still some shortcomings in the current research and further research should be carried out which includes the following: (1) Fish habitats not only provide living space for fish, but also provide all environmental factors such as water temperature, substrate, flow rate, pH value, and dissolved oxygen that satisfy the survival, growth, and reproduction of fish. The impact of various hydraulic factors on fish habitats should be considered in the future. (2) In addition to *C. idella* in the study area, other economically valuable local fish should be considered as research protection targets. Habitat models that consider multiple species should be established in the future. (3) The current research is only for the fish spawning period, and the whole life cycle of the fish should be considered in the future, including the hydraulic conditions of the feeding ground, the spawning ground, and the wintering ground.

5. Conclusion

Construction of water conservation projects changes natural flow regimes of rivers and causes adverse impacts on rivers such as obstruction of fish migration paths, as well as altering water quality and water temperature, and thereby causing profound impacts on the aquatic ecosystem. By coupling the fuzzy inference system, a two-dimensional laterally-averaged hydrodynamic model (CE-QUAL-W2), and a two-dimensional shallow water hydrodynamic model (MIKE21), the fuzzy logic-based habitat model is developed for evaluating the effect of inflow and outflow of the Xuanwei Reservoir on the spawning habitat suitability of *Ctenopharyngodon idella* (*C. idella*) in the Qingshui River. The model is constructed considering the reservoir and the downstream river channel, and explores the comprehensive effects of water temperature, velocity, and water depth on habitat suitability.

On the basis of the historical runoff data of the hydrological station, 64 different reservoir inflow and outflow scenarios are designed and simulated. The weighted usable area (WUA) and hydraulic habitat suitability (HHS) under each scenario are calculated and analyzed to obtain a suitable ecological flow range. The results show that the ecological flow for *C. idella* spawning can satisfy the basic demand when the reservoir inflow is greater than 60 m^3/s and the reservoir outflow is greater than 100 m^3/s. The habitat ecological suitability is the best when the reservoir outflow is 120 m^3/s. The suitable ecological flow range provides a reference and guidance for the ecological scheduling of reservoirs and plays an important role in conserving fish habitats. Further works should focus on data collection to consider other important physical variables such water quality, dissolved oxygen, and sediment in order to model a complex river ecosystem with more accuracy. Habitat simulation can evaluate river health and provide support for river management, and therefore it is needs to be quickly put into practice to improve local river ecology and environment.

Author Contributions: Data curation, L.Y.; Formal analysis, J.L.; Funding acquisition, J.Z.; Investigation, S.P.; Methodology, J.L.; Project administration, H.Q.; Software, S.P.; Supervision, L.Y. and L.T.; Validation, W.W.; Writing – original draft, J.L.; Writing – review & editing, H.Q.

Funding: This work is supported by the National Key R & D Program of China (2017YFC0405900), the National Natural Science Foundation of China (No. 91647114, 51979113, 91547208), and the National Public Research Institutes for Basic R & D Operating Expenses Special Project (CKSF2017061/SZ).

Acknowledgments: The authors wish to thank the anonymous reviewers and editors for their constructive comments.

References

1. Huang, L.; Li, X.; Fang, H.; Yin, D.; Si, Y.; Wei, J.; Liu, J.; Hu, X.; Zhang, L. Balancing social, economic and ecological benefits of reservoir operation during the flood season: A case study of the Three Gorges Project, China. *J. Hydrol.* **2019**, *572*, 422–434. [CrossRef]

2. Piper, A.T.; Rosewarne, P.J.; Wright, R.M.; Kemp, P.S. The impact of an Archimedes screw hydropower turbine on fish migration in a lowland river. *Ecol. Eng.* **2018**, *118*, 31–42. [CrossRef]

3. Zhang, M.; Wu, W.M. A two dimensional hydrodynamic and sediment transport model for dam break based on finite volume method with quadtree grid. *Appl. Ocean Res.* **2011**, *33*, 297–308. [CrossRef]

4. Wang, Y.; Lei, X.; Wen, X.; Fang, G.; Tan, Q.; Tian, Y.; Wang, C.; Wang, H. Effects of damming and climatic change on the eco-hydrological system: A case study in the Yalong River, southwest China. *Ecol. Indic.* **2018**, *7*, 1–12. [CrossRef]

5. Miranda, R.; Martinez-Lage, J.; Molina, J.; Oscoz, J.; Tobes, I.; Vilches, A. Effects of stress controlled loading of a reservoir on downstream fish populations in a Pyrenean river. *Environ. Eng. Manag. J.* **2012**, *11*, 1125–1131. [CrossRef]

6. Marques, H.; Dias, J.H.P.; Perbiche-Neves, G.; Kashiwaqui, E.A.L.; Ramos, I.P. Importance of dam-free tributaries for conserving fish biodiversity in Neotropical reservoirs. *Biol. Conserv.* **2018**, *224*, 347–354. [CrossRef]

7. Feng, Z.; Niu, W.; Cheng, C. China's large-scale hydropower system: Operation characteristics, modeling challenge and dimensionality reduction possibilities. *Renew. Energy* **2019**, *136*, 805–818. [CrossRef]

8. Huang, J.; Zhang, Y.; Arhonditsis, G.B.; Gao, J.; Chen, Q.; Wu, N.; Dong, F.; Shi, W. How successful are the restoration efforts of China's lakes and reservoirs? *Environ. Int.* **2019**, *123*, 96–103. [CrossRef]

9. Birk, S.; Bonne, W.; Borja, A.; Brucet, S.; Courrat, A.; Poikane, S.; Solimini, A.; Bund, W.; Zampoukas, N.; Hering, D. Three hundred ways to assess Europe's surface waters: An almost complete overview of biological methods to implement the Water Framework Directive. *Ecol. Indic.* **2012**, *18*, 31–41. [CrossRef]

10. Pander, J.; Geist, J. Ecological indicators for stream restoration success. *Ecol. Indic.* **2013**, *30*, 106–118. [CrossRef]

11. Bishop, M.; Eiler, J.H. Migration patterns of post-spawning Pacific herring in a subarctic sound. *Deep Sea Res. Part II Top. Stud. Oceanogr.* **2018**, *147*, 108–115. [CrossRef]

12. Frantzen, S.; Maage, A.; Duinker, A.; Julshamn, K.; Iversen, S.A. A baseline study of metals in herring (Clupea harengus) from the Norwegian Sea, with focus on mercury, cadmium, arsenic and lead. *Chemosphere* **2015**, *127*, 164–170. [CrossRef] [PubMed]

13. Benjankar, R.; Tonina, D.; McKean, J.A.; Sohrabi, M.M.; Chen, Q.; Vidergar, D. Dam operations may improve aquatic habitat and offset negative effects of climate change. *J. Environ. Manag.* **2018**, *213*, 126–134. [CrossRef] [PubMed]

14. He, W.; Lian, J.; Zhang, J.; Yu, X.; Chen, S. Impact of intra-annual runoff uniformity and global warming on the thermal regime of a large reservoir. *Sci. Total Environ.* **2019**, *658*, 1085–1097. [CrossRef] [PubMed]

15. Shen, Y.; Wang, P.; Wang, C.; Yu, Y.; Kong, N. Potential causes of habitat degradation and spawning time delay of the Chinese sturgeon (*Acipenser sinensis*). *Ecol. Inform.* **2018**, *43*, 96–105. [CrossRef]

16. Boskidis, I.; Kokkos, N.; Sapounidis, A.; Triantafillidis, S.; Kamidis, N.; Koutrakis, E.; Sylaios, G.K. Ecohydraulic modelling of Nestos River Delta under low flow regimes. *Ecohydrol. Hydrobiol.* **2018**, *18*, 391–400. [CrossRef]

17. Forbes, V.E.; Railsback, S.; Accolla, C.; Birnir, B.; Bruins, R.J.F.; Ducrot, V.; Galic, N.; Garber, K.; Harvey, B.C.; Jager, H.I.; et al. Predicting impacts of chemicals from organisms to ecosystem service delivery: A case study of endocrine disruptor effects on trout. *Sci. Total Environ.* **2019**, *649*, 949–959. [CrossRef] [PubMed]

18. Zhang, Q.; Zhang, Z.; Shi, P.; Singh, V.P.; Gu, X. Evaluation of ecological instream flow considering hydrological alterations in the Yellow River basin, China. *Glob. Planet. Chang.* **2018**, *160*, 61–74. [CrossRef]

19. Liu, C.; Zhao, C.; Xia, J.; Sun, C.; Wang, R.; Liu, T. An instream ecological flow method for data-scarce regulated rivers. *J. Hydrol.* **2011**, *398*, 17–25. [CrossRef]

20. Muñoz-Mas, R.; Martínez-Capel, F.; Alcaraz-Hernández, J.D.; Mouton, A.M. Can multilayer perceptron ensembles model the ecological niche of freshwater fish species? *Ecol. Model.* **2015**, *309–310*, 72–81. [CrossRef]

21. Wang, X.; Hao, G.; Yang, Z.; Liang, P.; Cai, Y.; Li, C.; Sun, L.; Zhu, J. Variation analysis of streamflow and ecological flow for the twin rivers of the Miyun Reservoir Basin in northern China from 1963 to 2011. *Sci. Total Environ.* **2015**, *536*, 739–749. [CrossRef] [PubMed]

22. Fu, X.; Lu, D.; Jin, G. Calculation of flow field and analysis of spawning sites for Chinese sturgeon in the downstream of Gezhouba Dam. *J. Hydrodyn.* **2007**, *19*, 78–83. [CrossRef]

23. Lange, C.; Schneider, M.; Mutz, M.; Haustein, M.; Halle, M.; Seidel, M.; Sieker, H.; Wolter, C.; Hinkelmann, R. Model-based design for restoration of a small urban river. *J. Hydro-Environ. Res.* **2015**, *9*, 226–236. [CrossRef]

24. Yi, Y.; Cheng, X.; Yang, Z.; Wieprecht, S.; Zhang, S.; Wu, Y. Evaluating the ecological influence of hydraulic projects: A review of aquatic habitat suitability models. *Renew. Sustain. Energy Rev.* **2017**, *68*, 748–762. [CrossRef]

25. Wang, P.; Shen, Y.; Wang, C.; Hou, J.; Qian, J.; Yu, Y.; Kong, N. An improved habitat model to evaluate the impact of water conservancy projects on Chinese sturgeon (*Acipenser sinensis*) spawning sites in the Yangtze River, China. *Ecol. Eng.* **2017**, *104*, 165–176. [CrossRef]

26. Yao, W.; Chen, Y. Assessing three fish species ecological status in Colorado River, Grand Canyon based on physical habitat and population models. *Math. Biosci.* **2018**, *298*, 91–104. [CrossRef] [PubMed]

27. Belgiorno, V.; Naddeo, V.; Scannapieco, D.; Zarra, T.; Ricco, D. Ecological status of rivers in preserved areas: Effects of meteorological parameters. *Ecol. Eng.* **2013**, *53*, 173–182. [CrossRef]

28. Akter, A.; Tanim, A.H. A modeling approach to establish environmental flow threshold in ungauged semidiurnal tidal river. *J. Hydrol.* **2018**, *558*, 442–459. [CrossRef]

29. Zingraff-Hamed, A.; Noack, M.; Greulich, S.; Schwarzwälder, K.; Pauleit, S.; Wantzen, K. Model-Based Evaluation of the Effects of River Discharge Modulations on Physical Fish Habitat Quality. *Water* **2018**, *10*, 374. [CrossRef]

30. Boavida, I.; Dias, V.; Ferreira, M.T.; Santos, J.M. Univariate functions versus fuzzy logic: Implications for fish habitat modeling. *Ecol. Eng.* **2014**, *71*, 533–538. [CrossRef]

31. Zhang, P.; Yang, Z.; Cai, L.; Qiao, Y.; Chen, X.; Chang, J. Effects of upstream and downstream dam operation on the spawning habitat suitability of *Coreius guichenoti* in the middle reach of the Jinsha River. *Ecol. Eng.* **2018**, *120*, 198–208. [CrossRef]

32. Mouton, A.M.; Alcaraz-Hernández, J.D.; de Baets, B.; Goethals, P.L.M.; Martínez-Capel, F. Data-driven fuzzy habitat suitability models for brown trout in Spanish Mediterranean rivers. *Environ. Model. Softw.* **2011**, *26*, 615–622. [CrossRef]

33. Fukuda, S.; de Baets, B.; Mouton, A.M.; Waegeman, W.; Nakajima, J.; Mukai, T.; Hiramatsu, K.; Onikura, N. Effect of model formulation on the optimization of a genetic Takagi–Sugeno fuzzy system for fish habitat suitability evaluation. *Ecol. Model.* **2011**, *222*, 1401–1413. [CrossRef]

34. Li, R.; Chen, Q.; Tonina, D.; Cai, D. Effects of upstream reservoir regulation on the hydrological regime and fish habitats of the Lijiang River, China. *Ecol. Eng.* **2015**, *76*, 75–83. [CrossRef]

35. Yi, Y.; Cheng, X.; Wieprecht, S.; Tang, C. Comparison of habitat suitability models using different habitat suitability evaluation methods. *Ecol. Eng.* **2014**, *71*, 335–345. [CrossRef]

36. Liu, F.; Lin, T.; Huang, D.; Perng, M.; Chiu, L. An automated system for egg collection, hatching, and transfer of larvae in a freshwater finfish hatchery. *Aquaculture* **2000**, *182*, 137–148. [CrossRef]

37. Zhao, Z.; Dong, S.; Wang, F.; Tian, X.; Gao, Q. Respiratory response of grass carp (*Ctenopharyngodon idellus*) to temperature changes. *Aquaculture* **2011**, *322*, 128–133. [CrossRef]

38. Feng, Z.; Niu, W.; Cheng, C. Optimization of hydropower reservoirs operation balancing generation benefit and ecological requirement with parallel multi-objective genetic algorithm. *Energy* **2018**, *153*, 706–718. [CrossRef]

39. Liu, Y.; Qin, H.; Mo, L.; Wang, Y.; Chen, D.; Pang, S.; Yin, X. Hierarchical flood operation rules optimization using multi-objective cultured evolutionary algorithm based on decomposition. *Water Resour. Manag.* **2019**, *33*, 337–354. [CrossRef]

40. Muñoz-Mas, R.; Marcos-Garcia, P.; Lopez-Nicolas, A.; Martínez-García, F.J.; Pulido-Velazquez, M.; Martínez-Capel, F. Combining literature-based and data-driven fuzzy models to predict brown trout (*Salmo trutta* L.) spawning habitat degradation induced by climate change. *Ecol. Model.* **2018**, *386*, 98–114. [CrossRef]

41. Kurup, R.G.; Hamilton, D.P.; Phillips, R.L. Comparison of two 2-dimensional, laterally averaged hydrodynamic model applications to the Swan River Estuary. *Math. Comput. Simul.* **2000**, *51*, 627–638. [CrossRef]

42. Zhang, Z.; Sun, B.; Johnson, B.E. Integration of a benthic sediment diagenesis module into the two dimensional hydrodynamic and water quality model—CE-QUAL-W2. *Ecol. Model.* **2015**, *297*, 213–231. [CrossRef]

43. Dong, L.; Liu, J.; Du, X.; Dai, C.; Liu, R. Simulation-based risk analysis of water pollution accidents combining multi-stressors and multi-receptors in a coastal watershed. *Ecol. Indic.* **2018**, *92*, 161–170. [CrossRef]

44. Hoque, M.A.; Perrie, W.; Solomon, S.M. Evaluation of two spectral wave models for wave hindcasting in the Mackenzie Delta. *Appl. Ocean Res.* **2017**, *62*, 169–180. [CrossRef]

45. Purba, J.H.; Tjahyani, D.T.S.; Ekariansyah, A.S.; Tjahjono, H. Fuzzy probability based fault tree analysis to propagate and quantify epistemic uncertainty. *Ann. Nucl. Energy* **2015**, *85*, 1189–1199. [CrossRef]

46. Liu, X.; Feng, X.; Pedrycz, W. Extraction of fuzzy rules from fuzzy decision trees: An axiomatic fuzzy sets (AFS) approach. *Data Knowl. Eng.* **2013**, *84*, 1–25. [CrossRef]

47. Van Broekhoven, E.; Adriaenssens, V.; de Baets, B.; Verdonschot, P.F.M. Fuzzy rule-based macroinvertebrate habitat suitability models for running waters. *Ecol. Model.* **2006**, *198*, 71–84. [CrossRef]

48. Fukuda, S. Consideration of fuzziness: Is it necessary in modelling fish habitat preference of Japanese medaka (*Oryzias latipes*)? *Ecol. Model.* **2009**, *220*, 2877–2884. [CrossRef]

49. Li, G.; Sun, S.; Zhang, C.; Liu, H.; Zheng, T. Evaluation of flow patterns in vertical slot fishways with different slot positions based on a comparison passage experiment for juvenile grass carp. *Ecol. Eng.* **2019**, *133*, 148–159. [CrossRef]

50. Šetlíková, I.; Bláha, M.; Edwards-Jonášová, M.; Dvořák, J.; Burianová, K. Diversity of phytophilous macroinvertebrates in polycultures of semi-intensively managed fishponds. *Limnologica* **2016**, *60*, 59–67. [CrossRef]

51. Xu, N.; Li, M.; Fu, Y.; Zhang, X.; Dong, J.; Liu, Y.; Zhou, S.; Ai, X.; Lin, Z. Effect of temperature on plasma and tissue kinetics of doxycycline in grass carp (*Ctenopharyngodon idella*) after oral administration. *Aquaculture* **2019**, *511*, 734204. [CrossRef]

Article

Environmental Flow Releases for Wetland Biodiversity Conservation in the Amur River Basin

Oxana I. Nikitina [1,*], Valentina G. Dubinina [2], Mikhail V. Bolgov [3], Mikhail P. Parilov [4] and Tatyana A. Parilova [4]

1   World Wide Fund for Nature (WWF-Russia), Moscow 109240, Russia
2   Central Directorate for Fisheries Expertise and Standards for the Conservation, Reproduction of Aquatic Biological Resources and Acclimatization, Moscow 125009, Russia; vgdu@mail.ru
3   Water Problems Institute of the Russian Academy of Sciences, Moscow 117971, Russia; bolgovmv@mail.ru
4   Khingan Nature Reserve, Arkhara 676748, Russia; mparilov@mail.ru (M.P.P.); tkuznetsova@mail.ru (T.A.P.)
*   Correspondence: onikitina@wwf.ru; Tel.: +7-910-462-90-57

Received: 31 August 2020; Accepted: 7 October 2020; Published: 10 October 2020

Abstract: Flow regulation by large dams has transformed the freshwater and floodplain ecosystems of the Middle Amur River basin in Northeast Asia, and negatively impacted the biodiversity and fisheries. This study aimed to develop environmental flow recommendations for the Zeya and Bureya rivers based on past flow rate records. The recommended floodplain inundation by environmental flow releases from the Zeya reservoir are currently impracticable due to technical reasons. Therefore, the importance of preserving the free-flowing tributaries of the Zeya River increases. Future technical improvements for implementing environmental flow releases at the Zeya dam would improve dam management regulation during large floods. The recommendations developed for environmental flow releases from reservoirs on the Bureya River should help to preserve the important Ramsar wetlands which provide habitats for endangered bird species while avoiding flooding of settlements. The results emphasize the importance of considering environmental flow during the early stages of dam planning and the need to enhance the role of environmental flow in water management planning.

Keywords: Amur; hydropower; dam; damage; environmental flow releases; biodiversity conservation; freshwater ecosystems; wetlands

---

1. Introduction

1.1. The Urgent Need for Restoring Freshwater Biodiversity

In recent decades, the biodiversity of freshwater ecosystems has been sharply decreasing. The 2020 Living Planet Index shows that the average abundance of freshwater populations monitored across the globe has declined by 84% since 1970 [1], and the extinction level of freshwater fish in the 20th century was the highest in the world among vertebrates [2]. The operation of dams has led to changes in flow regimes of many rivers in the world [3–6] including in Russia [7–11], and has depleted and changed the species composition of rivers, which makes freshwater ecosystems unstable to the effects of natural and anthropogenic factors. Irreversible water withdrawal and water pollution, water operation without effective means of protection of freshwater ecosystems, irrigation, and poaching have served as additional factors contributing to a sharp decline in freshwater biodiversity [12]. Urgent measures should be adopted to address freshwater biodiversity loss [13]. Studies have demonstrated that changes in freshwater ecosystems could be reversible and that ecosystems could retain sufficient potential for recovery [14]. The conservation and restoration of freshwater ecosystems should be a key component of sustainable water resources management, while biological productivity should be used as an indicator of the state of freshwater ecosystems [15]. A key element in the restoration of disturbed freshwater ecosystems is ensuring the hydrological regime for favorable environmental conditions within freshwater and floodplain ecosystems. The science and practice of environmental flow assessment enables identification and quantification of these attributes [1,16,17].

1.2. Environmental Value of the Amur River Basin

The Amur River, in Northeast Asia, is one of the ten largest rivers in the world. Its basin covers over 1.8 million square kilometers of Russian, Chinese, and Mongolian land [18] (see Figure 1). The state Sino-Russian border along the Amur River and its tributaries reaches 3500 km.

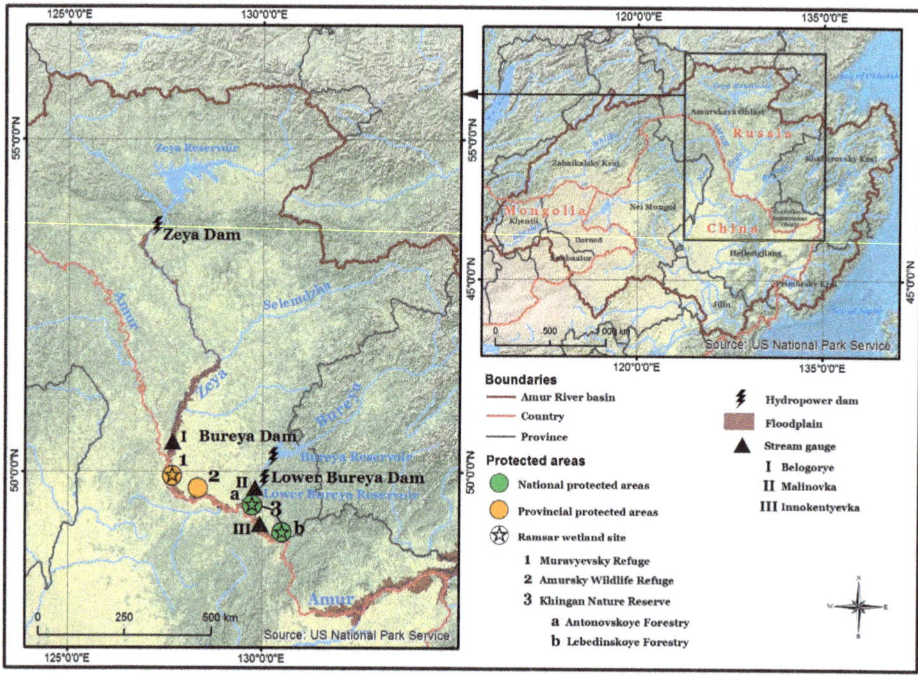

Figure 1. Large dams in the Russian part of the Amur River basin and the important wetlands, including Ramsar sites, located within protected areas under the dam impact.

The Amur River basin ecosystems maintain high levels of biodiversity [19], providing habitats for 130 fish species, 18 of which are endemic. The largest salmon and sturgeon populations of the Pacific Ocean live in the Amur River basin [20,21]. Endangered birds such as Oriental storks (*Ciconia boyciana*), red-crowned cranes (*Grus japonensis*), and white-naped cranes (*Grus vipio*) breed in the Amur wetlands. There are 320 terrestrial vertebrate fauna species that inhabit the floodplains of the Amur River. Flora assemblage of the riverside forests is estimated at more than 300 vascular plant species [22]. Despite the regulated flow of the main tributaries, the main channel of the Amur River remains free of dams, providing ecological integrity to the basin [23].

1.3. Amur River Basin and Floods

1.3.1. Natural Features

The river flow in the Amur River basin is influenced by rainfall, reaching 80% of the annual flow [24]. Frequent floods which are caused by the monsoons of Eastern Asia are common [25,26]. During the warm season from May to September, four to five floods occur, which have caused water level fluctuations of up to 6–8 m in large rivers. The maximum water levels are typical for July and August. Seasonal flow fluctuations in the Amur River basin are exceptionally large as compared with other regions of Russia.

Floods are formed within six major tributaries of the Amur River catchment, i.e., the Argun, Shilka, Zeya, Bureya, Sungari, and Ussuri rivers. Flood magnitude is determined by the maximum flow rates and flood levels on the tributaries as well as by flood volume [25,27]. Floods formed in a separate catchment area can also lead to a large flood on the Amur itself. Significant rises in water levels occur once every two to three years. Once every 10–15 years, floods on several large tributaries coincide, which lead to inundation of the Amur River floodplain and the adjacent areas [27].

1.3.2. The Role of Water Flow and Floods in Freshwater and Floodplain Ecosystem Maintenance and Function

Natural water flow provides ecosystem services such as maintaining specific hydraulic and geomorphological stream parameters (non-silting velocity, physical and chemical properties of water, etc.), and maintaining the area, timing, and duration of flooding of a floodplain. River flow which defines the thermal regime, water turbidity, as well as soil and vegetation cover, plays a vital role within freshwater and floodplain ecosystems. The permissible recurrence of ecologically unfavorable hydrological conditions should be determined by the characteristics of freshwater ecosystems and their inertia [15,28].

During high floods, the Amur River overflows its riverbanks for tens of kilometers and forms a backwater with its tributaries, bypassing their floodplain ecosystems and washing out oxbow lakes. Floods have a positive effect on fish reproduction, create favorable conditions for the migration of anadromous fish, and enhance the fertility of floodplain soils, creating a high potential for floodplain biological productivity [22,29]. Therefore, floods contribute to freshwater and floodplain ecosystem maintenance in the Amur River basin.

1.3.3. Socioeconomic Reasoning for Dam Building

Floods are the main natural disaster in the Amur River basin, negatively affecting the region's economy [30–33]. The most significant losses are common in the agriculture, industrial, and communal sectors [30,33].

Large dams regulate the flow of the following three main Amur tributaries: the Zeya and Bureya rivers in Russia, and the Songhua River in China. In Russia, three large dams (see Figure 1) have been built to produce electricity and protect areas from floods [25]. In the Zeya River basin, the Zeya dam (1330 MWt), located 640 km from the Zeya mouth, has been operating since 1984 [34]. The dam regulates 45% of the total Zeya flow [35]. On the Bureya River, the Bureya dam (2010 MWt), located 174 km from the Bureya mouth [36], has been operating since 2003. Since 2017, the Lower Bureya dam (320 MWt), located 85 km from the Bureya mouth [37] has also been operational. In general, the reservoirs successfully protect the human population from floods [25].

However, the catastrophic flood of 2013, which lasted over two months, became the largest flood in the Amur River basin for the last hundred years [24,38–40] and demonstrated the limitations of dams to protect the population. In Russia, tens of thousands of people were evacuated and many lost homes. In China, the flood also brought great disasters, including human casualties [33]. After this extraordinary flood, plans to build from four to ten new dams for additional flow regulation in Russia were announced [32,41,42], but have not yet been implemented.

1.4. Impact of Dams on the Regulated Zeya and Bureya Ecosystems

1.4.1. Long-Term Flow Regulation and Its Impact on Freshwater Ecosystems of the Zeya and Amur Rivers

As a result of flow regulation, maximum water flow rates of the Zeya River have decreased by more than 20%, and the frequency of floods has declined [43], while, as a result of irregular flooding, the settlement area within the Zeya floodplain has increased up to 1.5 times [44]. After construction of the Zeya dam, the Amur River water flows during summer reduced [45,46], which led to a deterioration of water exchange in floodplain wetlands and a gradual overgrowth of oxbow lakes [47,48]. The wetlands of the Muravyevsky Refuge (Ramsar Site, Wetland of International Importance) and the Amur Reserve in the Amur floodplain are affected by the flow regimes of the Amur and Zeya rivers. These wetlands are the habitat of endangered bird species such as the Oriental stork (*Ciconia boyciana*), red-crowned crane (*Grus japonensis*), white-naped crane (*Grus vipio*), and hooded crane (*Grus monacha*) [49].

Construction of the Zeya hydropower dam is an obstacle for fish migrations. The species composition of the Zeya reservoir, with an area of 2500 km^2, has decreased by a third; 12 out of 38 fish species have disappeared from the waterbody, including endangered species such as the kaluga (*Huso dauricus*) and Amur sturgeon (*Asipenser schrenckii*), while two invasive species have appeared (*Coregonus peled* and *Coregonus migratorius*) [50]. Species depletion and composition change make freshwater ecosystems unstable to the effects of natural and anthropogenic factors [28,51]. The Zeya reservoir was originally planned as a fishery reservoir; the main commercial catch was the Amur pike (*Esox reichertii*). However, the current pike catches are five to seven tons versus the expected 400 tons per year [52].

Fish stocks have decreased in all the water bodies located within the impacted area of the dam. In the 1980s, commercial fish productivity of the Zeya River was 20–25 kg per ha and the productivity of the floodplain lakes was 30–40 kg per ha. By 2008, the number had decreased to 0.34 and 0.22 kg per ha [53]. Before the dam construction, the main species of the Zeya river were fish laying eggs in the inundated terrestrial vegetation, i.e., phytophilic fish, representing up to 70–80% of the total freshwater fish stocks, including Amur catfish (*Silurus asotus*), Amur pike (*Esox reichertii*), the Prussian carp (*Carassius gibelio*), and the Amur carp (*Cyprinus carpio*). Due to decreased summer flows, the phytophilic fish spawning ground range and food supply has decreased, which has led to a decrease in their numbers [54]. The influence of cold water is found in the Zeya River section from the dam up to the confluence of the large free-flowing Selemdzha River, which has also changed the species composition towards lithophilic fish species [50,55].

Conversely, the maximum winter discharges and the water levels of the Middle and Lower Amur River basin have significantly increased [45,46]. This has led to intensified formation of intra-water ice, sludge, and congestion, which could threaten the stable operation of water intakes and other economic facilities [56]. Although an increase in the winter flow prevents fish suffocation, it does not contribute to a rise in their number [54].

1.4.2. Effects of Flow Regulation on Freshwater Ecosystems of the Bureya River Basin

The Khingano-Arkharinskaya Lowland (Ramsar Site, Wetland of International Importance) is impacted by the dams on the Bureya River (see Figure 1). The area provides breeding habitats for endangered bird species such as the Oriental stork (*Ciconia boyciana*), red-crowned crane (*Grus japonensis*), and white-naped crane (*Grus vipio*) (see Figure 2). Up to 10% of the world's Oriental stork population breeds here, and a large group of red-crowned cranes inhabit this area [57]. One of the main conditions for bird nesting is the location of nests in close proximity to weak-flowing water bodies inhabited by small fish [58,59]. The reduced fish abundance in lakes has worsened the living conditions for birds and lead to the loss of nesting area and a decrease in the number of nesting pairs [60].

Figure 2. The endangered bird species inhabiting wetlands under the impact of dams, and their food supply. (**A**) Oriental stork (*Ciconia boyciana*), by D. Korobov; (**B**) Red-crowned crane (*Grus japonensis*), by A. Sasin; (**C**) White-naped crane (*Grus vipio*), by A. Sasin; (**D**) Loaches (*Misgurnus mohoity*) and Amur sleeper (*Perccottus glenii*), typical food for storks and cranes, by E. Egidarev.

After the Bureya dam started operating in 2002, the magnitude of the high floods was reduced. The reduction in maximum water levels caused limited and infrequent water flow along small river channels and lakes, which resulted in their gradual overgrowth. The negative effect of the dam's operation has been reinforced during dry periods and has led to the loss of nesting sites for cranes [61]. At the peak of the drought in 2001–2003, the number of breeding cranes and storks decreased to a minimum. Under natural flow conditions, high floods have compensated for the adverse effects of droughts, restoring the productivity of wetlands [60].

Before dams were built, 36 fish species inhabited the Bureya River basin. After dam construction, the number of fish species in the Bureya reservoir decreased to 27 species of freshwater fish, while the number of fish species downstream from the dams decreased to 20 species [62].

1.5. Optimizing Reservoir Operating Rules to Meet Environmental Flows and Minimize the Ecological Effects of Flow Regulation

In order to reduce the negative impact of flow regulation and preserve freshwater and floodplain ecosystems, the flow regime in the warm season should mimic the natural flow by implementing environmental flow releases from reservoirs. These environmental flow releases should maintain river channels and spawning grounds which favorably affect the environment for fish, reconnect oxbow lakes and maintain a water regime of wetlands close to natural, and increase soil fertility within floodplains.

Russian official agencies responsible for assessing and limiting human impact on water bodies follow the Water Code when developing Standards for Permitted Limits of Impact on Water Bodies and Comprehensive Schemes for Water Bodies Management and Protection [63]. During the scheme's development, environmental flows are established for multiple river stretches in order to assess the water balance of the river basins, which are required in order to calculate the permitted water withdrawal for a particular water management area during the development of standards. In 2014, both the standards and scheme were developed for the Amur River basin and approved by the official agencies for the next decade [64,65]. However, these documents do not include sufficient requirements for maintaining the sustainable state of the freshwater and floodplain ecosystems [66]. Furthermore, in order to implement environmental flow releases from the reservoirs, such requirements ought to be outlined in the Reservoir Operating Rules, the underlying document on operational reservoir management [67].

Currently, environmental flow releases from the reservoirs on the Zeya and Bureya rivers have not been implemented, except for sanitary flows, to ensure compliance with water quality standards downstream from the dams. These sanitary flows are not enough to maintain freshwater or floodplain ecosystems. No previous recommendations for the implementation of environmental flow releases with broader objectives have been developed, although the nature conservation community and environmental experts have repeatedly pointed out their importance in resolving this issue [47,68–70].

Alongside the construction of the Lower Bureya Dam, an environmental compensation program was undertaken by the hydropower company RusHydro, the local government, the UNDP/GEF, and ecologists [70,71]. The project aimed to minimize impacts of the construction of the Lower Bureya dam on the biodiversity of the terrestrial ecosystem but did not include measures such as environmental flow releases aimed at conserving freshwater ecosystems or species [70,72].

2. Materials and Methods

2.1. Materials

The environmental flow requirements for the Zeya River basin were estimated according to the methodological approaches for determining surface water withdrawal and environmental flow (release) [73,74], which is the main approach applied in Russia for assessing environmental flow. For the calculations, the data on daily mean discharge for 1957–2017 at the Belogorye stream gauge was used. The stream gauge is located in the lower reaches of the Zeya River, i.e., 43 km from the river mouth, and 617 km from the dam (see Figure 1). The natural flow regime was analyzed using the data from 1957 to 1974, the period before the dam started operating.

To assess the environmental flow releases from the Bureya reservoirs, floods under natural and regulated flow regime were analyzed. The data on daily mean discharge from 1957 to 2017 at the Malinovka stream gauge, located 80 km from the mouth of the Bureya, downstream of the Bureya and Lower Bureya dams (see Figure 1), were used. The natural flow regime was analyzed using the data from 1957 to 1999, as the Bureya River was blocked for dam construction in 2000.

The flow rates for the water overflow from the channel into the adjacent floodplain were also determined from the Belogorye and Malinovka stream gauges.

2.2. Methodology for Assessment of Permitted Surface Water Withdrawal and Environmental Flow (Release)

The method is based on the principle of sustainable functioning of freshwater and floodplain ecosystems and preservation of natural breeding conditions.

The methodological approaches for determining the permitted surface water withdrawal and assessing environmental flow (release) are based on published materials [28,51,73,74]. The components of ecosystems in river basins depend on the ecologically significant elements of the hydrological regime that characterize their state. For rivers, an ecologically significant element of the hydrological regime is flow velocity and flow discharge. The volume of runoff describing the optimal and normal conditions should be determined, as well as water volume for critical conditions, when a sharp deterioration in living conditions and only minimal natural reproductive processes occur.

The river discharge which corresponds to the overflow of water from the channel into the floodplain can be used as a basis for establishing critical conditions [15,28,73,75]. The critical average daily water discharge was determined according to the hydrological data, as well as the corresponding discharge values when the water does not overflow into the floodplain. Then, the corresponding value of the critical annual runoff was determined.

The permitted surface water withdrawal is the maximum volume of water withdrawn from the river basin, which preserves the conditions for the stable and safe functioning of freshwater and floodplain ecosystems or their components. The value of the permitted surface water withdrawal from the river should preserve the intra-annual flow fluctuations, as close as possible to natural conditions without exceeding the limits of natural long-term fluctuations. In order to define the permitted surface water withdrawal, ecological criteria such as the conditions of natural reproductive processes of aquatic biological resources, the structure of the fish community, and the species diversity are used.

Environmental flow is calculated as the difference between the runoff volume and its permitted withdrawal. Therefore, this is the runoff with an admissible irreversible water withdrawal, which provides conditions for stable and safe ecosystem function and restoration.

The algorithm for defining the environmental flow is as follows:

1. On the basis of an analysis of the relationship among natural hydrological characteristics and freshwater ecosystem productivity or other indirect indicators, the critical discharge and volume of water, Q_{cr} and W_{cr} are defined, corresponding to the critical state of freshwater ecosystem.

2. The historically minimal runoff volume (W_{hist}) is taken as the restored minimum runoff per year with 99% probability of exceedance.

3. The difference between the critical runoff volume and the historically minimal runoff determines the volume which can be withdrawn from the river with minimal damage to the ecosystem in the long term. The average volume of permissible irrevocable withdrawal is determined by the formula:

$$W_{iw\ mean} = W_{cr} - W_{hist} \tag{1}$$

4. The intra-annual distribution of $W_{iw\ mean}$ is carried out according to the long-term natural flow regime.

5. The permitted water withdrawal is determined for the stream gauge located in the lower reaches of the river basin, mostly close to the river mouth. The withdrawal volume is determined for different water year types (e.g., normal, very wet, wet, dry, and very dry), considering intra-annual seasons of the water year:

$$W_{iw\ j} = W_{iw\ mean} \times (W_j / W_{mean}) \tag{2}$$

where W_j is the natural (or restored) runoff in the year with j% probability of exceedance and W_{mean} is the mean natural (restored) runoff in the lower reaches of the river basin.

6. The permitted water withdrawal value for the river sections located upstream is defined along the main river channel according to the formula:

$$W^{i}_{iw\ j} = K \times W_{iw\ j} \tag{3}$$

where $W_{iw\ j}$ is the permitted water withdrawal volume established for the whole basin for the year of j% probability of exceedance and K is the proportion between the total river runoff for the year with j probability of exceedance and the river runoff in a defined river section.

7. In dry years with a runoff below the critical volume, water withdrawal is allowed only for priority needs, such as drinking and domestic water supply.

8. According to the defined value of the permitted water withdrawal, the environmental flow (WiEFlow j) is calculated for the year of j% probability of exceedance:

$$W^i_{EFlow\,j} = W^i_j - W^i_{iw\,j} \tag{4}$$

9. The environmental flow release from the reservoir should consider the needs for fishery and ecosystem maintenance, channel-forming processes, and ensure proper sanitary requirements. If there is a lateral inflow downstream from the dam, the environmental flow release must ensure compliance with the defined requirements of environmental flow. When establishing recommendations for environmental flow releases, fishery needs can be taken as a basis, providing conditions for natural reproductive processes of commercially valuable and other fish species, as well as other aquatic flora and fauna inhabiting freshwater ecosystems downstream of the dam.

10. The intra-annual distribution of critical flow, as well as environmental flow and releases is carried out under the intra-annual flow distribution for a particular year.

When determining the permissible irrevocable water withdrawal, first, the calculation is carried out for the entire basin, and then for the upstream sections.

The permitted water withdrawal should not exceed 20% of the average annual runoff volume [76].

3. Results

3.1. Environmental Flow Assessment in the Zeya River Basin

On the Zeya River section from the hydropower dam up to the confluence of the Zeya and its tributary Selemdzha, the river course is relatively straight, with a small number of tributaries. Such a morphological structure determines the poverty of fish species. The negative effect of cold water on aquatic organisms is found here [55] and is an additional limiting factor in the reproduction of aquatic biological resources. After the confluence of the Zeya and Selemdzha rivers, the Zeya water is sufficiently heated. Its river course is meandering, forming channels, oxbow lakes, and grounds favorable for fish reproduction. A large number of channels and lakes are located within the wide floodplain on the right riverbank. Due to the thermal regime, this river section is comfortable for most freshwater fish living in the Zeya River basin. The Zeya's most important area for fish habitat is the channel and floodplain of the Zeya River below the confluence of the Selemdzha River and up to the Zeya River's lower reaches with a straight river course flowing into the Amur River. The river section between the confluence of the Selemdzha and Zeya rivers is the most ecologically valuable site of the Zeya River basin. The hydrological conditions of this area can be characterized by the data of the Belogorye stream gauge.

To define the environmental flow requirements, floods in the Lower Zeya River basin under the natural flow regime and the flow regulation were compared.

In order to inundate the Zeya floodplain and ensure conditions for natural reproduction processes for phytophilic fish, the river discharge at the Belogorye stream gauge should exceed 6500 m^3/s, which corresponds to the discharge of the overbank flow. According to hydrological observations under a natural flow regime between 1956 and 1974, the floodplain inundation at the Belogorye stream gauge has occurred annually and throughout the entire warm season. From early May to late July, during fish spawning, the duration of continuous periods of the floodplain inundation was 15–20 days, reaching 30–37 days some years (see Figure 3a–e). During August and September, at the end of the spawning and fattening periods, the floodplain was also annually inundated (see Figure 3f–h) for 5–32 days, causing water inflow into oxbow lakes. When river discharges exceeded 9000 m^3/s, the high rates would wash organic debris and macrophytes out of the lakes.

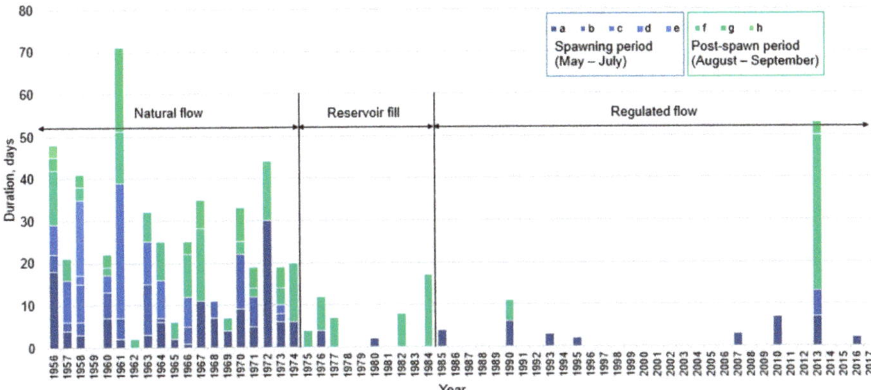

Figure 3. Under flow regulation, the frequency and duration of the Zeya River's bank overflow and floodplain inundation at the Belogorye stream gauge decreased, which negatively affects the reproduction of phytophilic fish.

To determine the environmental flow for the Zeya River at the Belogorye stream gauge, critical conditions for the freshwater ecosystem were determined, such as the flow rate and water volume which did not lead to bank overflow. The only relatively dry year in the short natural flow regime observation period of 1957–1974 was 1962 (78% probability of exceedance). Due to a short period of observations, the use of the 1962 data leads to an overestimation of the permitted water withdrawal, which is 20% more than the average flow rate. Such a large water withdrawal cannot contribute to the freshwater ecosystem's preservation. Therefore, the volume of critical flow (W_{cr}) is assumed to be equal to the runoff volume for a year with 90% probability of exceedance and equals 45 km^3. The difference between the volume of critical flow (W_{cr}) and the historically minimal runoff with 99% probability of exceedance (W_{hist}) is the average value of the permitted water withdrawal ($W_{iw\ mean}$):

$$W_{iw\ mean} = W_{cr} - W_{hist} = 45 - 35 = 10\ \text{km}^3. \tag{5}$$

Therefore, the average permitted water withdrawal is 10 km^3, equivalent to 17.5% of the Zeya's 57 km^3 average flow. The flow volumes and permitted water withdrawal were determined for wet (25% probability of exceedance), normal (50% probability of exceedance), dry (75% probability of exceedance), and very dry (95% probability of exceedance) water years.

For different water years, the environmental flow volumes are the following: 55 km^3 (P = 25%), 48 km^3 (P = 50%), 42 km^3 (P = 75%), 34 km^3 (P = 95%). The environmental flow volume is distributed between the warm and cold seasons in the ratios of 85–90% and 10–15% (see Table 1).

3.2. Flood Analysis in the Bureya River Basin for Environmental Flow Determination

The current surface water withdrawal from the Bureya River is 0.008% of the permitted withdrawal value for a very dry year [77] and the problem of excessive water withdrawal or the lack of water resources is not relevant at the present time or in the foreseeable future. Therefore, the volumes of irrevocable permitted water withdrawal have not been established.

We believe that flow regulation should also provide conditions for the preservation of freshwater ecosystems downstream from the dam, and the flows for periodic floodplain inundation should be maintained. To estimate the required river discharges, floods under natural flow regimes and under flow regulations were compared, and recommendations for the environmental flow releases were determined.

Table 1. Annual and intra-annual distribution of surface water withdrawal and environmental flow volumes of the Zeya River basin for different water years.

Characteristic	Annual Volume of Water km^3	Warm Season (May–September)		Cold Season (October–April)	
		Water Volume km^3	Proportion of Annual Flow %	Water Volume km^3	Proportion of Annual Flow %
W_{cr}	45.0	-	-	-	-
W_{hist}	35.0	-	-	-	-
$W_{iw\ mean}$	10.0	-	-	-	-
$W_{25\%}$ (wet year)	67.0	59.0	88	8.0	12
$W_{50\%}$ (normal year)	58.0	52.5	91	5.5	9
$W_{75\%}$ (dry year)	51.0	45.0	88	6.0	12
$W_{95\%}$ (very dry year)	41.0	34.5	84	6.5	16
$W_{iw\ 25\%}$ (wet year)	12.0	10.6	88	1.4	12
$W_{iw\ 50\%}$ (normal year)	10.0	9.1	91	0.9	9
$W_{iw\ 75\%}$ (dry year)	9.0	7.9	88	1.1	12
$W_{iw\ 95\%}$ (very dry year)	7.0	5.9	84	1.1	16
$W_{EFlow\ 25\%}$ (wet year)	55.0	48.4	88	6.6	12
$W_{EFlow\ 50\%}$ (normal year)	48.0	43.4	90	4.6	10
$W_{EFlow\ 75\%}$ (dry year)	42.0	37.1	88	4.9	12
$W_{EFlow\ 95\%}$ (very dry year)	34.0	28.6	84	5.4	16

3.2.1. Floods under Natural Flow Conditions

Under a natural flow regime, up to five to seven floods were observed during summer and early autumn, and the most significant level rises occurred during July and August. Large floods were typical for the lower reaches of the Bureya River, when the water level rise could reach 6–10 m above the pre-flood water levels, causing flooding of settlements and agricultural lands. Over the 48-year observation period at the time of 1966, the water level fluctuations measured at the Kamenka (Malinovka) stream gauge were 870 cm for the year with a 1% probability and 530 cm for the year with 50% probability of exceedance. Large floods reoccurred every 10–11 years [18]. High floods were observed in 1917, 1961, 1971, 1972, 1975, 1976, and 1984 [78].

The flushing water regime, during natural conditions, prevented the rapid overgrowth of floodplain lakes. The waters of the Bureya River entered the oxbow lakes and flowed south through them into the Amur River [61].

The water level rise, during the extraordinary flood of 1972, amounted to almost 9 m. The maximum discharge reached 17,700 m^3/s in July flow and rates above 10,000 m^3/s were observed for five days.

The last major flood under natural flow conditions occurred in August 1984. The high level of the Amur River created a backwater to the high flood on the Bureya River, thus the floodplains of the Amur, Bureya and Yarchikha rivers were flooded. The water levels of the floodplain lakes were very high, and the waters of the Bureya River merged with the waters of some oxbow lakes [79]. The flood of 1984 affected the high floodplain of the Bureya River and contributed to an increase in the fish productivity of its lakes after the drought, which reached its peak from 1980 to 1981. Maximum river discharges of the Bureya river at Kamenka (Malinovka stream gauge) reached 5270–7630 m^3/s (with equivalent water levels 529–560 cm above the zero gauge). Local precipitation was 170 mm in August with a climate average of 143 mm [80]. The watering of lakes in the Bureya floodplain occurred under the following simultaneous conditions:

1. High water level of the Amur River;
2. High water level (discharge) of the Bureya River;
3. Prolonged and abundant local precipitation.

During the flood, the lakes were washed out of macrophytes. It can be assumed that the washing of the lakes was ensured by the passage of a flood wave with a flow rate of more than 7000 m^3/s in combination with the high level of the Amur River and an excessive amount of precipitation. Thus, the probable value of the flushing flow rate for the Bureya river at the Malinovka stream gauge is 7000 m^3/s.

The years after the flood of 1984 were the most productive in terms of the number of breeding pairs of cranes and storks for all the observed years [60].

3.2.2. Floods under the Flow Regulation

Since the beginning of the Bureya dam operation in 2003, the maximum flow rate during summer floods has significantly reduced (see Figure 4).

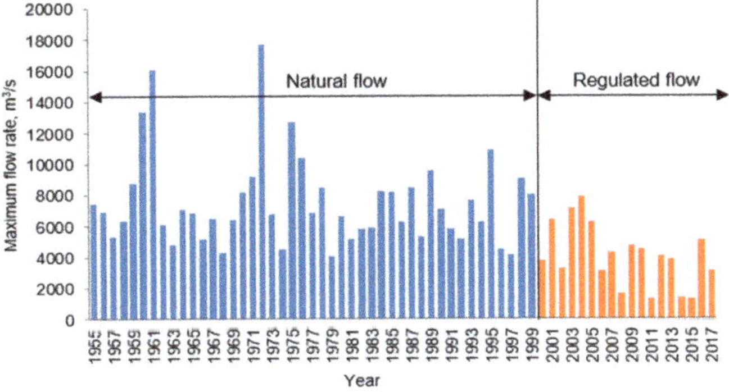

Figure 4. Maximum levels of the Bureya River at the Malinovka stream gauge decreased due to flow regulation by dams.

The summer flood of 2013 was not extraordinary in the Bureya River basin, yet the water levels were high as compared with an average summer. The maximum inflow to the Bureya reservoir reached 5175 m^3/s, while the maximum water release from the reservoir was 3670 m^3/s (see Figure 5), and the maximum flow rate at the Malinovka stream gauge reached 3800 m^3/s [81]. The provision of the maximum daily inflow is estimated at approximately 80% probability of exceedance; as a result of flow regulation, the maximum water level decreased at the Malinovka stream gauge by about 0.5 m [24].

The high-water level of the Amur River caused the backwater of the Bureya River and led to water overflow into the floodplain and hydrological connectivity of the Bureya River and the floodplain wetlands. Water of the Yarchikha River began to enter the Dolgoye Lake on August 19 (see Figure 6). On that day, the water level of the Bureya River at the Malinovka stream gauge was 395 cm, the river discharge was 3690 m^3/s, and the water release from the Bureya reservoir was 3670 m^3/s [81].

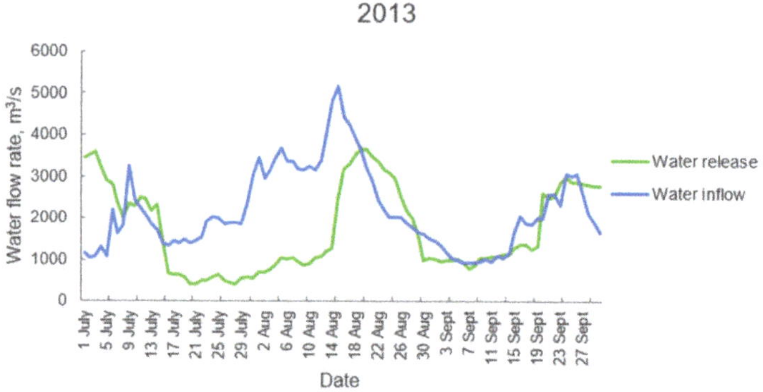

Figure 5. Water inflow to and releases from the Bureya reservoir during the flood of 2013.

Figure 6. On 20 August 2013, the water of the Bureya River entered the Dolgoye Lake through the Yarchikha River, ensuring the hydrological connectivity of the Bureya River and its floodplain wetlands.

A combination of the following conditions led to bank overflow and water access to the floodplain lakes:

- a high-water level of the Amur River (800 cm above the reference point of the Innokentyevka stream gauge) throughout August;
- excessive amount of precipitation (324 mm in August with a climate average of 143 mm for the Arkhara meteorological station); and
- the water release from the Bureya reservoir of 3700 m³/s, which provided the water levels of Bureya River at the Malinovka stream gauge of about 400 cm.

However, these conditions did not lead to the washing out of the lakes within the Bureya floodplain in the Antonovskoye Forestry (see Figures 1 and 7A). At the same time, high water levels of the Amur River, during the 2013 flood, washed out the lakes in the Amur floodplain in the Lebedinskoye Forestry, located downstream from the mouth of the Bureya River, as this area was less affected by flow regulation (see Figure 7B).

Figure 7. (**A**) Dolgoye Lake in the Bureya floodplain was not washed out by the 2013 flood, by O. Nikitina; (**B**) Lebedinoye Lake in the Amur floodplain was washed out by the 2013 flood of overgrowth, by T. Parilova.

In 2019, high water levels of more than 400 cm above the reference point of the Malinovka stream gauge were observed on the Bureya River, while the Amur River levels exceeded 800 cm above its reference point at Innokentyevka. The combination of the Bureya and Amur rivers' high-water levels along with abundant rainfall (225 mm in July as compared with the climate average of 130 mm) ensured the overbank flow and the floodplain inundation, as well as the water inflow into the floodplain lakes. The maximum discharges ranged from 5160 to 5600 m^3/s [81]; the lakes were not washed out due to the decrease in the maximum levels by the dam and the extension of the flood peak in time (see Figure 8).

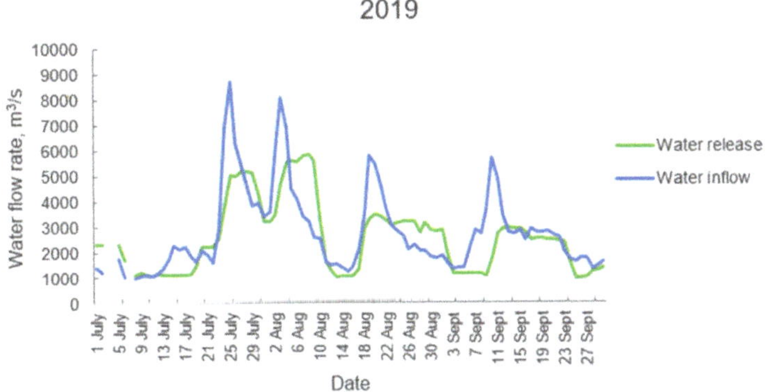

Figure 8. Water inflow to and releases from the Bureya reservoir during high-water period of 2019.

Comparison of hydrographs of 2013 and 2019 (Figures 5 and 8) reveals that, despite the higher flow rates of water released from the Bureya reservoir in 2019, the flood of 2013 was more severe because the water level of the Amur River during a major high-water period was higher than in 2019 (see Figure 9). This confirms the importance of the high-water level of the Amur River for defining the flooding and environmental flow conditions in the Lower Bureya River basin.

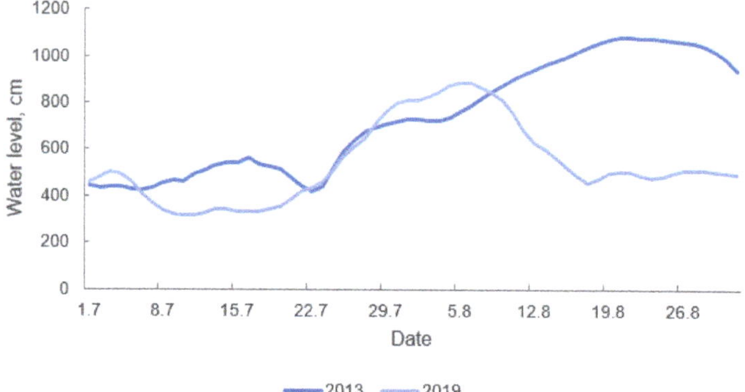

Figure 9. Water level of the Amur River at Innokentyevka stream gauge used to be higher for a major high-water period in 2013 as comparing with in 2019.

4. Discussion

Next, we interpret our findings in order to develop recommendations for reducing the negative impact of flow regulation and protecting freshwater and floodplain ecosystems in the Amur River basin.

4.1. Environmental Flow as a Basis for Freshwater Ecosystem Conservation

The construction of large dams, along with other negative factors, has led to the transformation of many rivers in Russia, including the Amur River, and their freshwater and floodplain ecosystems. For the Amur River basin, dam operation and flow regulation have led to the following: intra-annual flow redistribution and changes to fluvial processes, disturbance in the conditions for natural reproduction of aquatic organisms, a decrease in the area and period of floodplain flooding, loss of hydraulic connection between the river and its floodplain, habitat fragmentation and transformation, changes in species composition and diversity loss, a loss of fish spawning grounds and a decrease in fish catches, etc. [11,12,82].

Water management planning should consider environmental flow as a basis for preserving and restoring freshwater ecosystems. The recently released Emergency Recovery Plan for Freshwater Biodiversity promotes the accelerated implementation of environmental flows as a priority action to conserve and restore freshwater biodiversity. Other priority actions should also consider improving water quality, protecting and restoring critical habitats, sustainably managing the exploitation of freshwater species and river aggregates, preventing and controlling non-native species invasions, and safeguarding and restoring freshwater ecosystem connectivity, including the protection of the remaining free flowing rivers [13]. All these measures are highly relevant to the Amur River basin and should be further developed.

4.2. Interpretation of the Results of the Environmental Flow Assessment in the Zeya River Basin

For the Zeya River basin, the problem of exceeding the determined values of water withdrawal is not relevant currently and for the foreseeable future, although further assessments of critical conditions and water withdrawal are important for small rivers of the Middle Amur River basin actively being used for agriculture [83–85].

However, dam regulation of the intra-annual flow redistribution of the Zeya River, leads to water withdrawal during the warm season, which is important for reproduction of aquatic biological resources, as well as floodplain inundations. Under natural conditions, during the May to September warm season, the Zeya River flow at the Belogorye stream gauge has contributed up to 95% of the annual volume [18,45,46]. Due to flow regulation, the flow proportion in the Zeya River lower reaches, for the warm season, is 65–75% of the annual flow [45,46], which is 15–20% lower than the calculated environmental flow (see Table 1) and indicates a high level of Zeya River freshwater ecosystem transformation.

To ensure sustainable spawning of phytophilic fish and to maintain the floodplain ecosystems, the overbank flow should occur at the Lower Zeya reaches in wet years (25% probability of exceedance), at least once every four to five years. The average lifespan of the Amur pike (*Esox reichertii*) is three to five years, the Prussian carp (*Carassius gibelio*) lifespan lasts six to seven years, and the lifespan of the Amur carp (*Cyprinus carpio*) is more than 12 years [20]. In wet years, the environmental flow volume of high-water years should reach 48 km^3 during the May to September period (see Table 1). According to the natural flow regime, the Zeya River discharge at the Belogorye stream gauge should exceed 6500 m^3/s for 15–20 days in June and July and be formed by the releases from the Zeya reservoir and lateral inflow. Environmental flow release indicators for the Zeya basin could be fish stocks and catches under different water conditions, and the meadow vegetation productivity of the floodplain.

The analysis shows that bank overflow has occurred only once every five to ten years and for less than 10 days since the Zeya dam was fully completed in 1984. This is not enough to ensure proper conditions for the reproduction of aquatic organisms. In addition, environmental and fishing flow releases at the Zeya dam are unfeasible for technical reasons, as the opening of the spillway gates of the dam can only be carried out after the reservoir is filled to the 317.5 m mark. During the winter period, the reservoir level decreases to 309–310 m, and during the summer flood period, the level rises to 313–315 m, rarely reaching 317.5 m [86].

Changing the operating rules of the existing dams is important to restore freshwater and floodplain ecosystems and ecosystem services that have already been affected by the dam construction and operation. However, if the environmental flow releases were not initially included in early project stages, it may become technically impossible during operation, as evidenced by the Zeya hydropower dam. Furthermore, long-term dam operation reduces the likelihood of environmental flow releases being implemented since floodplains have already been developed by humans and floods and would have resulted in socioeconomic losses. Therefore, any construction within the floodplain that affects the state of the ecosystem should be strictly limited and thoroughly examined, especially within the area of a low floodplain, which should be noted in the Water Code.

In the future, technical restrictions for water releases from the Zeya reservoir below the 317.5 m mark should be solved, in order to allow environmental flow release implementation. For example, in 2011, an additional shore spillway of the Sayano-Shushenskaya hydropower dam was commissioned on the Yenisey River. The Sayano-Shushenskaya dam is the largest hydropower station in Russia in terms of installed capacity. The spillway was built, because of the need to improve the reliability and safety of hydraulic structures, to allow an additional passage of flow rates up to 4000 m^3/s during high-water periods and floods, and thereby reduce the load on the dam body [87]. Creation of a similar technical solution at the Zeya dam would allow implementation of environmental flow releases and also improve regulation of the Zeya dam management during severe and catastrophic floods. In turn, this would reduce the negative social and economic consequences of floods.

The large-scale flood of 2013 demonstrated that the floodplains of the Lower Zeya and the Amur rivers below their confluences were inundated, and the oxbow lakes were washed out, with high levels caused by the flow of unregulated tributaries of the Zeya River, such as the Selemdzha, Tom, Urkan, and Dep rivers. The average long-term value of the lateral inflow of the Zeya River, in the section from the Zeya hydropower dam to the Zeya mouth, is about twice as large as the inflow into the Zeya reservoir [35]. With their natural water regimes, the tributaries contribute to the preservation of the most ecologically valuable freshwater ecosystems of the Lower Zeya River basin. They also provide optimization of the thermal regime and other ecological functions [3], thereby improving conditions for reproduction of aquatic organisms in the Lower Zeya River basin. This indicates the importance of their preservation and the need for preventive protection from future flow regulation by the dams.

4.3. Determination of Environmental Flow Releases From the Bureya Reservoirs and Recommendations for Their Implementation

The interim operating rules for the Bureya reservoir include the option of implementing environmental flow releases to ensure environmental sustainability of the floodplain area located downstream from the Bureya dam [88,89]. The environmental flow releases should be aimed at periodic inundation of the floodplain, its wetlands, especially the floodplain lakes and channels. The releases should maintain the hydrological connectivity of the rivers and lakes and prevent the lakes overgrowth. The environmental flow release should be defined considering flood frequency and magnitude during the natural flow regime, which would ensure inundation of the floodplain and its wetlands and washing of the floodplain lakes.

The Bureya reservoir and the Bureya lower reaches have no commercial fishery value and the natural habitat for fish has been significantly disturbed by flow regulation [63]. Commercial fishing demands have been fulfilled by the construction of the Aniuysky fish hatchery in the Khabarovsky Krai as a compensation measure for the Bureya dam construction and operation. As agreed with the fishery authorities, releases from the Bureya reservoir to maintain conditions for fish should not be provided [77]. Therefore, the commercial fishery requirements were not considered when defining environmental flow releases.

Under flow regulation, wetlands inundation of the Bureya floodplain occurs under a combination of the following factors:

- a high-water level of the Amur River of more than 800 cm above the reference point at the Innokentyevka stream gauge;
- a water level of the Bureya River of 400 cm or more above the reference point at the Malinovka stream gauge, with water discharge from the Bureya and the Lower Bureya reservoirs of more than 3700 m^3/s;
- excessive rainfalls in July and August, providing a large amount of water in the floodplain lakes.

If the Amur River water level is not high, the Bureya River bank overflow occurs at a discharge rate of 6000 m^3/s at the Malinovka stream gauge. The bank overflows maintain the hydrological connectivity of the Bureya River with the valuable wetlands of its floodplain. To ensure sustainable conditions of the freshwater ecosystem, the environmental flow release should also wash out the oxbow lakes at a flow rate of 7000 m^3/s, while ensuring the conditions for non-flooding of settlements [90,91].

On the basis of the obtained results, the following recommendations for the environmental flow release implementation from reservoirs on the Bureya River are proposed:

1. To ensure the hydrological connectivity of the Bureya River and the wetlands of its floodplain in the lower reaches of the Bureya River, the water should be released at rates of 3700 to 7000 m^3/s for 10–15 days in July or August. The release rate should consider the Amur River level at the Innokentyevka stream gauge to ensure settlements are not flooded while providing the environmental flow for ecosystem conservation.
2. To halt the lakes overgrowth by cleaning the lakes of macrophytes, the duration of the maximum water releases of 6000–7000 m^3/s should last for two to three days in July or August. The value of the flushing flow rate can be adjusted during hydrological monitoring.
3. Environmental releases should be implemented at least once every six to seven years. This frequency exceeds the frequency of large floods under natural water conditions, which have occurred once every 10–11 years. Given the reduction in flood magnitude, a reduction in the time span is a compensation measure aimed at preserving wetlands under flow regulation.

Environmental flows should be released from the Bureya reservoir because it retains a large volume of water with sufficient capacity to provide such high volumes, unlike the smaller Lower Bureya reservoir to which it is coupled.

Indicators of the effectiveness of environmental flow releases could include the area of lakes and the rate of their overgrowth, the washing of channels and floodplain lakes from organic debris, the food supply for birds, i.e., the abundance of fish inhabiting wetlands, and the number of ichthyophagous birds, including storks and cranes.

The effectiveness of environmental flow releases can be assessed by hydrological monitoring. In 2019, we organized hydrological monitoring on the lakes of the Khingansky Nature Reserve with the support of the World Wide Fund for Nature in order to assess lake dynamics under the influence of flow regulation and climate changes. Subsequent expansion of the monitoring program should allow us to determine the water levels at which the water of the Bureya side enters the Dolgoye Lake through the Yarchikha River. Field observations could be supplemented with satellite imagery or terrain surveys using drones to identify the conditions under which the Bureya bank overflow occurs and hydrological connectivity is ensured.

Environmental flow release implementation in the Bureya River basin should help to preserve the valuable wetlands of the East Asian–Australasian Flyway, and therefore contribute to conservation of endangered bird species. In particular, the environmental releases should help to reduce the negative effect of flood regulation during droughts and to preserve the nesting grounds of cranes. In addition, these releases should indirectly contribute to the improvement of the conditions of natural fish reproduction in the Amur River, downstream of the confluence of the Bureya River, due to the spawning grounds inundation.

Furthermore, the filling of the Bureya reservoir in July and August should be limited by the reservoir to a level of 254.0 m in order to avoid the inundation of the Chekunda settlement located upstream from the dam [90]. This point is an additional incentive for environmental flow implementation. An additional effect of environmental flows would be the floodplain preservation of the Lower Bureya and the Amur rivers below the confluence of the Bureya River from anthropogenic development, primarily from agriculture and settlement development. Along with minimizing the negative impact of flow regulation on freshwater and floodplain ecosystems, this would contribute to the Amur River basin adaptation to floods.

In order to increase the effect of the environmental flows and to improve hydrological connectivity of the Bureya River and the floodplain lakes, it is necessary to expand culverts to allow the Yarchikha's water to easily flow under a road to the Dolgoye Lake in the Khingansky Nature Reserve.

The recommendations developed will be presented to the Federal Water Resources Agency for further discussion and clarification, in order to be included in the Operating Rules for reservoirs of the Bureya River and further implementation.

A relevant example of an environmental flow release is a dam re-operation of the Three Gorges Dam on the Yangtze River in China. The main functions of the dam include producing electricity and controlling floods downstream from the dam. Since 2011, environmental flow releases have been promoting Chinese carp spawning and propagation. However, although the environmental flow releases have improved environmental conditions, the number of fish is still far below the baseline values before the dam construction [17,88]. Another encouraging example is a water reservation for environmental purposes in the San Pedro Mezquital River Basin in Mexico. The wetlands of the large free-flowing river basin include mangrove forests belonging to the Marismas Nacionales Biosphere Reserve, which have been recognized as a Ramsar Site. The river is not water stressed in its middle and lower reaches, however, there have been concerns about future possible development in the river basin, including the Las Cruces hydropower dam construction. In 2014, an Environmental Water Reserve was established, with approximately 80% of its mean annual flow aimed at ensuring water and nutrient supply to the valuable wetlands, while the biological monitoring was focused on their mangroves [17,89].

4.4. The Legislative Changes Needed

Global experience demonstrates that the key factors for successful implementation of environmental flows include legislation on environmental flow and research on the impact of dams on the environment, as well as experimental monitoring of the efficiency of the environmental flow implementation [17,92]. For example, environmental flows have been incorporated into water legislation in South Africa and implemented through legally mandated catchment management agencies. The Crocodile River is an example of an environmental flow implementation. Another example of proactive reservation of flows for the environment is the National Water Reserves Program in Mexico [17,89]. This initiative sets sustainable water allocation limits for 189 rivers across the country, considering the social and economic benefits of environmental flows. The current water legislation in Russia does not include the requirements for environmental flows. The definition of terms for permitted surface water withdrawal, environmental flow, and environmental flow release should be introduced into the Water Code of the Russian Federation to indicate their key role in the freshwater ecosystem conservation.

4.5. Impacts of Climate Change on Flow Regime and Adaptation to Floods

Global climate changes have impacted the hydrological regime of the rivers in the Amur River basin [93], causing an increase in the frequency and power of floods, erosion of riverbanks, and instability of ice phenomena in the rivers [94]. Over time, climate-driven transformation of freshwater ecosystems can lead to a change in the habitats of flora and fauna [1,95]. Climate change and its associated risks should be assessed to develop appropriate adaptation measures for a river basin [33,94].

The studies conducted for the Zeya and Bureya rivers confirm that the maximum flows have an important ecosystem role in the maintenance of freshwater and floodplain ecosystems and their ecosystem services in the Amur River basin. At the same time, floods cause socioeconomic losses for the Amur region. Therefore, when determining the environmental flow regime, the allowable maximum flow reduction should be determined, as well as the magnitude, timing, frequency, and duration of maximum flows. A sufficient value for the maximum flow should both ensure the ecosystem functions of water and floodplain ecosystems, as well as consider the socioeconomic aspects, including the protection of territories from catastrophic floods. Further assessments of environmental flows should consider various climate change scenarios and their impacts on flow regimes.

Infrastructure development in floodplains has increased the number of people affected by floods in the Amur River basin. Following the results of the catastrophic flood of 2013, the recommendations for flood management have primarily included hard engineering measures such as construction of large dams and reservoirs, levee systems construction, and canal widening and deepening. Some reservoirs have been proposed on already modified tributaries such as the Zeya River, while other designs have targeted still free-flowing tributaries, such as the Selemdzha River [41,42,96,97]. In our opinion, construction of additional flood protection dams and reservoirs will not solve the problem of catastrophic floods. At the same time, the creation of reservoirs has been associated with a significant impact on ecosystems [3–7,9,11,12,72,82]. Therefore, the number of new dams should be strictly limited. Prior to the flood control reservoir project design, less harmful alternatives to the environment should be evaluated and carefully analyzed within the basin management plans. In the case of an urgent need to construct a dam, environmental flow releases should be included at the earliest stages of the project, taking into account basin planning and assessing the impact of a dam on the entire river basin instead of assessing the impact on the local area.

A broader approach to flood adaptation should emphasize the need to implement nature-based solutions, such as protection of flood retention capacities of the floodplains and their wetlands [33,96,97]. Additionally, improved land management at the watershed level is needed to find the optimal balance between the ecosystems' ability for water retention and regulation and economic development. These solutions can reduce the impact of extreme floods, while also helping to sustain floodplain ecosystems which are among the most productive and biodiverse habitats on the planet.

Economic, social, and environmental costs of floodplains properly managed for flood retention are lower than the costs of flood-retention reservoirs and other hard engineered measures. During the 2013 flood, in the Amur River basin, the volume of water accumulated by floodplains was higher than the volume of the existing and any planned hydropower reservoirs together [97,98]. In the future, recommendations should be prepared for establishing protected areas in floodplains that are particularly important in terms of accumulating floodwaters and preserving valuable natural ecosystems.

5. Conclusions

Floods are important for maintaining the biodiversity of freshwater and floodplain ecosystems in the Amur River basin. During high floods, the Amur River overflows for tens of kilometers and forms a backwater with its tributaries, inundating their floodplain ecosystems, washing out old lakes, providing a positive effect on fish reproduction, and increasing the fertility of floodplain soils. At the same time, floods cause socioeconomic losses for the Amur region. On the Zeya and Bureya rivers in Russia, three large dams have been built to generate electricity and protect the population from floods.

Flow regulation is one of the key factors determining the state of freshwater ecosystems in the Amur River basin. Decreased maximum water levels downstream of the Zeya dam have led to a reduction in fish spawning grounds and a decrease in their food supply, which have severely reduced fish numbers. Floodplain lakes and river channels have become overgrown. Rare inundation has caused changes in the land use of the floodplain areas of the Zeya River. Flow regulation has worsened the habitat for birds affected by dams in the Bureya floodplain, which has been especially important for the Ramsar site of the Khingano-Arkharinskaya Lowland, an important nesting area for endangered bird species such as the Oriental stork (*Ciconia boyciana*), red-crowned crane (*Grus japonensis*), and white-naped crane (*Grus vipio*).

To reduce the negative impact of flow regulation and to preserve freshwater and floodplain ecosystems, it is necessary to implement environmental flow releases, which would flush the channels, the spawning grounds, and the oxbow lakes; increase the fertility of floodplain soils; and ensure the proper water regime for wetlands, thereby preserving them.

The volumes of permitted surface water withdrawal and environmental flow for different water years and seasons have been determined for the entire Zeya River basin. The problem of surface water withdrawal is not relevant for the Zeya River basin, since the actual water withdrawal is less than 1% of the permitted volume. At the same time, the flow proportion during the warm season has decreased from 95 to 75%, which was 10–15% less than the calculated volume for environmental flow and has negatively impacted the state of freshwater water and floodplain ecosystems. It is recommended to inundate the floodplain at least once every five years for 15–20 days. However, environmental flow releases from the Zeya reservoir are impracticable due to technical reasons. Therefore, the importance of preserving the free-flowing tributaries of the Zeya River increases. With their natural water and thermal regimes, the tributaries contribute to the preservation of freshwater and floodplain ecosystems in the lower reaches of the Zeya basin. Future technical improvements at the Zeya hydropower dam would enable implementation of environmental flow releases, as well as improve the Zeya dam management regulation during severe and catastrophic floods and reduce negative social and economic consequences of floods.

To ensure hydrological connectivity between the river and wetlands and washing out of old lakes and channels in the Bureya floodplain, we recommend providing water discharges of 3700 to 7000 m^3/s from the Bureya reservoirs for 10–15 days in August. To flush the oxbow lakes, the high-water discharges of 6000–7000 m^3/s from the Bureya reservoirs should last for two to three days. Environmental flow releases should be implemented at least once every six to seven years. They should help preserve the important Ramsar wetlands of the Khingano-Arkharinskaya Lowland, which provide habitats for populations of endangered bird species while avoiding flooding of settlements. In addition, environmental flow releases should indirectly contribute to improved conditions for natural reproduction of fish in the Amur River, downstream from the confluence of the Bureya River.

An additional effect of the environmental flow releases would be the preservation of the Lower Bureya floodplain from development within flood-prone areas, which will contribute to the adaptation of the Amur River basin to floods.

Environmental flow legislation is the key factor for successful implementation of environmental flows. The definition of the terms for permitted surface water withdrawal, environmental flow, and environmental flow release should be introduced into the Water Code of the Russian Federation, indicating their key role in freshwater ecosystem conservation. Any construction within the floodplain that affects the state of the ecosystem should be strictly limited and thoroughly examined, especially within the area of a low floodplain, which should be noted in the Water Code.

The conducted studies confirm that maximum flows have an important ecosystem role in maintaining freshwater and floodplain ecosystems and their ecosystem services in the Amur River basin, but, at the same time, cause socioeconomic losses for the population. Therefore, when determining environmental flows, the allowable maximum flow reduction should be determined. A sufficient flow value should ensure the ecosystem functions of water and floodplain ecosystems while considering the protection of territories from catastrophic floods. Further environmental flow assessments should consider scenarios of climate change and their impact on the flow regime.

Global climate changes can lead to an increase in the frequency and power of floods. Prior to flood control reservoir project design, alternatives that are less harmful to the environment should be evaluated. Nature-based flood adaptation to floods should include measures such as protecting flood retention capacities of the floodplains and establishing protected areas in floodplains for simultaneously accumulating floodwaters and preserving ecosystems.

Author Contributions: Conceptualization, O.I.N., V.G.D., M.P.P., T.A.P. and M.V.B.; methodology, V.G.D.; validation, O.I.N., V.G.D., M.V.B., M.P.P. and T.A.P.; formal analysis, O.I.N.; investigation, O.I.N., V.G.D., M.P.P. and T.A.P.; resources, O.I.N., V.G.D., M.P.P., T.A.P. and M.V.B.; data curation, O.I.N.; writing—original draft preparation, O.I.N.; writing—review and editing, O.I.N., V.G.D., M.P.P. and M.V.B.; visualization, O.I.N.; supervision, V.G.D.; project administration, O.I.N.; funding acquisition, O.I.N. and V.G.D. All authors have read and agreed to the published version of the manuscript.

Funding: This research was funded by the World Wide Fund for Nature (WWF-Russia), grant number WWF001462.

Acknowledgments: We wish to thank our peer reviewers for their insightful comments and suggestions. We particularly thank Peter Osipov and Anna Serdyuk (Amur branch of WWF-Russia) for all their help and support throughout this research work. We would like to express our great appreciation to Oxana Erina (Moscow State University) for her enthusiastic encouragement and suggestions during the development of this research work. We are grateful to Yury Darman (advisor of the Amur branch of WWF-Russia) and Eugene Simonov (the Rivers without Boundaries Coalition) for their support and advice on the issue of environmental flows and freshwater ecosystem conservation.

Conflicts of Interest: The authors declare no conflict of interest.

References

1. Living Planet Report 2020: Bending the Curve of Biodiversity Loss: A Deep Dive into Freshwater. Available online: https://www.wwf.org.uk/sites/default/files/2020-09/LPR2020_freshwater.pdf (accessed on 30 September 2020).

2. Living Planet Report 2018: Aiming Higher. Available online: https://www.worldwildlife.org/pages/living-planet-report-2018 (accessed on 30 September 2020).

3. Grill, G.; Lehner, B.; Thieme, M.; Geenen, B.; Tickner, D.; Antonelli, F.; Babu, S.; Borrelli, P.; Cheng, L.; Crochetiere, H.; et al. Mapping the world's free-flowing rivers. *Nature* **2019**, *569*, 215–221. [CrossRef]

4. Gibson, L.; Wilman, E.N.; Laurance, W.F. How Green is "Green" Energy? *Trends Ecol. Evol.* **2017**, *32*, 922–935. [CrossRef] [PubMed]

5. World Commission on Dams. *Dams and Framework for Decision-Making*; Earthscan Publications Ltd.: London, UK; Sterling, VA, USA, 2000.

6. Ansar, A.; Flyvbjerg, B.; Budzier, A.; Lunn, D. Should we build more large dams? The actual costs of hydropower megaproject development. *Energy Policy* **2014**, *69*, 43–56. [CrossRef]

7.   Volovik, S.P.; Chikhachev, A.S. Anthropogenic transformation of the fish fauna of the Azov basin. In *Main Problems of Fisheries and the Protection of Fishery Reservoirs of the Azov-Black Sea Basin*; AzNIIRkh: Rostov-on-Don, Russia, 1998; pp. 7–22.

8.   Pavlov, D.S.; Savvaitova, K.A.; Sokolov, L.I.; Alekseev, S.S. *Rare and Endangered Animals—Fishes*; Higher school: Moscow, Russia, 1994; p. 334.

9.   Kirillov, V.V.; Kotsyuk, D.V.; Vizer, A.M.; Popov, P.A. Assessment of the impact of the constructed hydroelectric power plants on water and biological resources and their habitat in Siberia and the Far East. In *Fisheries Problems of Construction and Operation of Dams and Ways to Solve Them: Materials of the Meeting of the Thematic Community on the Problems of Large Dams and The Scientific Council of the Interdepartmental Ichthyological Commission*; WWF Russia: Moscow, Russia, 2010; pp. 19–32.

10. Ivanov, V.P.; Mazhnik, A.Y. *Fisheries of the Caspian Basin (White Book)*; TOO Fish Ind.: Moscow, Russia, 1997; p. 40.

11. Dubinina, V.G.; Zhukova, S.V. Assessment of the possible consequences of the construction of the Bagaevsky hydroelectric complex for the ecosystem of the Lower Don. *Fish Ind.* **2016**, *4*, 20–30.

12. Dubinina, V.G.; Nikitina, O.I. Considering environmental factor in water resources management of reservoirs. In Proceedings of the Reservoirs of the Russian Federation: Modern Environmental Problems, State, Management: Collection of Materials of the All-Russian Scientific and Practical Conference, Sochi, Russia, 23–29 September 2019; Lik: Novocherkassk, Russia, 2019.

13. Tickner, D.; Opperman, J.; Abell, R.; Acreman, M.; Arthington, A.; Bunn, S.; Cooke, S.; Dalton, J.; Darwall, W.; Edwards, G.; et al. Bending the Curve of Global Freshwater Biodiversity Loss: An Emergency Recovery Plan. *BioScience* **2020**, *70*, 330–342. [CrossRef] [PubMed]

14. Volovik, S.P.; Dubinina, V.G.; Semenov, A.D. Hydrobiology and dynamics of fisheries in the Azov Sea. In *General Fisheries Council for the Mediterranean. Studies and Reviews*; FAO: Rome, Italy, 1993; pp. 1–58.

15. Dubinina, V.G. Hydrological basis for increasing the scale of natural fish reproduction in the Azov-Don region. Fisheries research of the Sea of Azov. *Azniirkh Bull.* **1972**, *10*, 41–51.

16. Arthington, A.H.; Bhaduri, A.; Bunn, S.E.; Jackson, S.E.; Tharme, R.E.; Tickner, D.; Young, B.; Acreman, M.; Baker, N.; Capon, S.; et al. The Brisbane Declaration and Global Action Agenda on Environmental Flows. *Front. Environ. Sci.* **2018**, *6*, 45. [CrossRef]

17. Harwood, A.; Johnson, S.; Richter, B.; Locke, A.; Yu, X.; Tickner, D. *Listen to the River: Lessons from a Global Review of Environmental Flow Success Stories*; WWF-UK: Woking, UK, 2017; Available online: https://www.wwf.org.uk/sites/default/files/2017-09/59054%20Listen%20to%20the%20River%20Report%20download%20AMENDED.pdf (accessed on 30 September 2020).

18. Hydrometeorological publishing house. *Surface Water Resources of the USSR*; Hydrometeorological publishing house: Leningrad, Russia, 1966.

19. Simonov, E.A.; Dahmer, T.D. *Amur-Heilong River Basin Reader*; Ecosystems: Hong Kong, 2008.

20. Antonov, A.L.; Barabanshikov, E.I.; Zolotukhin, S.F.; Mikheev, I.E.; Shapovalov, M.E. *Fish of the Amur*; WWF: Vladivostok, Russia, 2019; p. 318.

21. Novomodny, G.V.; Zolotukhin, S.F.; Sharov, P.O. *The Amur Fish: Wealth and Crisis*; AVK Orange: Vladivostok, Russia, 2004; p. 51.

22. Sapaev, V.M. Regulation of the Amur River. Is it possible to optimize environmental conditions? *Sci. Nat. FE* **2006**, *2*, 86–95.

23. Zaletaev, V.S.; Dikareva, T.V.; Podolsky, S.A. Mega-ecosystem of a river basin and studies of intra-basin processes. In *Water: Ecology and Technology: Third International Congress. Ekvatek-98*; SIBIKO International: Moscow, Russia, 1998; pp. 49–50.

24. Shalygin, A.L.; Dugina, I.O. Catastrophic 2013 flood in the Amur River basin: Causes, features and consequences. In *Extreme Floods in the Amur River Basin: Hydrological Aspects. Papers on Hydrology*; Georgievsky, V.J., Ed.; FSBI "SHI": St. Petersburg, Russia, 2015; pp. 21–34.

25. Makhinov, A.N.; Kim, V.I.; Voronov, B.A. Flood of 2013 in the Amur basin: Causes and consequences. *FEB RAS Bull.* **2014**, *2*, 5–14.

26. Gartsman, B.I. *Rain-Induced Floods in Rivers in the Southern Far East: Methods for Calculation, Prediction, and Risk Assessment*; Dal'nauka: Vladivostok, Russia, 2008.

27. Kim, V.I. Flood features within the floodplain islands of the Lower Amur (on the example of Slavyansky Island). In *Formation of Surface Waters in the South of the Far East*; Far East Scientific Center of the Academy of Sciences of the USSR: Vladivostok, Russia, 1988; pp. 21–30.

28. Dubinina, V.G. *Methodical Foundations of the Environmental Regulation of the Water Irrevocable Withdrawals and the Establishment of Environmental Flow (Release)*; Economics and Informatics: Moscow, Russia, 2001; p. 120.

29. Fashchevsky, B.V. The ecological significance of the floodplain in the river ecosystems. *Sci. Notes Rshmu* **2007**, *5*, 118–119.

30. FGUP RosNIIVKh. *Scheme of the Integrated Use and Protection of Water Bodies in the Amur River Basin*; Fgup RosNIIVKh: Yekaterinburg, Russia, 2013.

31. Grek, E.A.; Ivanov, V.A.; Molchanova, T.G. Catastrophic floods on the Amur River in XIX–XX centuries. In *Extreme Floods in the Amur River Basin: Hydrological Aspects. Papers on Hydrology*; Georgievsky, V.J., Ed.; FSBI "SHI": St. Petersburg, Russia, 2015; pp. 4–20.

32. Gotvansky, V.I.; Sirotsky, S.E. Managing floods instead of fighting. *Natl. Inf. Agency Nat. Resour.* **2013**, *12*, 5.

33. Simonov, E.A.; Nikitina, O.I.; Osipov, P.O.; Egidarev, E.G.; Shalikovsky, A.V. *Amur Floods and Us: Lesson (Un)Learned*; Shalikovsky, A.V., Ed.; WWF-Russia: Moscow, Russia, 2016; p. 216. [CrossRef]

34. RusHydro. 2020 Zeya, H.P.P. Available online: http://www.zges.rushydro.ru/hpp/general/ (accessed on 13 August 2020).

35. Center for the Russian Register of Hydraulic Structures and State Water Cadastre. *Draft of the Rules for the Use of the Zeya Reservoir on the River Zeya—Report on the Topic M-10-02 "Modification of the Rules for the Use of Water Resources of the Zeya Reservoir"*; Center for the Russian Register of Hydraulic Structures and State Water Cadastre: Moscow, Russia, 2011.

36. RusHydro. 2020 Bureya, H.P.P. Available online: http://www.burges.rushydro.ru/hpp/general/ (accessed on 13 August 2020).

37. RusHydro. 2020 Lower Bureya, H.P.P. Available online: http://www.nbges.rushydro.ru/ (accessed on 13 August 2020).

38. Danilov-Danilyan, V.I.; Gelfan, A.N.; Motovilov, Y.G.; Kalugin, A.S. Disastrous flood of 2013 in the Amur basin: Genesis, recurrence assessment, simulation results. *Water Resour.* **2014**, *4*, 115–125. [CrossRef]

39. Bolgov, M.V.; Alekseevskii, N.I.; Gartsman, B.I.; Georgievskii, V.Y.; Shalygin, A.L.; Dugina, I.O.; Kim, V.I.; Makhinov, A.N. The 2013 extreme flood within the Amur basin: Analysis of flood formation, assessments and recommendations. *Geogr. Nat. Resour.* **2015**, *36*, 225–233. [CrossRef]

40. Bolgov, M.V.; Korobkina, E.A.; Osipova, N.V.; Filippova, I.A. The Analysis of Long-term Variability and Estimation of the Maximum Water Levels under Conditions of High Anthropogenic Impact for the Amur River. *Russ. Meteorol. Hydrol.* **2016**, *41*, 577–584. [CrossRef]

41. RusHydro. Flood Control Projects in the Amur Basin. 2013. Available online: http://blog.rushydro.ru/?p=9230 (accessed on 13 August 2020).

42. Motovilov, Y.G.; Danilov-Danilyan, V.I.; Dod, E.V.; Kalugin, A.S. Assessing the flood control effect of the existing and projected reservoirs in the Middle Amur basin by physically-based hydrological models. *Water Resour.* **2015**, *42*, 580–593. [CrossRef]

43. Gusev, M.N.; Pomiguev, Y.V. Channel activity of rivers of the Amurskaya region in conditions of current economic management. *Geogr. Nat. Resour.* **2008**, *29*, 137–141. [CrossRef]

44. Nikitina, O.I.; Bazarov, K.Y.; Egidarev, E.G. Application of remote sensing data for measuring freshwater ecosystems changes below the Zeya dam in the Russian Far East. *Proc. IAHS* **2018**, *379*, 49–53. [CrossRef]

45. Nikitina, O. Impacts of dams on freshwater ecosystems and flow regimes in the Middle Amur River basin. In *Uniting Europe for Clean Water: Cross-Border Cooperation of Old, New and Candidate Countries of EU for Identifying Problems, Finding Causes and Solutions*; Book of Abstracts: Budapest, Hungary, 2017; pp. 103–104.

46. Nikitina, O.I. *Transformation of Water and Floodplain Ecosystems of the Middle Amur as a Result of Flow Regulation—Scientific Graduate Report*; Water Problems Institute of the Russian Academy of Sciences: Moscow, Russia, 2018; p. 132.

47. Vlasova, S.N.; Voropayeva, I.S.; Dugina, I.O.; Dunayeva, I.M.; Kovtun, A.V.; Kramareva, L.S.; Krasnova, E.D.; Lobarev, S.A.; Mardashova, M.V.; Nikitina, O.I.; et al. *Giltchin River Watershed. History. Wetlands. Water Resources*; Smirenski, S.M., Smirenski, E.M., Eds.; Dalnauka: Vladivostok, Russia, 2016; p. 202.

48. Krasnova, E.D.; Gerasimova, O.V.; Belyaev, B.M. *Expedition Report on the Study of the Lakes in The Muravyevsky Park for Sustainable Nature Management*; MSU: Moscow, Russia, 2016; p. 42.

49. Wieland, H.; Smirenski, S.M. The importance of Muraviovka Park, Amur province, Far East Russia, for bird species threatened at regional, national and international level based on observations between 2011 and 2016. *Forktail* **2017**, *33*, 81–87.

50. Kotsyuk, D.V. Ichtyofauna structure and dynamics of the stock of basic food fish of the Zeya reservoir. In *Readings in Memory of S.M. Konovalov*; TINRO-Center: Vladivostok, Russia, 2008; pp. 133–137.

51. Dubinina, V.G.; Gargopa, Y.M.; Chebanov, M.S.; Katunin, D.N.; Fil, S.A. Methodological approaches to ecological regulation of anthropogenic runoff reduction. *Water Resour.* **1996**, *1*, 78–85.

52. Semenchenko, N.N.; (TINRO-Center, Vladivostok, Russia). Personal communication, 2018.

53. Golovko, V.I. *Justification of the Need to Provide Fishery Releases in the Regulation of the Operation of Hydroelectric Power Plants of the Amur Region—Appendix to the Letter to the Ministry of Agriculture and Agriculture*; Center for Fisheries Research: Blagoveshchensk, Russia, 2008.

54. Semenchenko, N.N. Hydrological regime of the Amur River and the number of commercial freshwater fish. In *The Current State of Aquatic Bioresources: Materials of Scientific Conference Dedicated to the 70th Anniversary of S.M. Konovalov*; TINRO-Center: Vladivostok, Russia, 2008; pp. 246–250.

55. Kotsuk, D.V. Ichtiological research in the Zeya River Basin. In *Hydroecological Monitoring of the Area Under the Zeya Hydropower Dam Impact*; FEB RAS: Khabarovsk, Russia, 2010; pp. 260–320.

56. Buzin, V.A.; Goroshkova, N.I. Features of formation and method of predicting highest ice-jam stages in the Amur River. In *Extreme Floods in the Amur River Basin: Hydrological Aspects. Papers on Hydrology*; Georgievsky, V.J., Ed.; FSBI "SHI": St. Petersburg, Russia, 2015; pp. 141–151.

57. Andronov, V.A.; Andronova, R.S.; Parilov, M.P.; Darman, Y.A. The current state of protection and knowledge of cranes in the South of the Far East. *Inf. Bull. Work Group Cranes Eurasia* **2001**, *2*, 13–14.

58. Winter, S.W. Diet of the Oriental White Stork (*Ciconia boyciana Swinhoe*) in the Middle Amur region, USSR. In *Biology and Conservation of the Oriental White Stork Ciconia Boyciana*; Savannah River Ecology Laboratory: Aiken, SC, USA, 1991; pp. 31–45.

59. Parilov, M.P. On the Ecology of the Japanese Crane in the Khingansky Reserve. Master's Thesis, Irkutsk State University, Irkutsk, Russia, 1996.

60. Ignatenko, S.Y.; Parilov, M.P.; Kastrikin, V.A.; Gusev, M.N. Impact of the Bureya and Zeya hydropower dams on nesting groups of cranes and storks within the Arkhara lowland. In *VII Far Eastern Conference on Conservation: Conference Materials*; ICARP FEB RAS: Birobidzhan, Russia, 2005; pp. 123–126.

61. Sapaev, V.M.; Voronov, B.A. *Composition and Forecast of the Downstream Changes in the Endangered Birds Habitat Due to the Creation of the Bureya Hydropower Dam*; Khabarovsk Complex Research Institute, USSR Academy of Sciences, Far Eastern Scientific Center: Khabarovsk, Russia, 1979; p. 123.

62. Ministry of Agriculture of the Russian Federation, Federal Agency for Fisheries, Khabarovsk Branch of the Pacific Research Fisheries Center, FGU "Amurrybvod", IVEP FEB RAS. *Scientific Socio-Ecological Monitoring and Databases of the Influence Zone of the Bureya Hydropower Dam. Ichthyological Monitoring of the Bureya Reservoir and the Bureya River Downstream from the Dam*; Ministry of Agriculture of the Russian Federation: Khabarovsk, Russia, 2008.

63. Federal Law. The Water Code of the Russian Federation of 03.06.2006 N 74-FZ. Available online: http://docs.cntd.ru/document/901982862 (accessed on 9 September 2020).

64. Finalization of the Comprehensive Schemes for Water Bodies Management and Protection Project for the Amur River basin. Book 4. Water Balances and Balances of Pollutants of Water Bodies of the Amur River. Available online: http://amurbvu.ru/576-skiovo-po-basseynu-reki-amur.html (accessed on 19 August 2020).

65. Standards for Permitted Limits of Impact on Water Bodies of Water Management Areas of the Amur River Basin. Approved by the Federal Water Resources Agency. Available online: http://amurbvu.ru/566-utverzhdennye-proekty-ndv.html (accessed on 19 August 2020).

66. Makhinov, A.N.; Kim, V.I. *Public Environmental Expertise of the Draft Scheme of Integrated Use and Protection of Water Bodies of the Amur River Basin*; Institute of Water and Ecological Problems FEB RAS: Khabarovsk, Russia, 2014; p. 26. Available online: https://amurinfocenter.org/directions/svobodno-tekushchiy-amur/obshchestvennaya-ekologicheskaya-ekspertiza-proekta-skhemy-kompleksnogo-ispolzovaniya-i-okhrany-vodn/ (accessed on 19 August 2020).

67. Ministry of Natural Resources and Environment of the Russian Federation. *Order of the Ministry of Natural Resources of the Russian Federation of January 26, 2011 No. 17 "On Approval of Guidelines for the Development of Rules for the Use of Reservoirs"*; Ministry of Natural Resources and Environment of the Russian Federation: Moscow, Russia, 2011.

68. Bureya, H.P.P. *The High-tension Area*; Podolsky, S.A., Ed.; WWF-Russia: Moscow, Russia, 2005; p. 80.

69. WWF. *WWF-Russia's Appeal on the Rules for the Use of Water Resources of the Zeya and Bureya Reservoirs*; World Wide Fund for Nature (WWF-Russia): Vladivostok, Russia, 2014.

70. Botha, M. Assessment of Biodiversity Mitigation for the Nizhne-Bureyskaya Hydroelectric Power Plant. Stage 1. Report to the UNDP GEF Project on "Mainstreaming Biodiversity into the Russian Energy Sector Policies and Operations"; GEF Agency: United Nations Development Programme. Available online: http://bd-energy.ru/documents/ENG%20Site/Reports/Assessment%20of%20Biodiversity%20Mitigation%20activities%20Nizhne-%20Bureyskaya%20HPP.pdf (accessed on 14 August 2020).

71. RusHydro. At the Lower Bureya HPP, the "Bureysky Compromise" Program is Being Completed. 2017. Available online: http://www.rushydro.ru/press/news/103941.html (accessed on 31 July 2020).

72. Simonov, E.A.; Nikitina, O.I.; Egidarev, E.G. Freshwater Ecosystems versus Hydropower Development: Environmental Assessments and Conservation Measures in the Transboundary Amur River Basin. *Water* **2019**, *11*, 1570. [CrossRef]

73. Dubinina, V.G.; Kosolapov, A.E.; Koronkevich, N.K.; Chebanov, M.S.; Skachedub, E.A. Methodological approaches to assessing environmental regulation of irrevocable water withdrawal and environmental flow (release). *Water Ind. Russ.* **2009**, *3*, 26–61.

74. Dubinina, V.G.; Kosolapov, A.E.; Koronkevich, N.I.; Chebanov, M.S. *Methodological Guidelines for the Irrevocable Water Withdrawal Regulation and Environmental Flow (Release) Establishment; State Contract No. M-08–18, 16 May 2008; Federal State Institution*; Interdepartmental Ichthyological Commission: Moscow, Russia, 2008; p. 40.

75. Dubinina, V.G.; Kosolapov, A.E.; Vuglinsky, V.S. Application of methodological documents on standardization of permitted irrevocable water withdrawal. In *Safety Problems in the Water Sector of Russia*; Avangard Plus LLC: Krasnodar, Russia, 2010; pp. 338–348.

76. Dubinina, V.G.; Nikitina, O.I. *Transboundary Water-Resource Distribution. Environmental Flow as the Basis for Freshwater Ecosystem Conservation (Russian Federation)*; Technical Report № 70/19-KAPE; KAPE: Moscow, Russia, 2020. [CrossRef]

77. Federal Agency for Water Resources of the Russian Federation; Amur Basin Water Administration. *Draft of the Permissible Exposure Limits for the Amur River Basin: Bureya River*; Federal Agency for Water Resources of the Russian Federation: Khabarovsk, Russia, 2012.

78. Afanasyev, P.Y. *Floods of the Upper Amur Region*; Bureyskaya HPP: Talakan, Russia, 2012; p. 48.

79. Darman, G.F. Waters. In *Nature Chronicle of the Khingan Reserve for 1985*; Khingan Reserve: Arkhara, Russia, 1986; p. 24.

80. Hydromet Centre of Russia. Climate Data for Cities Worldwide. Monthly Data. Arkhara. Available online: https://meteoinfo.ru/climatcities?p=1659 (accessed on 23 July 2020).

81. Hydrograph of Water Levels and Flow Rates. Discharge Record of the Bureya Dam. Available online: http://hgraph.ru/burges (accessed on 23 July 2020).

82. Simonov, E.A.; Egidarev, E.G.; Menshikov, D.A.; Khalyapin, L.E.; Korolyov, G.S. *Comprehensive Environmental and Socio-Economic Assessment of Hydropower Development in the Amur River Basin*; WWF-Russia, En+ Group: Moscow, Russia, 2015; p. 279. Available online: https://wwf.ru/resources/publications/booklets/comprehensive-environmental-and-socio-economic-assessment-of-hydropower-development-in-the-amur-rive/ (accessed on 14 August 2020).

83. Federal Agency for Water Resources of the Russian Federation; Amur Basin Water Administration. *Draft of the Permissible Exposure Limits for the Amur River Basin: Zeya River*; Federal Agency for Water Resources of the Russian Federation: Khabarovsk, Russia, 2012.

84. Bortin, N.N.; Gorchakov, A.M.; Milyaev, V.M. On the issue of permitted surface water withdrawal and environmental flow in water bodies of the Amur River basin. *Water Manag. Russ.* **2013**, *3*, 4–15.

85. Dubinina, V.G.; Nikitina, O.I.; Markov, M.L. Methodological approaches to determining the volumes of permitted irrevocable water withdrawal from poorly studied, unexplored and small rivers. *Water Manag. Russ.* **2015**, *4*, 80–97.

86. Hydrograph of Water Levels and Flow Rates. Discharge Record of the Zeya Dam. Available online: http://hgraph.ru/zeyges (accessed on 23 July 2020).

87. RusHydro. The Shore Spillway of the Sayano-Shushenskaya HPP. Available online: http://www.sshges. rushydro.ru/hpp/spillway/ (accessed on 23 July 2020).

88. Cheng, L.; Opperman, J.J.; Tickner, D.; Speed, R.; Guo, Q.; Chen, D. Managing the Three Gorges Dam to Implement Environmental Flows in the Yangtze River. *Front. Environ. Sci.* **2018**, *6*, 64. [CrossRef]

89. Barrios, O.J.E.; Rodríguez, S.; López, S.; Pérez, M.; Villón Bracamonte, R.; Rosales, A.F.; Guerra, G.A.; Sánchez Navarro, R. *National Water Reserves Program in Mexico: Experiences with Environmental Flows and the Allocation of Water for the Environment*; Interamerican Development Bank, Water and Sanitation Division: Washington, DC, USA, 2015.

90. Lenhydroproject. *Temporary Rules for the Use of Water Resources of the Bureya Reservoir on the Bureya River for the Period of May 2009–April 2010*; 1279–188t; Lenhydroproject: St. Petersburg, Russia, 2009; p. 61.

91. Set of rules and regulations № 2.07.01-89*. In *Urban Planning. Planning and Development of Urban and Rural Settlements*; FGUP TSPP: Moscow, Russia, 2007; p. 56. Available online: https://files.stroyinf.ru/Data2/1/4294854/4294854799.pdf (accessed on 9 September 2020).

92. Owusu, A.G.; Mul, M.; van der Zaag, P.; Slinger, J. Re-operating dams for environmental flows: From recommendation to practice. *River Res. Appl.* **2020**, 1–11. [CrossRef]

93. Kalugin, A.S. Assessment of the sensitivity of the Amur River flow to possible climate changes based on retrospective data. In Proceedings of the IX International Scientific Conference of Young Scientists and Talented Students, Water Resources, Ecology and Hydrological Safety, Moscow, Russia, 30 November–1 December 2015; pp. 46–49.

94. Makhinov, A.V.; Kim, V.I. Effect of climate changes on the hydrological regime of the Amur River. *Pac. Geogr.* **2020**, *1*, 30–39. [CrossRef]

95. World Bank Group. *Environmental Flows for Hydropower Projects: Guidance for the Private Sector in Emerging Markets*; International Finance Corporation: Washington, DC, USA, 2018; Available online: https://openknowledge.worldbank.org/handle/10986/29541 (accessed on 14 August 2020).

96. Nikitina, O.I.; Simonov, E.A.; Egidarev, E.G. River flood adaptation in the Amur Basin and nature conservation. *Use Prot. Nat. Resour. Russ.* **2015**, *3*, 15–24.

97. Egidarev, E.G.; Simonov, E.A.; Nikitina, O.I.; Osipov, P.E.; Shalikovsky, A.V. Wild Floods in the Amur River Basin. In *Heritage Dammed: Water Infrastructure Impacts on World Heritage Sites and Free Flowing Rivers*; World Heritage Watch: Berlin, Germany, 2019; pp. 80–83.

98. Egidarev, E.G. Mapping and evaluation of floodplains in the Amur River valley. *Vestnik DVO–FEB RAS Bull.* **2012**, *2*, 9–16.

Article

Species-Richness Responses to Water-Withdrawal Scenarios and Minimum Flow Levels: Evaluating Presumptive Standards in the Tennessee and Cumberland River Basins

Lucas J. Driver [1,*], Jennifer M. Cartwright [2], Rodney R. Knight [2] and William J. Wolfe [2]

[1] U.S. Geological Survey, Lower Mississippi-Gulf Water Science Center, Little Rock, AR 72211, USA
[2] U.S. Geological Survey, Lower Mississippi-Gulf Water Science Center, Nashville, TN 37211, USA; jmcart@usgs.gov (J.M.C.); rrknight@usgs.gov (R.R.K.); wmjwolfe@yahoo.com (W.J.W.)
* Correspondence: ldriver@usgs.gov; Tel.: +1-501-228-3600

Received: 19 March 2020; Accepted: 6 May 2020; Published: 8 May 2020

Abstract: Water-resource managers are challenged to balance growing water demand with protecting aquatic ecosystems and biodiversity. Management decisions can benefit from improved understanding of water-withdrawal impacts on hydrologic regimes and ecological assemblages. This study used ecological limit functions for fish groups within the Tennessee and Cumberland River basins to predict species richness responses under simulated constant-rate (CR) and percent-of-flow (POF) withdrawals and for different minimum flow level protections. Streamflow characteristics (SFC) and richness were generally less sensitive to POF withdrawals than CR withdrawals among sites, fish groups, and ecoregions. Species richness generally declined with increasing withdrawals, but responses were variable depending on site-specific departures of SFCs from reference conditions, drainage area, fish group, ecoregion, and minimum flow level. Under POF withdrawals, 10% and 20% daily flow reductions often resulted in loss of <1 species and/or ≤5% richness among fish groups. Median ecological withdrawal thresholds ranged from 3.5–31% for POF withdrawals and from 0.01–0.92 m^3/s for CR withdrawals across fish groups and ecoregions. Application of minimum flow level cutoffs often resulted in damping effects on SFC and richness responses, indicating that protection of low streamflows may mitigate hydrologic alteration and fish species richness loss related to water withdrawals. Site-specific and regionally summarized responses of flow regimes and fish assemblages under alternative withdrawal strategies in this study may be useful in informing water-management decisions regarding streamflow allocation and maintaining ecological flows.

Keywords: environmental flows; water management; water withdrawal; fish species richness; streamflow alteration; withdrawal threshold; minimum flow level; percent-of-flow; presumptive standard

1. Introduction

The natural flow regime is considered a master variable in determining lotic ecosystem processes [1]. It is well documented that alterations of streamflow regimes pose significant threats to riverine ecosystems and biodiversity worldwide [2–6]. Many studies have shown that aquatic communities can be negatively impacted by alteration of the magnitude, timing, and/or duration of flows to which they are adapted [2,4–8]. Water withdrawals for domestic, agricultural, and industrial purposes can directly reduce streamflow quantity and disrupt the natural timing and variability of flows [9]. With increasing human population and intensifying land-use practices, global water withdrawals have already increased dramatically over the past 50 years [10] and are expected to continue [11].

Effects of streamflow alteration on fish have been shown to be variable and context dependent but can have significant negative impacts on fish growth, reproduction, and survival, and ultimately community composition and diversity [5,8]. For example, altered streamflow variability has been associated with a loss of species richness [5], particularly among riffle-dwelling species [12,13], and reductions in streamflow via water withdrawals have been shown to impact the composition and structure of fish assemblages, specifically the occurrence and proportion of fluvial-dependent species [14,15].

As the global water demand continues to intensify, water-resource managers face increasing challenges to satisfy societal water needs while also protecting ecosystem flow requirements and biodiversity [6,16–18]. In the past several decades, much effort has been given to developing statistical and methodological frameworks to describe and quantify flow-ecology relationships to better inform resource managers and set conservation targets [3,18–23]. Despite increasing awareness of the importance of flow-ecology relationships [16,18,24] and continuing efforts to translate these relationships into decision support systems, the integration and implementation of ecological-flows management guidelines are continuing challenges at local and regional scales [25–27].

Resource managers seek accessible and easily implemented guidelines on water-withdrawal practices [18]. For example, minimum flow requirements and protection of minimum flow levels (MFL), which prohibit withdrawals when streamflow falls below specified critical levels, were some of the earliest ecological flow management strategies and are still commonly utilized [28]. However, it is well recognized that employment of MFLs alone are often insufficient to protect stream integrity across a range of hydrologic conditions [1,18,26,29]. Constant-rate (CR) withdrawal strategies are generally fixed withdrawal rates and are not related to changes in ambient flows. CR withdrawals are commonly used, proposed, or permitted for various surface and groundwater-withdrawal purposes based on the amount of water needed (i.e., consumer demand), the overall withdrawal capacity (i.e., supply and pumping capacity) of a facility, or basic rules-of-thumb, rather than on empirical, evidence-based ecological-flow relationships or dynamic environment and flow conditions. The potential for hydrologic and ecologic impacts of CR withdrawals can be quite high because CR withdrawals can strongly affect the streamflow regime by altering the natural (or baseline) variability of flows, particularly aspects of low flow. On the other hand, variable-rate water-withdrawal strategies based, for example, on set percentages of the natural or baseline flow regime (i.e., percentage-of-flow, POF) are increasingly recognized as attractive water-management strategies because they exert proportional withdrawals on the underlying flow regime and thereby are better suited to preserving natural variability of flows across the entire hydrograph [29,30]. More recently, presumptive POF water-withdrawal standards have been proposed at 10–20% of daily streamflow [30] which, in theory, inherently provide ecological protection. POF withdrawals and presumptive standards are conceptually simple strategies that can be easily adopted by water-resource managers, but implementation of POF withdrawals requires accurate streamflow measurements on a frequent time step (e.g., daily), and the ability to adjust withdrawal rates in real time. Although the vast majority of streams are ungauged, discrete or instantaneous streamflow can be measured using relatively simple methods [31], and prediction of steamflow at ungauged locations may be possible using basin-area/runoff models [32].

Possible disconnects exist between the cost, time, and amount of information needed to establish sound evidenced-based environmental flow requirements versus the type and urgency of information often sought by resource managers. These mismatches can prohibit or slow the practical implementation of flow-ecology guidelines, particularly under contentious social, political, and economic contexts [19,21,26]. In the meantime, water-resource managers and regulators are commonly left to make decisions based on generalized rules-of-thumb and experience rather than evidence-based approaches [18,21,26,33].

Water-management decisions associated with permitting and allocation of water withdrawals can benefit from a better understanding of how different streamflow-withdrawal strategies impact ecological communities and the establishment of evidence-based guidelines to balance water withdrawals and protection of ecological assemblages [18,27,30]. Specifically, current understanding is limited regarding the range of potential ecological responses along a gradient of hydrologic alteration scenarios (i.e., so-called "top-down" approach; [19]) using plausible water-management strategies (e.g., water-withdrawal rates and MFL protections). Additionally, there is little consensus regarding what might be considered 'acceptable' limits of ecological change (e.g., species loss) or biological condition with which to establish potential ecological-flow thresholds [21,22,34,35]. Despite the growing number of ecological-flow studies investigating ecological responses in relation to various types of flow alteration (e.g., related to climate change, impoundments, withdrawals, and land-use changes) [5,17,26,34,36], few have explored these relationships in an incremental fashion across relatively long alteration gradients or set out to identify ecological or hydrologic thresholds.

This study expands on a series of previously developed hydrologic and ecological relationships within the Cumberland River and Tennessee River basins [37–41]. We used published ecological-limit functions (ELFs) [39,41] for fish groups within Blue Ridge (BR), Ridge and Valley (RV), Interior Plateau (IP), and Cumberland Plateau (CP) Level III ecoregions [42] to model incremental changes in streamflow characteristics (SFCs) and species richness responses along gradients of simulated withdrawal scenarios. We ran these simulations using simplified CR and POF withdrawal strategies and different MFL protections. The primary objectives of this study were to (1) simulate water withdrawals and characterize potential hydrologic and ecological impacts of CR and POF scenarios and (2) identify potential withdrawal thresholds below which conservative levels of species richness loss may be minimized.

We investigated the following general hypotheses related to hydrologic and ecological impacts of CR compared to POF withdrawals: (1) SFCs should be less sensitive to POF withdrawal scenarios than to CR withdrawals, particularly SFCs associated with flow ratios, frequency, variability, and timing of flows; (2) resulting predicted fish species richness should generally decrease with increased water withdrawals regardless of CR or POF withdrawals; and (3) application of MFLs should generally dampen patterns of change of SFCs, species richness, and withdrawal threshold responses regardless of CR or POF withdrawals.

2. Materials and Methods

Data analyses and prediction of ecological effects of water-withdrawal scenarios were performed using flow-ecology relationships (i.e., ELFs) that were developed in a series of previously published studies conducted on stream sites within the Tennessee and Cumberland River basins [37–39,41]. In [37], a large suite of SFCs were statistically related to metrics of fish diversity at select free-flowing gauged stream sites and a subset of ecologically significant SFCs were identified for the BR, RV, CP, and IP ecoregions (see Table 1).

Table 1. Definitions of streamflow characteristic (SFC) used in ecological limit functions and the respective ecoregion(s) where they were found to be statistically significant; modified from [39].

Flow Category	Streamflow Characteristic (SFC)	Definition (Units)	Eco-Region
Magnitude	MA41: mean annual runoff	Annual mean streamflow divided by the drainage area ($ft^3s^{-1}mi^{-2}$)	RV, CP
	AMH10: maximum October streamflow	Maximum October flow across period of record divided by watershed area ($ft^3s^{-1}mi^{-2}$)	BR, CP, IP
	LRA7: rate of streamflow recession	Log of the median change in log of flow for days that the change is negative across the entire flow record (flow units per day)	IP
Ratio	LDH13: average 30-day maximum	Log of the average over period of record of annual maximum 30-day moving average flows divided by median for entire record	CP
	ML20: base flow	Divide daily flow record into 5-day blocks. Assign minimum (min.) flow for the block as a base flow if 90% of that min. flow is less than the min. flows for blocks on either side; otherwise, set to zero. Fill in zero values using linear interpolation. Compute total flow and total base flow for entire record. ML20 is total flow: total base flow (ratio)	CP
	TA1: constancy	Measure the stability of flow regimes by dividing daily flows into predetermined flow classes	RV, CP, IP

Table 1. *Cont.*

Flow Category	Streamflow Characteristic (SFC)	Definition (Units)	Eco-Region
Frequency	FH6: frequency of moderate flooding	Average number of high-flow events per year ≥ 3 times the median annual flow for the period of record (number per year)	IP
Variability	LDL6: variability of annual minimum daily average streamflow	Log of the standard deviation for the annual minimum daily average streamflow. Multiply by 100 and divide by the mean streamflow for the period (%)	CP
	LDH16: variability in high-pulse duration	Log of the standard deviation for the yearly average high-flow pulse durations (daily flow > 75th percentile) (%)	RV
	FL2: variability in low-pulse count	Coefficient of variation for the number of annual occurrences of daily flows < 25th percentile	RV
Timing	TL1: annual min. flow	Date of annual mininum (min.) flow occurrence (Julian day)	CP, IP

Hydrological reference profiles (i.e., interquartile ranges; IQR) of SFCs were then defined for reference streams within BR, RV, and IP [38,39] and CP [41]. Finally, ELFs were developed using quantile regression that described the upper limits (≥85th quantile) of species richness responses among different fish groups as a function of the hydrologic departure from reference of SFC(s) among streams within BR, RV, and IP [39] and CP [41].

2.1. Study Area

The geographic scope of this study includes the Tennessee and Cumberland River basins, which together drain approximately 150,000 km^2 in Tennessee, Virginia, Georgia, Alabama, Mississippi, and Kentucky in the USA (Figure 1). Population growth, intensification of land-use practices, and growing demand for surface water and groundwater throughout this region pose increasing risks for hydrologic alteration and degradation of water quality [10]. Further, with more than 250 species of fish, 115 crayfish, 141 freshwater mussels, and 160 aquatic snails, including more than 100 taxa identified as being at risk, the Tennessee and Cumberland River basins collectively represent a major concentration of North American freshwater biodiversity and one of the most diverse temperate freshwater systems in the world [43,44].

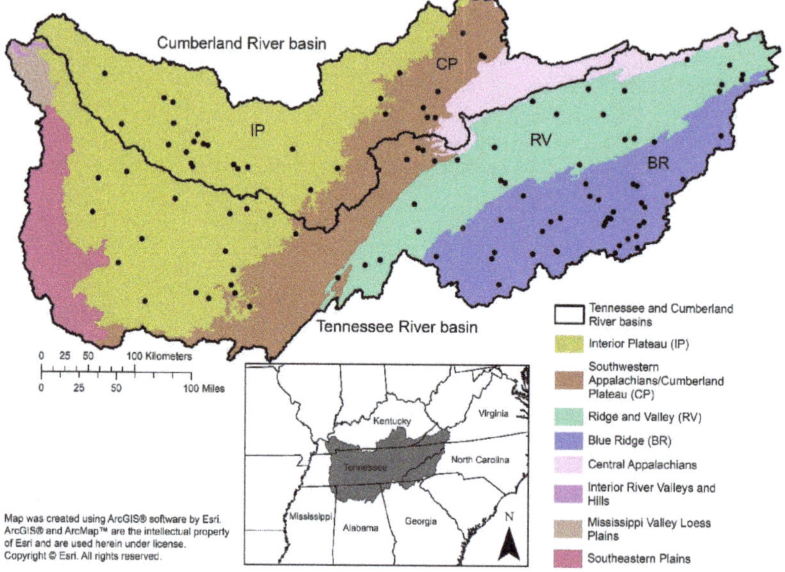

Figure 1. Distribution of selected stream sites within Level III ecoregions of the Tennessee and Cumberland River basins.

The Tennessee and Cumberland River basins transect multiple Level III ecoregions [42], including the Blue Ridge, Ridge and Valley, Interior Plateau, and Southwestern Appalachians. The Southwestern Appalachians Level III ecoregion generally corresponds with the 'Cumberland Plateau' physiographic section [45] (Figure 1). In order to maintain consistency with adjacent ecoregion boundaries used in [39], as well as maintain consistency of ecoregion terminology used in [41], we refer to the Southwestern Appalachians Level III ecoregion as the Cumberland Plateau (CP) throughout this paper.

2.2. Site Selection

U.S. Geological Survey (USGS) streamgauge sites within the Cumberland River basin (4-digit hydrological unit code, HUC4: 0513) and Tennessee River basin (HUC4s: 0601, 0602, 0603, 0604) and within BR, RV, CP, and IP Level III ecoregions [42] were initially screened for inclusion. USGS sites with drainage areas ranging from 26–3890 km^2 and with daily streamflow records for a minimum of 15 complete water years (following recommendations in [46]) between 1975 and 2016 were retained for this study. Prior or existing hydrologic alteration was not considered for site selection. Web-based satellite imagery was used to identify water-control structures and streamgauge sites within close downstream proximity were excluded from analysis. Sites located outside the geographic extent of stream sites used in prior model development [38,39,41] were also excluded. For example, development of ELFs for CP streams was based on sites and datasets from the northern portion of the ecoregion [41] and thus USGS streamgauge sites from the southern portion of CP were not used in this analysis (see Figure 1). Out of 641 USGS daily streamgauge sites across the HUC4s, a total of 112 USGS sites were retained across the four ecoregions: IP, 41 sites; CP, 11 sites; RV, 25 sites; and BR, 35 sites (Figure 1). To assess potential patterns associated with stream size, sites were categorized according to drainage area: 25–130 km^2, 130–260 km^2, 260–780 km^2, 780–1300 km^2, 1300–2600 km^2, and 2600–3890 km^2.

2.3. Water-Withdrawl Models

Mean daily streamflow data for each of the retained stream sites were accessed from USGS National Water Information System (NWIS) database [47]. Hydrologic data sets were queried using the "dataRetrieval" package [48] and R statistical software [49]. For each site, water-withdrawal scenarios were simulated from these observed (baseline) streamflow datasets. For this study, simulated water withdrawals were assumed to be the only source of streamflow alteration during the period of record for each site and previous or existing alterations (e.g., upstream withdrawals or effluent discharges) were assumed to be negligible.

2.3.1. Constant-Rate (CR) Withdrawals

The CR withdrawal model simulated the application of a constant-rate of water withdrawal across the entire period of record for each site, regardless of variability in ambient daily streamflow. A series of incremental CR withdrawal scenarios, ranging from 0.003 to 0.028 m^3/s by 0.003 m^3/s increments and 0.028 to 1.416 m^3/s by 0.028 m^3/s increments, were individually simulated for each site by subtracting simulated withdrawals from the baseline daily streamflow data set. CR withdrawal scenario increments were initially constructed using cubic feet per second (ft^3/s) units and then later converted to metric units. A total of 60 CR withdrawal scenarios (i.e., hydrographs), including the baseline dataset, were simulated for each site and for each of 4 MFLs (detailed in Section 2.3.3). Because the amount of daily withdrawal was held constant during high or low flows, CR withdrawals were anticipated to disproportionately affect periods of low streamflow (Figure 2a).

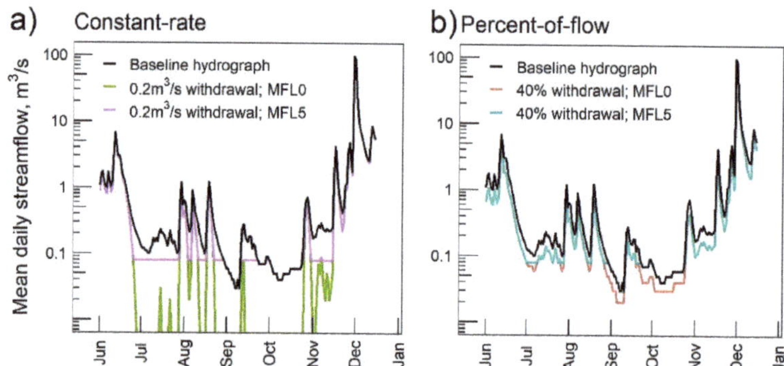

Figure 2. Example daily hydrographs of (**a**) constant-rate and (**b**) percent-of-flow water-withdrawal scenarios and minimum flow levels (MFLs). MFL0 hydrographs represent the absence of any MFL protection; MFL5 represents a MFL set at 5% of the mean annual flow. Daily streamflow values extending below y-axis represent flows at or near 0 m³/s on log10 scale.

2.3.2. Percent-of-Flow (POF) Withdrawals

The POF model simulated the withdrawal of set percentages of measured daily streamflow for each site, ranging from 1% to 40% daily streamflow withdrawal at 1% increments (i.e., total of 41 POF withdrawal scenarios, including the baseline data set, for each site and for each MFL). For the POF model, the absolute amount of simulated water withdrawal during high flows is greater than during low-flow periods, although the relative proportion of withdrawals is held constant. In general, hydrographs simulated under incremental POF withdrawal scenarios were anticipated to be susceptible to an overall reduction in streamflow magnitude but, in contrast to CR withdrawals, POF scenarios should theoretically better preserve aspects of timing, ratios, and variability of streamflows in relation to the baseline hydrograph for each site (Figure 2b). To prevent simulation of unrealistic daily withdrawal rates during higher flows (i.e., withdrawals beyond the pumping capacity of most facilities), a maximum instantaneous withdrawal limit was set at 1.41 m³/s, which was approximately the 99th percentile of all reported annual mean monthly withdrawals from permitted water-withdrawal facilities across Tennessee during 2010 [50].

2.3.3. Minimum Flow Level (MFL) Protection

Four MFLs were applied to each incremental withdrawal scenario to investigate the potential effects of low-flow protections on simulated CR and POF withdrawals. MFLs were defined as set daily streamflow values below which additional withdrawal was prohibited. MFL scenarios were applied at 0%, 5%, 10%, and 30% (MFL0, MFL5, MFL10, MFL30, respectively) of the mean annual daily flow calculated for the period of record of each site. MFL0 represents the absence of minimum-flow protection such that streamflow was allowed to be withdrawn to 0 m³/s under certain circumstances (see Figure 2a). Withdrawal scenarios using MFL0 were used as the basis for comparison of the hydrologic and ecologic effects of simulated water withdrawals. Importantly, for both CR and POF models, water withdrawals were treated as 'consumptive', meaning that water returns (i.e., effluent discharges) from upstream withdrawals were not accounted for.

2.4. Data Processing and Analyses

2.4.1. Calculation of SFCs

Prior to simulation of water withdrawals, baseline daily streamflow datasets were formatted for compatibility with the EflowStats R package [51]. The core functions of EflowStats calculate a series of SFCs from a daily streamflow data set, using water years defined as 365 daily flow values beginning on 1 October of one year and ending on 30 September of the following year(s). Gaps in the daily hydrograph were identified and incomplete water years were removed. The final baseline data set for each site consisted of either one continuous or several non-continuous periods that together totaled a minimum of 15 complete water years.

For each site, SFCs were calculated for each data period for (1) the baseline (observed) hydrograph and (2) hydrographs simulated under each incremental withdrawal scenario using CR and POF withdrawal models. For sites with multiple data periods (i.e., sites with gaps in the streamflow record), average SFC values were taken across periods for each withdrawal scenario. For example, for a site with two data periods—a 5-year period and a 10-year period—SFCs were calculated for each period and a weighted average was calculated using the number of water years as the weighting variable. SFC values for each withdrawal scenario were then transformed to standardized (unitless) values using published mean and standard deviation values of SFCs and formulae previously reported in [38].

2.4.2. SFC Sensitivity

Patterns of changes in standardized SFCs in response to incremental water-withdrawal scenarios were visually assessed by plotting SFC values for each site across withdrawal scenarios and for each MFL in relation to reference IQRs of SFCs for each ecoregion (e.g., Figure 3a and Figures S1–S8). The mean sensitivity of each SFC under CR and POF withdrawals was determined by calculating the mean absolute change in SFC values across all incremental withdrawal scenarios relative to the baseline hydrograph for each site and then taking the mean across all sites for each ecoregion. SFCs were considered generally sensitive to water withdrawals if the mean sensitivity for MFL0 was greater than 0. The presence of damping effects of MFLs was indicated by relatively lower mean sensitivity of SFCs at MFL5, MFL10, and/or MFL30 in relation to MFL0. The relative magnitudes of change in mean sensitivity as well as the appearance of 'dampened' SFC withdrawal responses based on MFLs (i.e., decreasing slope of site-specific responses with increasing MFLs; see Figure 3a) were used to compare the relative strength and consistency of damping effects for some SFCs among ecoregions as well as generally between CR and POF withdrawals. For example, SFCs that exhibited relatively small decreases in mean sensitivity and subtle or indistinguishable responses to MFLs were considered to have relatively weak damping effects. Damping effects on SFCs were considered absent if an SFC showed no change in mean sensitivity across MFLs or if mean sensitivities of MFL5, MFL10, and MFL30 were each greater than MFL0 (i.e., considered an anomalous or unpredicted increase in sensitivity due to application of MFLε).

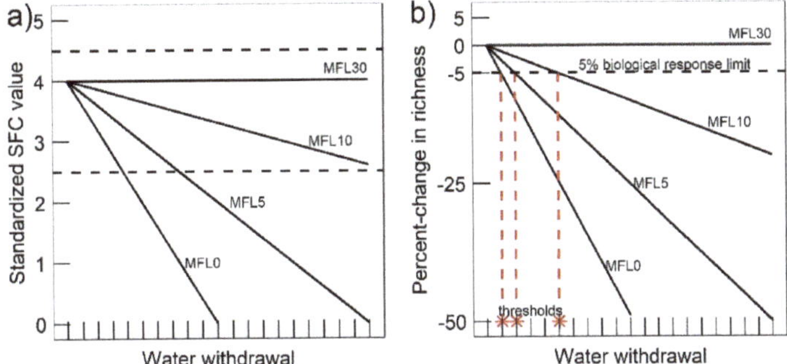

Figure 3. Hypothetical responses of (**a**) SFC values and (**b**) percent-changes in species richness plotted for a single stream site across increasing water-withdrawal scenarios and for different minimum flow levels (solid lines). (**a**) Dashed horizontal lines represent the pre-defined interquartile range of SFC responses among reference sites. (**b**) Dashed horizontal line represents an arbitrary biological response limit; Dashed red vertical lines represent intersection points at which an ecological withdrawal threshold occurs, i.e., at which a 5% loss of species richness occurs.

2.4.3. Departure from Reference

Departures of SFCs from hydrologic reference conditions were quantified for each incremental withdrawal scenario for each site by calculating the absolute difference of standardized SFC values from reference IQRs of SFCs for each ecoregion. For example, calculated SFC values for each withdrawal scenario that were within the reference IQR (i.e., >25th and <75th percentile values) (e.g., Figure 3a) received a hydrologic departure value of 0, whereas values outside of the IQR (i.e., <25th or >75th percentiles) received a departure value > 0. We note that by using absolute departures values, neither the initial relative position nor the direction of change in departure for each withdrawal scenario in relation to reference IQR are specifically known. Absolute departures of SFCs were then summed across the relevant SFC(s) that defined the ELF for each fish group (Table S1; [39,41]) to arrive at a cumulative hydrologic departure value for each withdrawal scenario for each site for each fish group.

2.4.4. Predicted Species Richness Responses to Withdrawal Scenarios

The ELF for each fish group defines the upper limit (≥85th percentile) of observed fish species richness as a function of the cumulative hydrologic departure of SFCs from reference hydrologic condition [39,41]. Predicted richness under each water-withdrawal scenario, including the original baseline streamflow data set, was calculated for each stream site by solving the following equation for each ELF:

$$SR = m \times x + b \tag{1}$$

where SR is predicted species richness, x is cumulative hydrologic departure and m and b are the slope and y-intercept of the ELF for each fish group [39,41] (see Table S1). Because the slope of each ELF is negative (Table S1), cumulative departure values > 0 result in predicted species richness less than the theoretical maximum richness for each fish group (i.e., y-intercept). It is important to note that predicted species richness for any given stream site may increase, decrease, or remain unchanged among withdrawal scenarios depending on the relative change in cumulative departure among relevant SFC(s) between scenarios. For example, predicted species richness will decrease if increasing water withdrawals causes one or more SFC(s) to move farther from the reference condition (i.e., increase in departure). Conversely, predicted richness would increase if departures decreased.

To summarize overall patterns and direction of change in predicted species richness among fish groups, first the mean change in richness was calculated for each withdrawal scenario relative to richness predicted for the baseline hydrologic regime for each site, and then an overall mean was taken across sites for each fish group. Predicted species richness values were also converted to the percent change in richness relative to the richness predicted for the baseline hydrologic regime for each site and plotted to allow more consistent comparison of change in richness among streams and ecoregions with potentially different richness values (e.g., small versus large streams). Negative and positive values represent the predicted relative loss or gain, respectively, of species richness. It is important to note that large positive values of percent change in predicted richness in relation to baseline can result from relatively small increases in predicted species richness among withdrawal scenarios. We recognize that increases in species richness in response to increased water withdrawals are possible under certain circumstances, for example, streams with higher than normal flows (e.g., due to dam release) and/or rivers where reduced velocities and water depths could create additional habitat suitable to more species. However, actual gains in fish species richness are likely to be modest. As such, large gains in predicted richness (absolute or percent change) were considered mathematical artifacts rather than plausible real-world outcomes to increased water withdrawals.

Further, to compare the magnitude of species-richness loss under POF withdrawals with increasingly utilized "presumptive" percent-of-flow standards [30], mean absolute loss and mean percent loss of species richness were calculated across stream sites for 10 and 20 POF withdrawal scenarios for MFL0 for each fish group and ecoregion. Stream sites that did not result in any loss in richness among withdrawal scenarios were excluded from this analysis.

2.4.5. Biological Response Limit and Maximum Withdrawal Thresholds

A 5% loss of species richness (i.e., −5% change relative to richness predicted under baseline hydrologic conditions) was set as an arbitrary biological response limit at which to identify potential withdrawal thresholds for each stream site. Specifically, for any given stream site modeled for each fish group and MFL, an ecological withdrawal threshold (also referred to hereafter as 'withdrawal threshold' or simply 'threshold') was identified at the maximum incremental water-withdrawal scenario that resulted in less than 5% loss of species richness. Figure 3b illustrates hypothetical examples of monotonic percent changes in species richness along a withdrawal gradient and for different MFLs for a single stream site and identifies different ecological withdrawal thresholds at which richness loss reaches −5%. Damping effects of MFLs on percent change in richness should generally appear as a decreased slope of species richness responses when plotted in relation to MFL0 (e.g., solid lines in Figure 3b), whereas damping effects of MFLs on withdrawal thresholds should appear as (a) higher threshold values with higher MFLs (e.g., asterisks in Figure 3b) or (b) a shift to no threshold in relation to MFL0 (e.g., MFL30 in Figure 3b).

Because species richness varies naturally among streams and because each fish group contains varying subsets of species, using percent loss of predicted species richness to inform a biological response limit—rather than absolute richness values—provides a consistent and scalable metric of biological change across stream sites, fish groups, and ecoregions. A 5% biological response limit is protective of 95% of the relative species richness of each fish group and should buffer against the loss of some sensitive species while still allowing for some water withdrawals to meet human needs. While similar biological response limits have been previously proposed [13,52], there is little consensus (among ecologists, resource managers, citizens, policy makers, or other stakeholders) regarding the degree of change in biological condition (i.e., absolute or percent species-loss) that could/should be tolerated in response to hydrologic alteration [22,34].

Withdrawal thresholds identified under the CR model are reported as daily streamflow withdrawal rates (m^3/s) and POF thresholds are reported as percentages of daily streamflow withdrawals (%). Boxplots were used to summarize the distributions and patterns of thresholds among fish groups and by drainage area for each MFL scenario. Median values calculated among all thresholds for each fish group, and overall regional mean withdrawal thresholds calculated across these median values were used to summarize central tendencies among sites, fish groups, and ecoregions. Additionally, the number of sites that reached an ecological withdrawal threshold was calculated for each fish group, and the cumulative number of sites that reached a threshold across all fish groups was summed for each ecoregion. Because each stream site was modeled independently for each fish group, sites that reached a threshold in multiple fish groups were counted more than once.

2.4.6. Model Archive and Bulk Data

A model archive containing detailed R scripts with reproducible procedures as well as bulk data tables containing site-specific results for standardized SFCs, cumulative departures, predicted species richness, percent-change in richness, and withdrawal thresholds for each MFL scenario for both CR and POF withdrawals are available from [53]. Supporting information and large format data summary tables are provided as supplementary tables (Tables S1–S4). Plotted responses among all stream sites for each SFC (Figures S1–S8) and percent-changes in richness (Figures S9–S16) and ecological withdrawal thresholds (Figures S17 and S18) for each fish group MFL and ecoregion are provided as supplementary figures.

3. Results

3.1. Sensitivity of Streamflow Characteristics

3.1.1. Sensitivity of SFCs to CR Withdrawals

Each SFC showed some sensitivity to CR withdrawals (i.e., mean values greater than 0) for MFL0, although the relative degree of sensitivity (Table 2) and underlying patterns of response varied within and among SFCs (e.g., Figure 4 and Figures S1–S4).

Table 2. Mean absolute change (i.e., sensitivity) of streamflow characteristics (SFC; standardized unitless values) across water-withdrawal scenarios and for each minimum flow level (MFL) under constant-rate and percent-of-flow withdrawals. MFLs represent low-flow protections set at 0, 5, 10, and 30% of mean annual flow.

Eco-Region	Flow Category	SFC	Constant-Rate Withdrawal				Percent-of-Flow Withdrawal			
			MFL0	MFL5	MFL10	MFL30	MFL0	MFL5	MFL10	MFL30
BR	Magnitude	AMH10	0.04	0.04	0.04	0.04	0.06	0.06	0.06	0.06
RV	Magnitude	MA41	0.32	0.30	0.28	0.21	0.28	0.28	0.28	0.24
	Ratio	TA1	0.60	0.44	0.30	0.16	0.16	0.15	0.13	0.09
	Variability	LDH16	0.01	0.01	0.01	0.02	0.00	0.00	0.00	0.00
	Variability	FL2	0.02	3.17	3.09	0.24	0.00	0.00	0.03	0.23
CP	Magnitude	MA41	0.16	0.14	0.13	0.10	0.20	0.19	0.19	0.16
	Magnitude	AMH10	0.01	0.01	0.01	0.01	0.02	0.02	0.02	0.02
	Ratio	LDH13	0.29	0.34	0.30	0.14	0.23	0.23	0.23	0.17
	Ratio	ML20	0.08	0.18	0.15	0.07	0.16	0.15	0.14	0.09
	Ratio	TA1	0.27	0.05	0.03	0.02	0.12	0.03	0.03	0.02
	Variability	LDL6	1.16	0.05	0.00	0.00	0.00	0.05	0.00	0.00
	Timing	TL1	3.80	0.02	0.00	0.00	0.00	0.01	0.00	0.00
IP	Magnitude	AMH10	0.03	0.03	0.03	0.02	0.03	0.03	0.03	0.03
	Magnitude	LRA7	1.33	0.95	0.68	0.19	0.11	0.13	0.14	0.10
	Ratio	TA1	0.63	0.35	0.23	0.11	0.15	0.11	0.08	0.06
	Frequency	FH6	0.17	0.15	0.14	0.13	0.05	0.05	0.05	0.07
	Timing	TL1	4.59	1.34	0.46	0.00	0.00	0.10	0.12	0.01

Responses of some SFCs showed variability among streams of different drainage area size, in which the initial SFC values at baseline conditions were often stratified according to size (e.g., lower values among sites with relatively smaller drainage area for TA1 in Figure 4), and/or SFCs of sites with smaller drainage areas were often more strongly influenced by incremental water withdrawals relative to larger streams (e.g., greater rates of change across withdrawals among drainage-area sizes under MFL0 for TL1 in Figure 4). Among SFCs, TL1 and LDL6 (associated with the timing and variability of low flows, respectively) and LRA7 (magnitude of flow recession rate) were particularly sensitive to CR withdrawals (Table 2). These SFCs exhibited consistent and strong responses to incremental withdrawals for MFL0 among sites for their respective ecoregions (CP and/or IP) (e.g., TL1 in Figure 4). In general, SFCs associated with flow magnitude (e.g., AMH10 and MA41) showed consistent and/or monotonic responses to increasing CR withdrawals among sites and MFLs across ecoregions (e.g., relatively weak monotonic responses of AMH10 across drainage-area sizes in Figure 4). In contrast, SFCs associated with flow ratios, frequency, variability, and timing of flows (see categories in Table 1) tended to have more variable and sometimes non-monotonic responses to increasing withdrawals and among MFLs (e.g., shifts between decreasing and increasing values among some sites for TA1 in Figure 4; also see examples in Figures S2–S4).

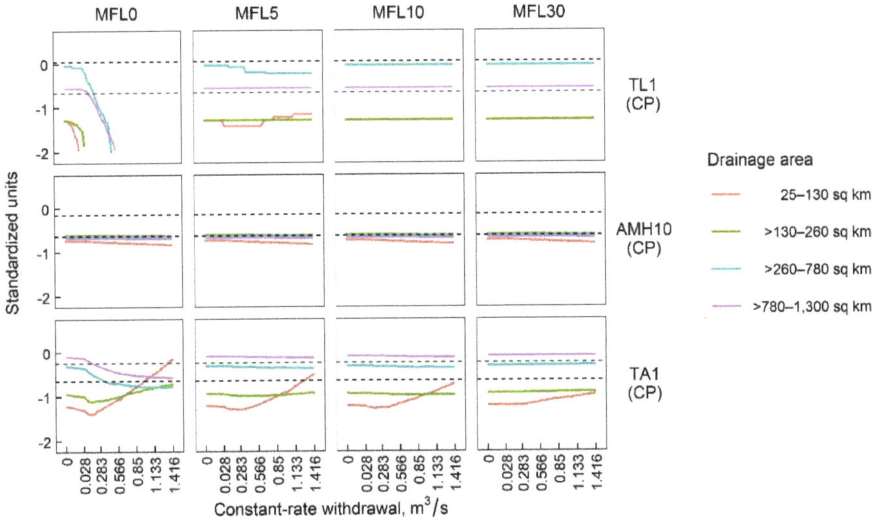

Figure 4. Mean responses of select streamflow characteristics (SFCs) to incremental constant-rate withdrawal scenarios and minimum flow levels (MFL) within the Cumberland Plateau (CP). Solid lines represent the mean SFC response across sites for each drainage-area size; horizontal dashed lines are the interquartile range (IQR) of SFCs from reference sites within each ecoregion [38,39,41].

Relatively strong and consistent damping effects were observed among TL1, TA1, LDL6, and LRA7 among respective ecoregions, in which mean sensitivity values were lower among MFLs in relation to MFL0 (Table 2) and slopes of SFC responses were also generally lower for MFL5, MFL10, and/or MFL30 (e.g., TL1 and TA1 in Figure 4 and Figures S2–S4). Slight decreases in mean sensitivities among MFLs also indicated the presence of relatively weak damping effects for MA41 and FH6 (Table 2). LDH13 and ML20 showed inconsistent responses in mean sensitivity among MFL scenarios (Table 2) but plots suggested the presence of some damping effects (Figure S3). Damping effects of MFLs were largely absent among AMH10 (e.g., Figure 4), LDH16, or FL2 (Table 2). Increases in the mean sensitivity observed for some SFCs across MFLs, particularly FL2, were counter to our general expectations and are most likely attributed to prolonged periods of reduced variability of low flows introduced by MFL cutoffs (e.g., illustrated in Figure 2a as periods of flat lines in the MFL5 hydrograph).

3.1.2. Sensitivity of SFCs to POF Withdrawals

Mean sensitivities of SFC for MFL0 indicated that some SFCs were generally less sensitive to POF withdrawals than CR withdrawals, particularly TL1 (timing), LDH13 and TA1 (flow ratio), LDL6, LDH16 and FL2 (flow variability), FH6 (frequency), and LRA7 (magnitude) (Table 2; Figures S5–S8). In contrast to CR withdrawals, TL1 and LDL6 were not sensitive to POF withdrawals under MFL0 (Table 2). Additionally, underlying patterns of most SFC responses to POF withdrawals were largely monotonic across sites, ecoregions, and MFLs. Responses of streams with relatively small drainage area were also often similar and/or proportional to those with larger drainage area (e.g., TA in Figure 5 and Figures S5–S8).

Decreases in mean sensitivity of TA1, ML20, MA41, LDH13, and LRA7 (Table 2) among MFLs indicate the presence of some damping effects under POF withdrawals, in which TA1 (Figure 5) appeared to have relatively stronger and more consistent damping effects compared to others SFCs across respective ecoregions. However, compared to CR withdrawals, damping effects under POF withdrawals were generally weaker and less consistent among most SFCs. Damping effects appeared to be absent among TL1, LDL6, FH6, LDH16, FL2, and AMH10 (Table 2, Figures S5–S8). Similar to CR withdrawals, some SFCs showed some greater sensitivity at higher MFLs under POF withdrawals, specifically FL2 among RV streams and TL1 among IP streams (Table 2; see example of FL2 in Figure S6 and TL1 in Figure S8).

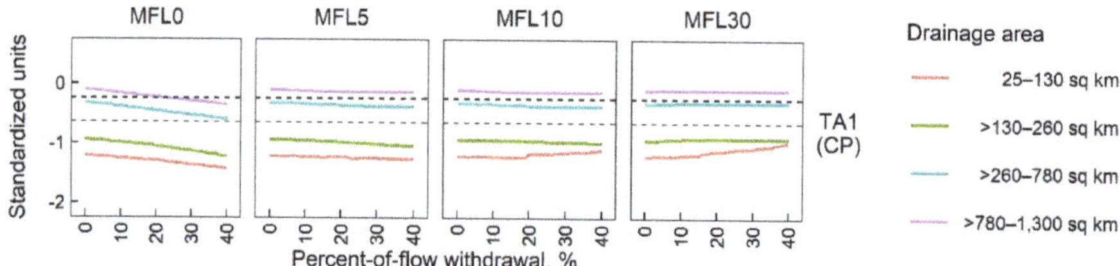

Figure 5. Mean responses of select streamflow characteristics to incremental percent-of-flow withdrawal scenarios and minimum flow levels (MFL) within the Cumberland Plateau (CP). Solid lines represent mean values across sites for each drainage-area size; horizontal dashed lines are the interquartile range (IQR) of SFCs from reference sites within the ecoregion [38,39,41].

3.2. Species-Richness Responses

Mean changes in richness (averaged among withdrawal scenarios and across stream sites) were negative across most fish groups and MFLs. Compared to CR withdrawals, overall mean loss of richness predicted under POF withdrawals was generally lower across fish groups and ecoregions (Table 3 and Table S2).

Table 3. Mean change in predicted fish species richness across all stream sites and water-withdrawal scenarios for select fish groups under constant-rate and percent-of-flow withdrawals.

Eco-Region	Fish Group [1,2,3]	Streamflow Characteristic (s)	Constant-Rate Withdrawal Mean Change in Richness				Percent-of-Flow Withdrawal Mean Change in Richness			
			MFL0	MFL5	MFL10	MFL30	MFL0	MFL5	MFL10	MFL30
BR	All species	AMH10	−0.28	−0.28	−0.28	−0.26	−0.40	−0.40	−0.40	−0.39
RV	All species	FL2	−0.10	−4.89	−5.57	−1.01	0.00	0.00	−0.19	−0.85
	Specialized insectivores	TA1 + LDH16	−0.71	0.94	1.11	0.15	0.51	0.67	0.81	−0.29
CP	All species	TL1	−12.69	−0.02	0.00	0.00	0.00	−0.09	0.00	0.00
	Specialized insectivores	AMH10 + TA1 + TL1	−5.54	0.16	0.01	0.04	−0.25	−0.16	−0.07	0.03
IP	All species	AMH10 + TL1	−21.50	−8.13	−2.97	−0.05	−0.09	−0.52	−0.91	−0.08
	Specialized insectivores	AMH10 + TA1	−0.12	0.53	0.59	0.31	−0.02	0.11	0.21	−0.01

[1] Maximum possible species richness for each fish group is represented by the 'intercept' coefficient in Table S1; [2] Mean change in richness for all fish groups modelled in this study are presented in Table S2; [3] The identities of fish species within each group were not considered and membership of an individual species to a fish group is not mutually exclusive.

However, patterns of change in predicted species richness along the withdrawal gradients were highly variable among sites for different fish groups, ecoregions, MFLs, and streams of different drainage-area sizes under both CR (Figures S9–S12) and POF withdrawals (Figures S13–S16), in which richness responses were related to the relative individual sensitivity and cumulative departures from reference conditions of SFC(s) that defined the ELF for each combination of fish group and ecoregion.

Under CR withdrawals, overall mean changes in predicted richness ranged from −21.5 species (loss) to 0.02 species (gain) across fish groups for MFL0 (Table S2). The relative magnitude of richness loss was greatest among fish groups influenced by more sensitive SFCs (Table 3 and Table S2). For example, the strongest patterns of both mean change in richness and damping effects of MFLs among stream sites were associated with fish groups influenced by TL1 for CP (10 of 11 fish groups) and IP (6 of 10 fish groups) (e.g., Table 3; see examples of the 'All species' fish groups for CP and IP in Figure 6), in which predicted richness losses were generally greater at MFL0 than MFL5–MFL30. In contrast, fish groups that were predominantly (or solely) influenced by SFC(s) that were only relatively weakly sensitive to CR withdrawals and/or MFLs showed correspondingly weak patterns in richness responses across withdrawal scenarios (e.g., BR sites for the 'All species' group in Table 3 and Figure 6).

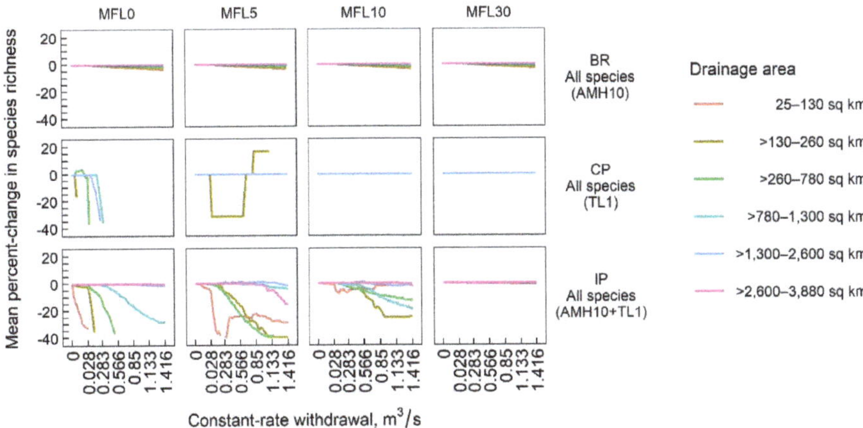

Figure 6. Patterns of mean percent change in fish species richness in response to incremental constant-rate withdrawal scenarios and minimum flow levels (MFL) for the "All Species" fish group for Blue Ridge (BR), Cumberland Plateau (CP), and Interior Plateau (IP) ecoregions. Solid lines represent individual stream sites. SFC(s) associated with the ecological limit function for each fish group and ecoregion are in parentheses.

Under POF withdrawals, overall mean changes in predicted richness (averaged across all withdrawal scenarios for each site) ranged from −0.44 to 0.51 species among fish groups for MFL0 (Table S2). Patterns of richness loss under POF withdrawals were less strongly associated with a single SFC (Table 3 and Table S2), compared to CR. With the exception of relatively small decreases in the mean predicted richness loss among some fish groups influenced by TA1 among CP and IP (e.g., specialized insectivores from CP in Table 3), patterns of change in richness among MFLs were largely inconsistent among fish groups and ecoregions (Table S2, Figures S13–S16). These inconsistencies indicated generally lesser damping effects of MFLs on outcomes of species richness under POF withdrawals.

In relation to 10- and 20-percent presumptive standards [30], mean loss in richness across all stream sites for each fish group and ecoregion for MFL0 ranged from 0.01 to 0.66 species at 10% withdrawals and from 0.02 to 1.34 richness loss at 20% daily water withdrawal. The overall mean predicted richness loss collectively among ecoregions was 0.17 and 0.33 species, respectively (Table 4 and Table S3). Mean percent loss in species richness among fish groups ranged from 0.15% to 12.95% at 10% withdrawal and from 0.22% to 18.89% at 20% withdrawals, with an overall mean percent richness loss of 2.57% and 4.96%, respectively, across all fish groups and ecoregions (Table 4 and Table S3).

Table 4. Mean loss and mean percent-loss in predicted fish species across stream sites at 10 and 20% (POF) withdrawal scenarios for MFL0. Means calculated only among streams and fish groups that exhibited declines in richness.

Eco-Region	Number of Fish Groups [1]	Mean Loss in Richness (Number of Species)				Mean Percent Loss in Richness (%)			
		10% Flow Withdrawal		20% Flow Withdrawal		10% Flow Withdrawal		20% Flow Withdrawal	
		Range	Mean	Range	Mean	Range	Mean	Range	Mean
BR	3	0.11–0.32	0.19	0.16–0.43	0.27	0.85–0.99	0.93	1.20–1.34	1.29
RV	2	0.34–0.43	0.39	0.61–0.81	0.71	3.86–10.59	7.23	7.31–13.19	10.25
CR	9	0.02–0.31	0.13	0.05–0.61	0.30	0.75–12.95	3.45	1.28–18.89	7.67
IP	10	0.01–0.66	0.15	0.02–1.34	0.29	0.15–7.12	1.34	0.22–13.00	2.57
Overall Mean			0.17		0.33		2.57		4.96

[1] Mean values for each fish group presented in Table S3.

3.3. Ecological Withdrawal Thresholds

3.3.1. CR Withdrawal Scenarios

Box plots of ecological withdrawal thresholds from select fish groups under CR withdrawals show a high degree of variability among stream sites both within and among drainage-area sizes, MFLs, and ecoregions (Figure 7 and Figure S17). Despite considerable variability of thresholds among streams of the same drainage-area size, relatively smaller-size streams often were associated with lower thresholds compared to larger-size streams across fish groups (Figure 7 and Figure S17). For MFL0, median thresholds ranged from 0.01–0.92 m^3/s across all fish groups and ecoregions, and median thresholds were often higher among BR and RV fish groups compared to CP and IP (Table 5). For MFL0, the relatively higher number of stream sites that reached an ecological withdrawal threshold and the relatively lower threshold values among CP and IP fish groups (compared to BR and RV) (Table 5) were associated with the prevalence and influence of TL1 among fish groups (Table S4). In partial support of our hypothesis, the reduced number of sites that reached a threshold among MFL5–MFL30 in relation to MFL0 among most fish groups for CP and IP indicated damping effects and suggested that some sites became less sensitive to loss of species richness as MFLs increased under CR withdrawals (Table 5 and Table S4; Figures S11 and S12). For example, the cumulative number of ecological withdrawal thresholds reached among stream sites across fish groups for IP decreased from 267 under MFL0 to 187, 118, and 33 under MFL5, MFL10, MFL30, respectively (Table 5 and Table S4). Counter to predictions, however, the number of thresholds reached among most RV fish groups tended to increase and threshold values tended to decrease with increasingly protective MFLs, related to the influence of FL2 among RV fish groups, in which FL2 showed greater sensitivity to CR withdrawals at higher MFLs (Table 5 and Table S4; Figures S2 and S10).

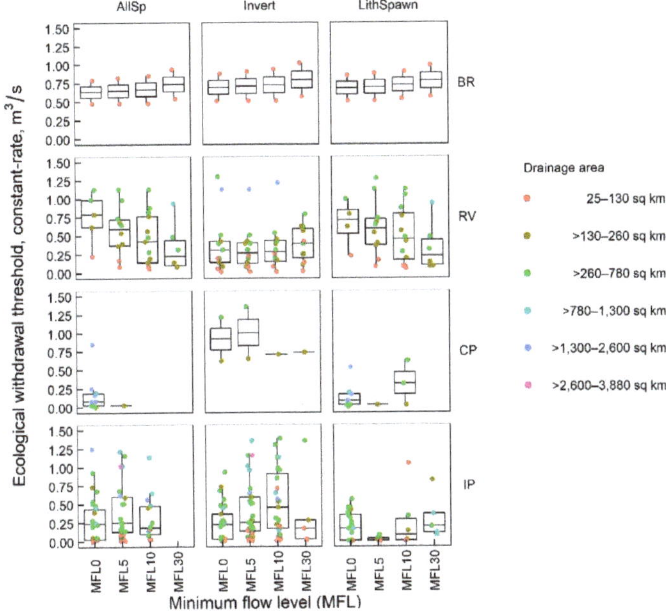

Figure 7. Boxplots of predicted ecological withdrawal thresholds under constant-rate withdrawal scenarios among select fish groups and minimum flow levels (MFLs) for Blue Ridge (BR), Ridge and Valley (RV), Cumberland Plateau (CP), and Interior Plateau (IP) ecoregions. Boxes represent the 25th and 75th quantile; center line represents the median (50th percentile). Points are threshold values for individual stream sites and are color coded by drainage-area size. Results for all fish groups are illustrated in Figure S17.

Table 5. Range and mean of median values (across fish groups) of predicted ecological withdrawal thresholds under constant-rate (m^3/s) and percent-of-flow (%) withdrawal scenarios for each ecoregion. "n" = cumulative number (sum) of sites that reached an ecological withdrawal threshold across all fish groups for each ecoregion.

Constant-Rate (CR) Ecological Withdrawal Thresholds (m^3/s) [1]											
	MFL0			**MFL5**			**MFL10**			**MFL30**	
Eco-Region	**n** [2]	**Range**	**Mean**	**n**	**Range**	**Mean**	**n**	**Range**	**Mean**	**n**	**Range**	**Mean**
BR	6	0.64–0.68	0.67	6	0.64–0.68	0.68	6	0.67–0.72	0.70	6	0.74–0.79	0.77
RV	47	0.20–0.81	0.62	73	0.24–0.68	0.51	80	0.06–0.64	0.39	52	0.23–0.57	0.34
CP	98	0.01–0.92	0.16	25	0.03–1.01	0.23	23	0.01–0.92	0.35	11	0.03–0.85	0.40
IP	267	0.17–0.37	0.23	187	0.10–0.37	0.20	118	0.17–0.37	0.26	33	0.17–1.33	0.67

Percent-of-flow (POF) ecological withdrawal thresholds (%) [1]											
	MFL0			**MFL5**			**MFL10**			**MFL30**	
Eco-region	**n** [2]	**Range**	**Mean**	**n**	**Range**	**Mean**	**n**	**Range**	**Mean**	**n**	**Range**	**Mean**
BR	6	23.5–27.0	25.5	6	23.5–27.0	25.5	6	23.5–27.0	25.5	6	25.0–28.0	26.8
RV	17	6.0–6.5	6.3	16	6.5–14.5	10.5	25	1.0–32.5	22.9	49	6.0–16.0	13.0
CP	33	4.0–16.0	9.4	29	4.0–28.0	15.7	22	4.0–29.0	13.7	14	2.0–17.0	7.6
IP	63	3.5–31.0	18.8	102	9.0–31.0	22.6	76	2.0–33.5	18.6	23	1.0–37.0	14.7

[1] Median threshold values and the number of sites (n) for each fish group are presented in Table S4; [2] Sites that reached a threshold in multiple fish groups were counted more than once, meaning cumulative "n" can be greater than the number of individual stream sites for each ecoregion.

3.3.2. POF Withdrawal Scenarios

Under POF scenarios, patterns of ecological withdrawal thresholds were also variable among drainage-area sizes, MFLs, and ecoregions and spanned the entire gradient, from 0 to 40% of the mean daily streamflow (Figure 8 and Figure S18). For MFL0, median thresholds ranged from 3.5 to 31% of the daily withdrawal among all fish groups and ecoregions (Table 5), with most fish group median thresholds < 20% daily withdrawal (Table S4). Compared to CR withdrawals, the cumulative number of stream sites that reached an ecological withdrawal threshold under POF withdrawals was consistently lower for MFL0 among fish groups and ecoregions (Table 5), indicating generally reduced sensitivity of stream sites to species richness loss (i.e., 5% biological response limit). Decreases in the number of thresholds reached at MFL5–MFL30 in relation to MFL0 among fish groups influenced predominantly by TA1 among RV, CP, and IP ecoregions indicate some damping effects of MFLs (Table S4).

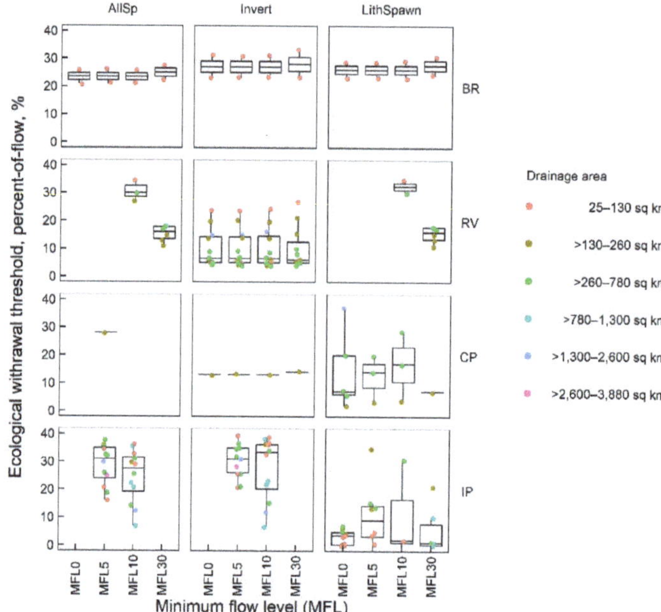

Figure 8. Boxplots of predicted ecological withdrawal threshold under percent-of-flow (POF) withdrawal scenarios among select fish groups and minimum flow levels (MFLs) for Blue Ridge (BR), Ridge and Valley (RV), Cumberland Plateau (CP), and Interior Plateau (IP) ecoregions. Boxes represent the 25th and 75th quantile; center line represents the median (50th percentile). Points are threshold values for individual stream sites and are color coded by drainage-area size. Results for all fish groups are illustrated in Figure S18.

4. Discussion

Despite some variability among ecoregions, fish groups, and SFCs, our findings generally indicate that POF withdrawals better preserved aspects of the baseline flow regime and had overall less negative impacts on fish species richness compared to CR withdrawals. In general, CR withdrawals resulted in greater sensitivity of SFCs across several ecoregions compared to POF withdrawals, particularly for metrics associated with low streamflows. Similarly, CR withdrawals produced greater sensitivity of SFCs to the protective effects of MFLs. CR withdrawals also resulted in more frequent instances of predicted declines in species richness with greater simulated withdrawals. This included greater overall numbers of sites that reached an ecological withdrawal threshold corresponding to loss of at least 5% species richness at some point along the simulated water-withdrawal gradient. In contrast, POF withdrawals resulted in lower sensitivity, or in some cases, a lack of sensitivity of SFCs related to variability and timing of streamflows, lower sensitivity of SFCs and species richness to MFLs, and relatively fewer ecological withdrawal thresholds reached among streams across fish groups and ecoregions. These findings support the notion that different water-withdrawal strategies can have important hydrologic and ecologic consequences [5,30] and our findings are consistent with other studies demonstrating the potential benefits of POF-type water-management strategies [1,30].

Under both CR and POF withdrawal strategies, the sensitivity of SFCs to withdrawals and the resulting changes in predicted species richness showed apparent relations to drainage area, with smaller streams showing greater sensitivity to water withdrawals than larger streams. Hydrologic and ecologic structuring along stream-size gradients is well documented [54] and classifying streams by size and/or hydrologic regime is often a critical step in developing ecological-flow relationships [22]. A similar study conducted recently among streams in Virginia, suggests that stream size (e.g., drainage area) may be one of the most important factors influencing patterns of change in species richness associated with alteration of streamflows (personal communication, J. Rapp, USGS Virginia/West Virginia Water Science Center, Richmond, VA, USA). Similarly, a study that investigated potential impacts of water withdrawals among streams within the Marcellus Shale region in West Virginia and Pennsylvania, indicates that streams with a drainage area greater than 1000 km^2 may be less susceptible to changes in hydrologic indices and potential negative ecological responses [52]. The relatively small number of streams sites within each drainage-area size among ecoregions for our study precluded more detailed comparison of patterns among sizes, but general patterns suggest that stream size is an important factor when considering potential withdrawal thresholds at local and regional scales.

Water-withdrawal simulations in this study generally support the idea that MFLs can provide important protection to ecological flow regimes. Specifically, the damping effects of MFLs on the sensitivities of select SFCs and patterns of change in species richness and ecological withdrawal thresholds under both CR and POF water withdrawals imply that MFLs can help protect against richness loss [1,29,30] and, in some circumstances, may allow for greater water withdrawals with relatively lower ecological impacts. Although other SFCs were also sensitive to the damping effects of MFLs, the prevalence of TL1 and TA1 in ELFs among fish groups and ecoregions and the high sensitivity of these SFCs to withdrawals, particularly under CR withdrawals, appear to be the primary drivers of patterns of ecological withdrawal thresholds among many fish groups, MFLs, and ecoregions. Because low flows are particularly vulnerable to water withdrawals [7,52,55], the high sensitivity of TL1 (as well as LDL6) to CR withdrawals for MFL0 and the limited changes in richness loss observed for MFL5–MFL30 are not surprising and highlight the potential influence of even modest MFLs on mitigating fish species richness loss under CR withdrawal scenarios. However, variable (or lack of) sensitivity of some SFCs to MLFs support the notion that minimum flow cutoffs alone may not be sufficient for protecting against richness loss [1,6,30]. Damping effects of MFLs were less prevalent under POF withdrawals, but the general influence of higher MFLs on TA1 and corresponding fish groups across multiple ecoregions further argues in favor of the potential ecological benefits of applying some form of MFL, regardless of whether CR or POF withdrawals are employed.

Our findings show that increasing CR or POF water withdrawals can result in reductions in fish-species richness within some fish groups at some sites. However, the high degree of variability in predicted species richness responses and the range of ecological withdrawal thresholds identified across local (sites specific) and regional (ecoregion) scales suggests that application of these results for management decisions should be implemented with caution. For example, median withdrawal thresholds predicted under CR withdrawals for MFL0 (Table 5 and Table S4) were substantially higher than a large majority of annual mean-monthly withdrawal rates reported for nearly 800 permitted water-withdrawal facilities within the State of Tennessee during 2010 [50], of which the median withdrawal value was approximately 0.013 m^3/s. Although the 2010 withdrawal data used by [50] is now somewhat dated, and reported withdrawals may often be less than permitted withdrawals [56], these data suggest that many of the withdrawal thresholds predicted under CR withdrawals in this study would greatly exceed the reported withdrawal rates of many water-use facilities. As a site-specific example, the lowest ecological withdrawal threshold predicted for Yellow Creek at Ellis Mills, TN (ecoregion = Interior Plateau; USGS station number = 03436690; drainage area = 237 km^2), was 0.396 m^3/s [53], whereas the annual mean monthly withdrawal rate reported by a public supply facility that withdraws surface water from Yellow Creek near the USGS gauge was several orders of magnitude lower: 0.0008 m^3/s [50]. Additionally, site-specific ecological withdrawal thresholds were often much greater than 7Q10 flows (annual 7-day minimum flow with a 10-year recurrence interval) [53], a metric commonly used by water-management agencies for assessing water-withdrawal limits. On the one hand, these results could be interpreted as general evidence that some existing withdrawal rates could be increased; however, a more cautious interpretation, and more likely, is that the CR withdrawal model may not fully capture the potential ecological responses to increasing withdrawals among many stream sites, particularly at lower CR withdrawal rates. Overall, the magnitudes of ecological withdrawal thresholds predicted under CR withdrawals for many stream sites and for most fish groups exceeded real-world water-withdrawal rates. This indicates that the applicability of specific CR withdrawal thresholds for management decisions is limited and should be interpreted with caution.

Regarding POF withdrawal scenarios, the range of median thresholds summarized among fish groups and ecoregions is largely consistent with the range of recommended limits to daily POF streamflow reductions (approximate 6–35%) reported among several empirical case studies (e.g., [57–59], and summarized in [30]. The mean percent loss in species richness among fish groups for across ecoregions at 10 and 20 POF withdrawal scenarios (Table 4 and Table S3) are generally supportive of Richter's [30] presumptive flow standards, and are also in agreement with the use of 5% loss in richness as a conservative biological response limit. However, richness losses greater than 5% among some fish groups indicate that POF withdrawal rates less than 10 or 20% of daily streamflow may be necessary to protect particularly sensitive fish groups or species (e.g., specialized insectivores (RV, CP), intolerant species (CP), and lithophilic spawners (IP) in Table S3). Further, overall averages of mean percent loss and mean absolute loss in species richness for 10 and 20 POF withdrawal scenarios were similar to or lower than those predicted for streams throughout North Carolina [36] and are in agreement with findings for streams in Virginia (personal communication, J. Rapp, USGS Virginia/West Virginia Water Science Center, Richmond, VA, USA). Specifically, [36] predicted a loss of 0.49 and 1.0 fish species for flow reduction scenarios of 15% and 25% among streams in North Carolina, whereas the predicted loss of richness at 20% withdrawals among fish groups and ecoregions in our study was often much less than 0.5 species and averaged only 0.33 species across groups (Table 4). Although the methods and criteria for developing flow-ecology relationships differed between our study and others [36], similarities in predicted ecological outcomes at 10 and 20% flow reductions suggest that these POF withdrawal rates may be generally applicable among basins and ecoregions of the southeastern U.S.

Our data indicated that protection of low stream flows, specifically timing of low flows, was particularly important for steams and fish groups within CP and IP ecoregions. Others have shown that low-flow metrics may be particularly sensitive to water withdrawals [52]. Water withdrawals can exacerbate low-flow conditions and alteration of the timing, magnitude, duration, and frequency of low flows are known to have negative impacts on aquatic biota, stream habitat, and water quality [2,5,7,55]. Intensification of low-flow conditions can impact fish species and communities at local and regional scales through increased mortality and disruption of reproduction, recruitment, and dispersal [2,7,60]. Nuisance and/or harmful algal blooms (HABs) and associated decreases in dissolved oxygen often occur during low flows, and cyanobacteria HABs and toxicity in lotic ecosystems are an increasing management concern in some regions [61,62]. Use of percentage-of-flow and presumptive standards are increasingly popular among water-management agencies within the US and elsewhere [30,57,63,64]. Despite high variability in the ecological withdrawal thresholds among stream sites and variability in responses among fish groups, our data provide support for water-management strategies using POF withdrawals and the implementation of MFLs to help mitigate hydrologic alteration and negative impacts on fish species richness.

Considerations and Caveats of Water-Withdrawal Models

The importance of SFC selection when characterizing ecologically relevant aspects of the flow regime is well recognized [1,65], as is the potential variability in sensitivities of SFCs to hydrologic alteration [46,52]. Although the ecological and statistical relevance of specific SFCs used in this study have been previously described for streams within the Cumberland and Tennessee River basins [37–39,41], the potential sensitivities of these SFCs to CR and POF water withdrawals and MFL scenarios were not specifically considered in previous works. SFCs that had greater individual sensitivity to withdrawals and are of demonstrated importance to fish communities may be particularly important targets for water-management decisions. On the other hand, SFCs with limited sensitivity to CR or POF withdrawals in this study may still be important and warrant attention from resource managers under different flow alteration conditions than considered in this study. In the Tennessee and Cumberland River basins, TL1 and TA1 were of particular ecological importance, but other SFCs may be more important in other regions or other climate contexts [8].

Ecological withdrawal thresholds summarized across sites according to drainage area, fish groups, or cumulatively for each ecoregion using central tendencies (e.g., median or means), may be as or more valuable than site-specific thresholds for use in decision making at local and regional scales. Specifically, summarized thresholds may inform decisions regarding water withdrawals among streams. Although site-specific withdrawal thresholds [53] may be useful for specific stream systems, in which species richness responses and thresholds modeled independently for each fish group and MFL represent multiple lines of evidence to guide management decisions, high variability at the individual site level may confound decision making.

Modelled hydrologic and ecological responses to increasing CR and POF water-withdrawal scenarios were based on underlying flow-ecology relationships that assume that patterns of fish species richness are determined solely by streamflow and that responses would be generally linear. However, non-linear responses are common in natural systems [64], and although [39] found some evidence of internal structuring of species-richness responses presumed to be associated with factors other than streamflow (e.g., water quality, physical habitat), the influence of additional environmental factors in determining fish species richness were not considered in this study.

Daily streamflow data used to calculate water-withdrawal scenarios from USGS streamgauge sites were assumed to represent the baseline (i.e., unaltered) flow regime for each site, and the simulated water withdrawals modelled in this study were assumed to be the only source of streamflow alteration during the period of record for each site. However, we acknowledge that this assumption ignores additional real-world sources of streamflow alteration, including effects of previous or existing flow alterations and cumulative effects of upstream withdrawals or effluent discharges on downstream sites.

Quantile regression results used in prior studies to develop the ELFs used in this study provide regional expected upper-bound predictions of fish species richness in relation to the departure of SFCs from defined hydrologic reference conditions. Some stream sites may fit regional ELF equations better than others, depending on the relative position of specific SFCs under the initial baseline flow regime to that of the defined reference condition. For example, when subjected to water withdrawals, a calculated baseline SFC value that was initially greater than the so-called reference condition may move closer to the reference condition (i.e., decreased departure), and can result, mathematically, in an increase in predicted species richness. While it is ecologically plausible that some species may benefit from reductions in streamflow under certain conditions, large increases in predicted fish species richness with increasing water withdrawals were not considered realistic outcomes and thus were largely ignored in this study. Actual gains in species richness would likely be limited to a modest number of potential colonists by natural abiotic (i.e., physical and chemical habitat limitations), biotic (e.g., competition), and spatial mechanisms (e.g., distance or dispersal barriers) [66]. A review among 165 empirical studies by [5] found that only 13% of the studies reported some positive association between ecological metrics and flow alteration. In this study, some sites exhibited no change in predicted richness under any simulated water-withdrawal scenario. These results should also be interpreted with caution because it is unrealistic to assume that fish species richness would be categorically unaffected by increasingly large withdrawals.

Predicted loss (or gain) of fish species at any given stream site was relative to the predicted richness under baseline streamflow (i.e., simulated withdrawal = 0 m^3/s) which was produced by a regionalized quantile regression curve, assuming >85% probably of species richness at 'optimal' or reference conditions [39]. If the initial baseline flow conditions are far from reference conditions, the actual fish species richness that exists at any given stream may be different than what is predicted using the given ELF equations. Additionally, stream size was not considered during the initial development of ELFs, which means that the same ELF equation is used to predict richness among streams regardless of stream size. For example, a stream with a drainage area size of 10 km^2 could be predicted to have the same or even greater richness than a stream with a drainage area of 1,000 km^2 in the same ecoregion, depending on the hydrologic departure from reference of the subset of SFCs that define each ELF.

5. Conclusions

Results of this study indicate that increasing water-withdrawal rates, regardless of the withdrawal strategy, will likely result in increased loss of fish species richness across a variety of streams and ecoregions. However, the alteration of hydrologic regimes and the resulting predicted richness losses were generally less severe under the POF water withdrawals compared to CR withdrawals, suggesting potential benefits of POF withdrawal scenarios in maintaining aquatic biodiversity. Richness losses and patterns among ecological withdrawal thresholds under POF withdrawals showed general concurrence with Richter's [30] presumptive standards and suggest that limiting water withdrawals to 10–20% of daily streamflow may indeed be generally protective of species richness among most stream drainage area sizes and ecoregions. The sensitivity of low-flow SFCs to CR withdrawals and the prevalence of the low-flow metrics among ELFs across ecoregions suggests that protocols for protection of low flows, such as MFL cutoffs, are likely critical when CR water-withdrawal strategies are employed. Our findings also showed that application of MFL cutoffs, regardless of the withdrawal strategy, were generally effective at damping hydrologic alteration and species richness responses to both withdrawal strategies and indicate that employing MFLs may allow for larger permitted water withdrawals under some circumstances. Development of specific and practical guidelines that limit water withdrawals within ecological response boundaries have been difficult to achieve. This study provides a general framework of comparative and alternative potential outcomes under plausible water-withdrawal strategies. Localized and regionalized estimates of predicted hydrologic and ecologic responses generated in this study could be useful for informing resource management and policy decisions. Lastly, more detailed prioritization of sensitive ecoregions, geographical areas, sub-watersheds, and streams of a certain size or drainage area is an important next step in guiding more targeted management and conservation strategies and actions.

Supplementary Materials: The following are available online at http://www.mdpi.com/2073-4441/12/5/1334/s1, Table S1: Streamflow characteristics and slope and intercept coefficients of ecological limit functions (ELF) for fish groups among ecoregions, Table S2: Mean change in species richness across all stream sites and all water-withdrawal scenarios for each fish group and minimum flow level (MFL), Table S3: Mean loss and mean percent-loss in predicted fish species across stream sites at 10 and 20% water-withdrawal scenarios under MFL0, Table S4: Median values of predicted ecological withdrawal thresholds under constant-rate and percent-of-flow withdrawals for each fish group and overall means of median thresholds calculated for each ecoregion, Figure S1: Responses of SFCs among Blue Ridge (BR) streams to constant-rate (CR) withdrawals, Figure S2: Responses of SFCs among Ridge and Valley (RV) streams to constant-rate (CR) withdrawals, Figure S3: Responses of SFCs among Cumberland Plateau (CP) streams to constant-rate (CR) withdrawals, Figure S4: Responses of SFCs among Interior Plateau (CP) streams to constant-rate (CR) withdrawals, Figure S5: Responses of SFCs among Blue Ridge (BR) streams to percent-of-flow (POF) withdrawals, Figure S6: Responses of SFCs among Ridge and Valley (RV) streams to percent-of-flow (POF) withdrawals, Figure S7: Responses of SFCs among Cumberland Plateau (CP) streams to percent-of-flow (POF) withdrawals, Figure S8: Responses of SFCs among Interior Plateau (CP) streams to percent-of-flow (POF) withdrawals, Figure S9: Predicted percent-change (%) in fish species richness among Blue Ridge (BR) streams to constant-rate (CR) withdrawals, Figure S10: Predicted percent-change (%) in fish species richness among Ridge and Valley (RV) streams to constant-rate (CR) withdrawals, Figure S11: Predicted percent-change (%) in fish species richness among Cumberland Plateau (CP) streams to constant-rate

(CR) withdrawals, Figure S12: Predicted percent-change (%) in fish species richness among Interior Plateau (CP) streams to constant-rate (CR) withdrawals, Figure S13: Predicted percent-change (%) in fish species richness among Blue Ridge (BR) streams to percent-of-flow (POF) withdrawals, Figure S14: Predicted percent-change (%) in fish species richness among Ridge and Valley (RV) streams to percent-of-flow (POF) withdrawals, Figure S15: Predicted percent-change (%) in fish species richness among Cumberland Plateau (CP) streams to percent-of-flow (POF) withdrawals, Figure S16: Predicted percent-change (%) in fish species richness among Interior Plateau (CP) streams to percent-of-flow (POF) withdrawals, Figure S17: Boxplots of predicted ecological withdrawal thresholds under constant-rate (CR) withdrawals, Figure S18: Boxplots of predicted ecological withdrawal thresholds under percent-of-flow (POF) withdrawals.

Author Contributions: Conceptualization, funding acquisition, writing—review and editing, R.R.K., J.M.C., W.J.W.; methodology, R.R.K., J.M.C., W.J.W., and L.J.D.; formal analysis, visualization, writing—original draft preparation, L.J.D. All authors have read and agreed to the published version of the manuscript.

Funding: This research was funded by the Tennessee Department of Environment and Conservation (TDEC) and U.S. Geological Survey (USGS).

Acknowledgments: This study was supported by the Tennessee Department of Environment and Conservation and the U.S. Geological Survey Cooperative Water Program. We thank Jennifer Rapp (USGS VA/WV Water Science Center) and anonymous reviewers for comments on earlier drafts of this manuscript. Any use of trade, firm, or product names is for descriptive purposes only and does not imply endorsement by the U.S. Government.

Conflicts of Interest: The authors declare no conflict of interest. The funders had no role in the design of the study; in the collection, analyses, or interpretation of data; in the writing of the manuscript, or in the decision to publish the results.

References

1. Poff, N.L.; Allan, J.D.; Bain, M.B.; Karr, J.R.; Prestegaard, K.L.; Richter, B.D.; Sparks, R.E.; Stromberg, J.C. The natural flow regime. *BioScience* **1997**, *47*, 769–784. [CrossRef]

2. Bunn, S.E.; Arthington, A.H. Basic principles and ecological consequences of altered flow regimes for aquatic biodiversity. *Environ. Manag.* **2002**, *30*, 492–507. [CrossRef] [PubMed]

3. Tharme, R.E. A global perspective on environmental flow assessment: Emerging trends in the development and application of environmental flow methodologies for rivers. *River Res. Appl.* **2003**, *19*, 397–441. [CrossRef]

4. Dudgeon, D.; Arthington, A.H.; Gessner, M.O.; Kawabata, Z.-I.; Knowler, D.J.; Lévêque, C.; Naiman, R.J.; Prieur-Richard, A.-H.; Soto, D.; Stiassny, M.L.J.; et al. Freshwater biodiversity: Importance, threats, status and conservation challenges. *Biol. Rev.* **2006**, *81*, 163–182. [CrossRef]

5. Poff, N.L.; Zimmerman, J.K.H. Ecological responses to altered flow regimes: A literature review to inform the science and management of environmental flows. *Freshw. Biol.* **2010**, *55*, 194–205. [CrossRef]

6. Annear, T.; Chisholm, I.; Beecher, H.; Locke, A.; Aarrestad, P.; Coomer, C.; Estes, C.; Hunt, J.; Jacobson, R.; Jobsis, G.; et al. *Instream Flows for Riverine Resource Stewardship*, Revised ed.; Instream Flow Council: Cheyenne, WY, USA, 2004.

7. Rolls, R.J.; Leigh, C.; Sheldon, F. Mechanistic effects of low-flow hydrology on riverine ecosystems: Ecological principles and consequences of alteration. *Freshw. Sci.* **2012**, *31*, 1163–1186. [CrossRef]

8. Walters, A.W. The importance of context dependence for understanding the effects of low-flow events on fish. *Freshw. Sci.* **2016**, *35*, 216–228. [CrossRef]

9. Doll, P.; Fiedler, K.; Zhang, J. Global-scale analysis of river flow alterations due to water withdrawals and reservoirs. *Hydrol. Earth Syst. Sci.* **2009**, *12*, 2413–2432. [CrossRef]

10. Robinson, J. *Public-Supply Water Use and Self-Supplied Industrial Water Use in Tennessee, 2010*; U.S. Geological Survey Scientific Investigation Report 2018–5009; U.S. Geological Survey: Reston, VA, USA, 2018; p. 30.

11. Shen, Y.; Oki, T.; Utsumi, N.; Kanae, S.; Hanasaki, N. Projection of future world water resources under SRES scenarios: Water withdrawal. *Hydrolog. Sci. J.* **2008**, *53*, 11–33. [CrossRef]

12. Meador, M.R.; Carlisle, D.M. Relations between altered streamflow variability and fish assemblages in eastern USA streams. *River Res. Appl.* **2012**, *28*, 1359–1368. [CrossRef]

13. Phelan, J.; Cuffney, T.; Patterson, L.; Eddy, M.; Dykes, R.; Pearsall, S.; Goudreau, C.; Mead, J.; Tarver, F. Fish and invertebrate flow-biology relationships to support the determination of ecological flows for North Carolina. *J. Am. Water Resour. As.* **2017**, *53*, 42–55. [CrossRef]

14. Freeman, M.C.; Marcinek, P.A. Fish assemblage responses to water withdrawals and water supply reservoirs in Piedmont streams. *Environ. Manag.* **2006**, *38*, 435–450. [CrossRef] [PubMed]

15. Kanno, Y.; Vokoun, J.C. Evaluating effects of water withdrawals and impoundments on fish assemblages in southern New England streams, USA. *Fish. Manag. Ecol.* **2010**, *17*, 272–283. [CrossRef]

16. Baron, J.S.; Poff, N.L.; Angermeier, P.L.; Dahm, C.N.; Gleick, P.H.; Hairston, N.G.; Jackson, R.B.; Johnston, C.A.; Richter, B.D.; Steinman, A.D. Meeting ecological and societal needs for freshwater. *Ecol. Appl.* **2002**, *12*, 1247–1260. [CrossRef]

17. Arthington, A.; James, C.; Mackay, S.; Rolls, R.; Sternberg, D.; Barnes, A. *Hydro-Ecological Relationships and Thresholds to Inform Environmental Flow Management*; Science Report; International WaterCentre: Brisbane, Australia, 2012; p. 224.

18. Arthington, A.H.; Bunn, S.E.; Poff, N.L.; Naiman, R.J. The challenge of providing environmental flow rules to sustain river ecosystems. *Ecol. Appl.* **2006**, *16*, 1311–1318. [CrossRef]

19. Arthington, A.H.; Brizga, S.; Kennard, M. *Comparative Evaluation of Environmental Flow Assessment Techniques: Best Practice Framework*; LWRRDC Occasional Paper 25/98; Land and Water Resources Research and Development Corporation: Canberra, Australia, 1998; p. 26.

20. Wheeler, K.; Wenger, S.J.; Freeman, M.C. States and rates: Complementary approaches to developing flow-ecology relationships. *Freshw. Biol.* **2018**, *63*, 906–916. [CrossRef]

21. Acreman, M.C.; Overton, I.C.; King, J.; Wood, P.J.; Cowx, I.G.; Dunbar, M.J.; Kendy, E.; Young, W.J. The changing role of ecohydrological science in guiding environmental flows. *Hydrol. Sci. J.* **2014**, *59*, 433–450. [CrossRef]

22. Poff, N.L.; Richter, B.D.; Arthington, A.H.; Bunn, S.E.; Naiman, R.J.; Kendy, E.; Acreman, M.; Apse, C.; Bledsoe, B.P.; Freeman, M.C.; et al. The ecological limits of hydrologic alteration (ELOHA): A new framework for developing regional environmental flow standards: Ecological limits of hydrologic alteration. *Freshw. Biol.* **2010**, *55*, 147–170. [CrossRef]

23. Pastor, A.V.; Ludwig, F.; Biemans, H.; Hoff, H.; Kabat, P. Accounting for environmental flow requirements in global water assessments. *Hydrol. Earth Syst. Sci.* **2014**, *18*, 5041–5059. [CrossRef]

24. Naiman, R.J.; Bunn, S.E.; Nilsson, C.; Petts, G.E.; Pinay, G.; Thompson, L.C. Legitimizing fluvial ecosystems as users of water: An overview. *Environ. Manag.* **2002**, *30*, 455–467. [CrossRef]

25. Poff, N.L.; Allan, J.D.; Palmer, M.A.; Hart, D.D.; Richter, B.D.; Arthington, A.H.; Rogers, K.H.; Meyer, J.L.; Stanford, J.A. River flows and water wars: Emerging science for environmental decision making. *Front. Ecol. Environ.* **2003**, *1*, 298–306. [CrossRef]

26. Davies, P.M.; Naiman, R.J.; Warfe, D.M.; Pettit, N.E.; Arthington, A.H.; Bunn, S.E. Flow–ecology relationships: Closing the loop on effective environmental flows. *Mar. Freshw. Res.* **2014**, *65*, 133–141. [CrossRef]

27. Cartwright, J.; Caldwell, C.; Nebiker, S.; Knight, R. Putting flow–ecology relationships into practice: A decision-support system to assess fish community response to water-management scenarios. *Water* **2017**, *9*, 196. [CrossRef]

28. Tennant, D.L. Instream Flow Regimens for Fish, Wildlife, Recreation and Related Environmental Resources. *Fisheries* **1976**, *1*, 6–10. [CrossRef]

29. Richter, B.; Baumgartner, J.; Wigington, R.; Braun, D. How much water does a river need? *Freshw. Biol.* **1997**, *37*, 231–249. [CrossRef]

30. Richter, B.D.; Davis, M.M.; Apse, C.; Konrad, C. A presumptive standard for environmental flow protection. *River Res. Appl.* **2012**, *28*, 1312–1321. [CrossRef]

31. U.S. Environmental Protection Agency Water: Monitoring and Assessment, Section 5.1 Stream Flow. Available online: https://archive.epa.gov/water/archive/web/html/vms51.html (accessed on 2 January 2020).

32. Farmer, W.H.; Over, T.M.; Kiang, J.E. Bias correction of simulated historical daily streamflow at ungauged locations by using independently estimated flow duration curves. *Hydrol. Earth Syst. Sci.* **2018**, *22*, 5741–5758. [CrossRef]

33. Cook, C.N.; Hockings, M.; Carter, R. (Bill) Conservation in the dark? The information used to support management decisions. *Front. Ecol. Environ.* **2010**, *8*, 181–186. [CrossRef]

34. Kendy, E.; Apse, C.; Blann, K. *A Practical Guide to Environmental Flows for Policy and Planning, with Nine Case Studies in the United States*; The Nature Conservancy: Arlington, VA, USA, 2012.

35. Davies, S.P.; Jackson, S.K. The biological condition gradient: A descriptive model for interpreting change in aquatic ecosystems. *Ecol. Appl.* **2006**, *16*, 1251–1266. [CrossRef]

36. Hain, E.F.; Kennen, J.G.; Caldwell, P.V.; Nelson, S.A.C.; Sun, G.; McNulty, S.G. Using regional scale flow-ecology modeling to identify catchments where fish assemblages are most vulnerable to changes in water availability. *Freshw. Biol.* **2018**, *63*, 928–945. [CrossRef]

37. Knight, R.R.; Brian Gregory, M.; Wales, A.K. Relating streamflow characteristics to specialized insectivores in the Tennessee River Valley: A regional approach. *Ecohydrology* **2008**, *1*, 394–407. [CrossRef]

38. Knight, R.R.; Gain, W.S.; Wolfe, W.J. Modelling ecological flow regime: An example from the Tennessee and Cumberland River basins. *Ecohydrology* **2012**, *5*, 613–627. [CrossRef]

39. Knight, R.R.; Murphy, J.C.; Wolfe, W.J.; Saylor, C.F.; Wales, A.K. Ecological limit functions relating fish community response to hydrologic departures of the ecological flow regime in the Tennessee River basin, United States. *Ecohydrology* **2014**, 1262–1280. [CrossRef]

40. Murphy, J.C.; Knight, R.R.; Wolfe, W.J.; S. Gain, W. Predicting ecological flow regime at ungaged sites: A comparison of methods. *River Res. Appl.* **2012**, *29*, 660–669. [CrossRef]

41. Knight, R.R.; Cartwright, J.M.; Ladd, D.E. *Streamflow and Fish Community Diversity Data for Use in Developing Ecological Limit Functions for the Cumberland Plateau, Northeastern Middle Tennessee and Southwestern Kentucky, 2016*; U.S. Geological Survey data release; U.S. Geological Survey: Reston, VA, USA, 2016. Available online: http://dx.doi.org/10.5066/F7JH3J83 (accessed on 1 September 2018).

42. Omernik, J.M. Map Supplement: Ecoregions of the Conterminous United States. *Ann. Assoc. Am. Geogr.* **1987**, *77*, 118–125. [CrossRef]

43. Abell, R.; Olson, D.; Dinerstein, E.; Hurley, P.; Diggs, J.; Eichbaum, W.; Walters, S.; Wettengel, W.; Allnutt, T.; Loucks, C.; et al. *Freshwater Ecoregions of North America: A Conservation Assessment*; Island Press: Washington, DC, USA, 2000.

44. Elkins, D.C.; Sweat, S.C.; Hill, K.S.; Kuhajda, B.R.; George, A.L.; Wenger, S.J. *The Southeastern Aquatic Biodiversity Conservation Strategy. Final Report*; University of Georgia River Basin Center: Athens, Greece, 2016; p. 237.

45. Fenneman, N.M. Physiographic Divisions of the United States. *Ann. Assoc. Am. Geogr.* **1928**, *18*, 261. [CrossRef]

46. Kennard, M.J.; Mackay, S.J.; Pusey, B.J.; Olden, J.D.; Marsh, N. Quantifying uncertainty in estimation of hydrologic metrics for ecohydrological studies. *River Res. Appl.* **2010**, 137–156. [CrossRef]

47. U.S. Geological Survey (USGS) USGS Water Data for the Nation: U.S. *Geological Survey National Water Information System Database*. Available online: http://dx.doi.org/10.5066/F7P55KJN (accessed on 1 September 2018).

48. Hirsch, R.; De Cicco, L. *User Guide to Exploration and Graphics for RivEr Trends (EGRET) and dataRetrieval: R Packages for Hydrologic Data*; U.S. Geological Survey Techniques and Methods book 4, chap. A10; U.S. Geological Survey: Reston, VA, USA, 2015; p. 93.

49. *R Core Team R: A Language and Environment for Statistical Computing*; R Foundation for Statistical Computing: Vienna, Austria, 2019.

50. Robinson, J.A. *Water Use in Tennessee, 2010*; U.S. Geological Survey data release; U.S. Geological Survey: Reston, VA, USA, 2017. Available online: https://doi.org/10.5066/F7V9868K (accessed on 1 September 2018).

51. Mills, J.; Blodgett, D. *EflowStats: Hydrologic Indicator and Alteration Stats. R Package Version 5.0.0*; 2017; Available online: https://github.com/USGS-R/EflowStats (accessed on 1 September 2017).

52. Buchanan, B.; McManamay, R.A.; Auerbach, D.; Fuka, D.; Walter, M. *Environmental Flow Analysis for the Marcellus Shale Region*; Appalachian Landscape Conservation Cooperative: Shepherdstown, WV, USA, 2015; p. 170.

53. Driver, L.J. *Ecological Flow Analyses Results: Streamflow Characteristics, Predicted Fish Responses, and Ecological Withdrawal Thresholds for Select Stream Sites within the Cumberland and Tennessee River Basins*; U.S. Geological Survey data release; U.S. Geological Survey: Reston, VA, USA, 2019. Available online: https://doi.org/10.5066/F7Q23Z4B (accessed on 1 September 2019).

54. Vannote, R.L.; Minshall, G.W.; Cummins, K.W.; Sedell, J.R.; Cushing, C.E. The River Continuum Concept. *Can. J. Fish. Aquat. Sci.* **1980**, *37*, 130–137. [CrossRef]

55. Poff, N.L.; Ward, J.V. Implications of Streamflow Variability and Predictability for Lotic Community Structure: A Regional Analysis of Streamflow Patterns. *Can. J. Fish. Aquat. Sci.* **1989**, *46*, 1805–1818. [CrossRef]

56. Shank, M.K.; Stauffer, J.R. Land use and surface water withdrawal effects on fish and macroinvertebrate assemblages in the Susquehanna River basin, USA. *J. Freshw. Ecol.* **2015**, *30*, 229–248. [CrossRef]

57. Flannery, M.S.; Peebles, E.B.; Montgomery, R.T. A percent-of-flow approach for managing reductions of freshwater inflows from unimpounded rivers to Southwest Florida estuaries. *Estuaries* **2002**, *25*, 1318–1332. [CrossRef]

58. Acreman, M.C.; Ferguson, A.J.D. Environmental flows and the European Water Framework Directive: Environmental flows and WFD. *Freshw. Biol.* **2010**, *55*, 32–48. [CrossRef]

59. Zorn, T.G.; Seelbach, P.W.; Rutherford, E.S.; Wills, T.; Cheng, S.; Wiley, M. *A Regional-Scale Habitat Suitability Model to Assess the Effects of Flow Reduction on Fish Assemblages in Michigan Streams*; Michigan Department of Natural Resources: Ann Arbor, MI, USA, 2008; pp. 871–895.

60. Driver, L.J.; Hoeinghaus, D.J. Spatiotemporal dynamics of intermittent stream fish metacommunities in response to prolonged drought and reconnectivity. *Mar. Freshw. Res.* **2016**, *67*, 1667–1679. [CrossRef]

61. Loftin, K.A.; Clark, J.M.; Journey, C.A.; Kolpin, D.W.; Van Metre, P.C.; Carlisle, D.; Bradley, P.M. Spatial and temporal variation in microcystin occurrence in wadeable streams in the southeastern United States: Microcystins in southeastern US streams. *Environ. Toxicol. Chem.* **2016**, *35*, 2281–2287. [CrossRef] [PubMed]

62. Graham, J.L.; Dubrovsky, N.M.; Foster, G.M.; King, L.R.; Loftin, K.A.; Rosen, B.H.; Stelzer, E.A. Cyanotoxin occurrence in large rivers of the United States. *Inland Waters* **2020**, *10*, 109–117. [CrossRef]

63. North Carolina Ecological Flows Science Advisory Board. *Recommendations for Estimating Flows to Maintain Ecological Integrity in Streams and Rivers in North Carolina*; North Carolina Department of Environmental Quality: Raleigh, NC, USA, 2013. Available online: https://files.nc.gov/ncdeq/Water%20Resources/files/eflows/sab/EFSAB_Final_Report_to_NCDENR.pdf (accessed on 4 January 2019).

64. Rosenfeld, J.S. Developing flow-ecology relationships: Implications of nonlinear biological responses for water management. *Freshw. Biol.* **2017**, *62*, 1305–1324. [CrossRef]

65. Olden, J.D.; Poff, N.L. Redundancy and the choice of hydrologic indices for characterizing streamflow regimes. *River Res. Appl.* **2003**, *19*, 101–121. [CrossRef]

66. Jackson, D.A.; Peres-Neto, P.R.; Olden, J.D. What controls who is where in freshwater fish communities–the roles of biotic, abiotic, and spatial factors. *Can. J. Fish. Aquat. Sci.* **2001**, *58*, 157–170. [CrossRef]

MDPI

St. Alban-Anlage 66

4052 Basel

Switzerland

Tel. +41 61 683 77 34

Fax +41 61 302 89 18

www.mdpi.com

Water Editorial Office

E-mail: water@mdpi.com

www.mdpi.com/journal/water